Number Theory
in
Digital Signal Processing

PRENTICE-HALL SIGNAL PROCESSING SERIES

Alan V. Oppenheim, Editor

ANDREWS and HUNT *Digital Image Restoration*
BRIGHAM *The Fast Fourier Transform*
CASTLEMAN *Digital Image Processing*
HAMMING *Digital Filters*
OPPENHEIM and SCHAFER *Digital Signal Processing*
OPPENHEIM, et al. *Applications of Digital Signal Processing*
RABINER and GOLD *Theory and Applications of Digital Signal Processing*
RABINER and SCHAFER *Digital Processing of Speech Signals*

Advanced Monographs

MCCLELLAN and RADER *Number Theory in Digital Signal Processing*
TRIBOLET *Seismic Applications of Homomorphic Signal Processing*

James H. McClellan
M. I. T.

Charles M. Rader
M. I. T.
Lincoln Laboratory

Number Theory in Digital Signal Processing

Prentice-Hall, Inc. Englewood Cliffs, New Jersey 07632

Library of Congress Cataloging in Publication Data

McClellan, James H (date)
Number theory in digital signal processing.

(Prentice-Hall signal processing series)
Bibliography: p.
Includes index.
1. Signal processing—Digital techniques.
2. Numbers, Theory of. 3. Convolutions (Mathematics)
4. Fourier transformations. I. Rader, Charles M.,
joint author. II. Title.
TK5102.5.M216 621.38′.043 78-12313
ISBN 0-13-627349-1

Printed in the United States of America

10 9 8 7 6 5 4 3 2 1

Prentice-Hall International, Inc., *London*
Prentice-Hall of Australia Pty. Limited, *Sydney*
Prentice-Hall of Canada, Ltd., *Toronto*
Prentice-Hall of India Private Limited, *New Delhi*
Prentice-Hall of Japan, Inc., *Tokyo*
Prentice-Hall of Southeast Asia Pte. Ltd., *Singapore*
Whitehall Books Limited, *Wellington, New Zealand*

Contents

Preface

The usefulness of an engineering method becomes clear only when that method is evaluated by the wide community of potential users, who consider it in relation to their needs. A complicated method, or one unfamiliar to its potential users, may never be properly evaluated—is it widely useful or merely a footnote in engineering history? At present several algorithms, intended to be used for digital signal processing, can be understood and programmed only with a knowledge of elementary number theory. Since so many engineers and programmers are not familiar with number theory, the ultimate usefulness of these algorithms is an open question. We have compiled this book in order to make available to the engineering community the background in number theory needed to understand these new algorithms, to teach the principles of the algorithms in digital signal processing courses, and to apply the ideas of number theory to engineers' work.

The book is primarily a collection of papers representative of the applications of number theoretic ideas to the development of algorithms for the two most basic signal processing computations: convolution and the discrete Fourier transform (DFT). The algorithms presented in this book fall into two classes. In one class of algorithms, the finite field transforms, the number theoretic properties of the data values are exploited. In the other, the number theoretic properties of the addresses of the data are exploited. The most highly developed example of the latter type of algorithm is the Winograd Fourier transform algorithm, which is the first fundamentally different approach to computing the DFT since the development of the fast Fourier transform (FFT). A program which implements the Winograd Fourier transform is included in the book. (We are indebted to Mr. H. Nawab for the use of the program.)

We have included an introductory section about elementary number theory as it applies to digital signal processing. This section introduces only the most elementary concepts in the study of integers and polynomials, principally the properties of divisibility and congruence as they have been applied to digital signal processing in the papers in this collection. It might be interesting to mention some of the important topics we have not discussed.

Number theory is divided into elementary number theory and analytic number theory. The latter is far beyond the scope of this book, but it has attracted the attention of most of the great mathematicians, and the methods developed in pursuing questions of analytic number theory have been of great importance in the other branches of mathematics. For example, Jacobi's development of the elliptic functions, which serves as the basis of important filter design methods, was motivated by his interest in analytic number theory. Even within elementary number theory, our introductory material ignores Farey sequences and quadratic residues, two topics that have recently been applied to signal processing problems; it also ignores the Pythagorean triplets and many similar subjects which might well have applications of their own. We have written only one section, and we do not pretend that it alone will compete with the many excellent textbooks on the theory of numbers. In fact, we hope that some readers will be motivated to study further.

Our material is sprinkled with simple problems meant to reinforce the material immediately with concrete examples. Working out these little problems should help the reader get a feeling for number theoretic principles, rather than leaving him with only memorized definitions and theorems. None of these problems is very difficult or tricky.

We have wrestled with the problem of notation and have finally adopted, for ourselves, two conventions that, compared with the more usual notations, simplify notation and improve the clarity and appearance of many of the equations. One of these notations has to do with sequences, where the domain of definition is the integers. For this, we use the notation $f[n]$ to represent the nth term of the sequence $\{f[n]\}$. This prevents confusion with the continuous function $f(t)$ from which we might have produced $\{f[n]\}$ by sampling at times $t = nT$; yet it spares us the need to introduce confusing subscripts, superscripts, and other ways of doctoring the "f" when it is the domain which has been treated differently.

The other notation is $\langle x \rangle_n$ for the residue of x modulo n. Most books on number theory deal with *congruence*, a relation between classes of numbers. To implement algorithms, however, we must locate data with addresses, or store numbers in registers, and we cannot be content with number classes in this sort of application. A residue is a number and can unambiguously be used in a computational recipe. A number of equations, before adopting these two conventions, required superscripts and subscripts on seven distinct levels.

We appreciate the aid and encouragement which a great many people have given us in our work on this collection. We would like to express particular thanks to Dr. Alan V. Oppenheim, who convinced us that the book would fill a real need. For their critical suggestions and aid we would like to acknowledge the contributions of Drs. Thomas Parks, Ramesh Agarwal, C. Sidney Burrus, James Cooley, Shmuel Winograd, Harvey Silverman, and L. Robert Morris. For her aid in preparing the manuscript we are extremely grateful to Ms. Monica Edelman.

Acton, Massachusetts

Charles M. Rader
James H. McClellan

Part One

I

the discrete Fourier transform and circular convolution

This section introduces the discrete Fourier transform and circular convolution of sequences. Since we assume that the reader is thoroughly familiar with these topics our coverage is brief. Its main purpose is providing the basis for a consistent notation throughout this section. A more detailed coverage of these topics, however, can be found in the text by Alan V. Oppenheim and Ronald W. Schafer—*Digital Signal Processing*.

Processing signals with a digital computer or with special-purpose digital hardware involves the implementation of computational schemes on sequences of numbers. Thus a *discrete-time signal* $x[n]$ is a sequence, that is, a function in the mathematical sense that assigns to each integer a unique real or complex number. If the range of values of the sequence is a finite subset of the real or complex numbers, the signal is termed a *digital signal*. This is the case when representing numbers in a digital computer.

As a practical matter it is not possible to process an entire (infinitely long) sequence, although it is common practice to analyze many processing systems as though this were the case. In our study of signal-processing algorithms we shall concentrate on the case of *finite length sequences*, that is, discrete-time signals that are identically zero outside some finite interval of the integers. Typically, we shall take this interval to be the first N non-negative integers $\{n : 0 \leq n \leq N - 1\}$.

The class of *linear shift-invariant* (LSI) systems contains many of the useful signal-processing systems that are widely employed. The input-output relationship for a LSI system is that of convolution

$$y[n] = x[n] * h[n] = \sum_{k=-\infty}^{\infty} x[k]h[n - k] \tag{1.1}$$

where $x[n]$ is the input signal, $y[n]$ the output, and $h[n]$ the *impulse response* of the LSI system. The LSI system is completely characterized by its response $h[n]$ to just one input, the unit sample input signal $\delta[n]$. Accordingly, many properties of the LSI system can be expressed in terms of the impulse response sequence $h[n]$. For example, the system is *stable* if and only if the impulse response is absolutely summable; the system is *causal* if and only if $h[n] = 0$ for $n < 0$, etc. When the duration of the impulse response is finite, the infinite sum in (1.1) reduces to a finite sum†

†Recall that convolution is a commutative operation:

$$x[n] * h[n] = h[n] * x[n].$$

For other properties of convolution consult Oppenheim and Schafer, *Digital Signal Processing*, Prentice-Hall, Englewood Cliffs, N.J., 1975.

$$y[n] = \sum_{k=0}^{N-1} h[k]x[n-k] \tag{1.2}$$

Efficient implementation of (1.2) is the central topic of this monograph. We shall discuss methods for the "direct" implementation of (1.2) in the time domain as well as "transform" methods.

Another common combination of signals is the *cross-correlation* $R_{xy}[n]$ of $x[n]$ and $y[n]$

$$R_{xy}[n] = \sum_{k=-\infty}^{\infty} x[k]y^*[n+k] \tag{1.3}$$

The autocorrelation sequence $R_{xx}[n]$ is used extensively in power spectrum estimation of random signals. The cross-correlation sequence can be expressed in terms of the convolution operator†

$$R_{xy}[n] = x[-n] * y^*[n] \tag{1.4}$$

The computation of R_{xy} requires that one signal be time reversed before being convolved with the complex conjugate of the other signal. Thus the discussion of efficient algorithms for convolution will extend naturally to efficient algorithms for correlation.

Perhaps the most powerful tools for the analysis of signals and LSI systems are those known as *transform methods.* We begin our discussion with the z-transform. The sequence $\{x[n]\} = \{c^n\}$, $-\infty < n < \infty$, plays a very special role with respect to LTI systems. When $\{c^n\}$ is the input to a LSI system the output is easily calculated to be $\{\lambda c^n\}$†† where λ is the constant $H(c) = \sum_k h[k]c^{-k}$. If an arbitrary input can now be decomposed into a linear combination of such functions, then the linearity of the system will allow one easily to construct the output.

The appropriate representation of a general sequence is given by the inverse z-transform

$$x[n] = \frac{1}{2\pi j} \oint_C X(z)z^{n-1}\,dz \tag{1.5}$$

where C is a closed counterclockwise contour that encircles the origin and lies in the region of convergence of the analytic function $X(z)$ which is the z-transform of $x[n]$.

$$X(z) = \sum_{n=-\infty}^{\infty} x[n]z^{-n} \tag{1.6}$$

The fact that eigenfunctions underlie z-transform analysis manifests itself in the convolution property for z-transforms:

If $\qquad y[n] = x[n] * h[n]$ then $Y(z) = X(z)H(z)$

Thus, convolution in the time domain is converted to polynomial multiplication in the transform domain.

†An asterisk in superscript means *complex conjugate.* An asterisk as a dyadic operator means *convolution.*

††A sequence for which the output of the system is proportional to the input is called an *eigenfunction* of the system and λ is the corresponding *eigenvalue.* Note that λ depends on the complex number c but is independent of n.

PROBLEM Show that cross-correlation of $x[n]$ and $y[n]$ is converted to $X(z^{-1})Y^*(z^*)$ in the z-transform domain.

An important class of LSI systems is those possessing rational z-transforms. In this case $H(z) = \frac{B(z)}{A(z)}$ where $A(z)$ and $B(z)$ are polynomials (of finite degree). Since $Y(z) = H(z)X(z)$ we obtain $A(z)Y(z) = B(z)X(z)$. In the time domain the input-output relation of such a system is the difference equation

$$\sum_{k=0}^{P} a_k y[n-k] = \sum_{k=0}^{Q} b_k x[n-k] \tag{1.7}$$

We may assume, without loss of generality, that $a_0 = 1$. Then solving for $y[n]$, we obtain a recursive formula for obtaining the output from the input and initial conditions on $y[n]$.

$$y[n] = -\sum_{k=1}^{P} a_k y[n-k] + \sum_{k=0}^{Q} b_k x[n-k] \tag{1.8}$$

In general, the impulse response of a rational system is indefinitely long and (1.8) offers a practical way to implement the system. It is interesting to note, incidentally, that each sum in (1.8) is itself a convolution sum. This fact might be exploited in direct form implementations that use special convolution algorithms.

An important subclass of the set of rational z-transforms are those in which the denominator $A(z) = 1$; thus $H(z)$ is a polynomial in z. In this case the difference equation implementation of (1.8) contains no feedback terms and the output $y[n]$ is merely the convolution of the input $x[n]$ with the coefficients, b_k, of the polynomial $B(z)$. The response of this system to an impulse is the finite length output sequence $\{b_0, b_1, \ldots, b_Q\}$. Such systems are often called *finite impulse response* (FIR) systems or nonrecursive systems.

In general a rational z-transform will possess both poles (values of z such that $|H(z)| = \infty$) and zeros (values of z where $H(z) = 0$). If the system is FIR then $H(z)$ will have no poles except possibly at $z = 0$ or $z = \infty$.

The evaluation of the z-transform along the unit circle ($z = e^{j\omega}$) in the z-plane yields the Fourier transform relationships

$$x[n] = \frac{1}{2\pi}\int_{-\pi}^{\pi} X(e^{j\omega})e^{j\omega n}\, d\omega \tag{1.9a}$$

$$X(e^{j\omega}) = \sum_{n=-\infty}^{\infty} x[n]e^{-jn\omega} \tag{1.9b}$$

The Fourier transform pair is well defined for all sequences that have finite energy, $\sum_n |x[n]|^2 < \infty$. Parseval's theorem shows that the energy in the time domain equals the energy in the frequency domain.

PARSEVAL'S THEOREM

$$\sum_n |x[n]|^2 = \frac{1}{2\pi}\int_{-\pi}^{\pi} |X(e^{j\omega})|^2\, d\omega \tag{1.10}$$

Since (1.9a) describes the decomposition of the signal $x[n]$ in terms of a linear combination of the eigenfunctions $e^{j\omega n}$, Parseval's theorem implies that $X(e^{j\omega})$ is a measure of the signal's energy distribution as a function of frequency.

Since the Fourier transform can be viewed as a special case of the z-transform the convolution property still holds. Namely, if $y[n] = x[n]*h[n]$, then $Y(e^{j\omega}) = X(e^{j\omega})H(e^{j\omega})$, provided that the Fourier transforms exist.

PROBLEM Calculate the Fourier transform of the autocorrelation of $x[n]$, $R_{xx}[m]$.

The Fourier transform is a useful analytical tool, hence we would like to have efficient computational methods for calculating this transform. The definition of $X(e^{j\omega})$ given in (1.9b) involves an infinite summation. Furthermore, $X(e^{j\omega})$ is a function of the continuous frequency variable ω. Three concessions must be made in order to perform a finite calculation: (1) $x[n]$ must be a finite length signal; (2) we must settle for samples of $X(e^{j\omega})$; (3) only a finite number of samples $\{X(e^{j\omega_k})\}$ of $X(e^{j\omega})$ can be computed. Thus we are led to consider the finite Fourier transform

$$X[k] = X(e^{j(2\pi k/L)}) = \sum_{n=0}^{N-1} x[n]e^{-j(2\pi nk/L)} \tag{1.11}$$

where $k = 0, 1, 2, \ldots, L-1$. We restrict our attention to the special case $L = N$ in which case we obtain the *discrete Fourier transform* (DFT)

$$X[k] = \sum_{n=0}^{N-1} x[n]W_N^{nk} \quad k = 0(1)N-1 \tag{1.12†}$$

where $W_N = e^{-j(2\pi/N)}$. Concerning the DFT defined in (1.12) we would like to know whether an inverse transform exists and what sort of convolution property applies.

The inverse DFT (or IDFT) can be obtained by first proving the identity

$$\sum_{n=0}^{N-1} W_N^{nk} = \begin{cases} N & \text{if } k = 0, \pm N, \pm 2N, \ldots \\ 0 & \text{otherwise} \end{cases} \tag{1.13}$$

Proof of (1.13) breaks into two cases. If k is an integer multiple of N, then $W_N^{nk} = 1$ and the sum becomes $\sum_{n=0}^{N-1}(1)$ and is equal to N. If k is not a multiple of N, then the sum in (1.13) can be recognized as the sum of the first N terms of a geometric sequence with ratio W_N^k and we obtain

$$\sum_{n=0}^{N-1} W_N^{nk} = \frac{1 - W_N^{Nk}}{1 - W_N^k}$$

Since $W_N^{Nk} = 1$, the sum is zero and (1.13) is thus verified. Now suppose that the inverse DFT is of the form

$$x[n] = N^{-1} \sum_{k=0}^{N-1} X[k]W_N^{-nk} \tag{1.14}$$

†The notation $k = r(s)t$ is an abbreviation for $k = r, r+s, r+2s, \ldots, t$.

Substituting the definition of $X[k]$ from (1.12) into (1.14) we obtain

$$N^{-1}\sum_{k=0}^{N-1}\sum_{l=0}^{N-1} x[l]W_N^{lk}W_N^{-nk}$$

Interchanging the order of summations we obtain

$$N^{-1}\sum_{l=0}^{N-1} x[l]\sum_{k=0}^{N-1} W_N^{(l-n)k}$$

The inner sum can be recognized as the form of (1.13) and is nonzero only for $l = n$. Hence, the right side of (1.14) reduces to

$$N^{-1}x[n]N = x[n]$$

Equations (1.12) and (1.14) constitute the discrete Fourier transform pair:

$$X[k] = \sum_{n=0}^{N-1} x[n]W_N^{nk} \tag{1.15a}$$

$$x[n] = N^{-1}\sum_{k=0}^{N-1} X[k]W_N^{-nk} \tag{1.15b}$$

One of the most successful applications of the DFT is in the high-speed implementation of convolution. In order to use the DFT in this application we must develop a convolution property for the DFT. Therefore, we consider the inverse DFT of the product of the two DFT's $X[k]H[k]$

$$y[n] = N^{-1}\sum_{k=0}^{N-1} H[k]X[k]W_N^{-nk}$$

Substituting the definition of $X[k]$ and interchanging the order of summations, we obtain

$$y[n] = \sum_{l=0}^{N-1} x[l]N^{-1}\sum_{k=0}^{N-1} H[k]W_N^{-k(n-l)}$$

which yields

$$y[n] = \sum_{l=0}^{N-1} x[l]h[n-l] \tag{1.16}$$

In order that the sum in (1.16) make sense, the signal $h[n-l]$ must be periodically extended to have a period of N. That means that, if $n-l$ is not in the range $0 \leq n-l \leq N-1$, we must add or subtract the unique multiple of N so that $0 \leq n-l+rN \leq N-1$, and use $h[n-l+rN]$ in (1.16). We can rewrite (1.16) to express this periodic extension of $h[n]$.

$$y[n] = \sum_{l=0}^{N-1} x[l]h[n-l \bmod N] \tag{1.17}$$†

The sums of (1.16), (1.17) are usually called N-point *circular* convolution and will be denoted as $y[n] = x[n] \circledN h[n]$.

†See Section III for a discussion of arithmetic modulo N.

If circular convolution is to be useful for computing the aperiodic (or linear) convolution of (1.2), we must somehow compensate for the periodic nature of the circular convolution. Recall that the linear convolution of an L-point sequence $x[n]$ with an M-point sequence $h[n]$ will yield an N-point sequence $y[n]$ where $N = L + M - 1$. In order to calculate all of $\{y[n]\}$ using circular convolution, the length of the circular convolution must be at least N points. Furthermore, $x[n]$ and $h[n]$ must be extended to create N-point sequences. This is accomplished by appending zeroes to both $x[n]$ and $h[n]$ to form $\hat{x}[n]$ and $\hat{h}[n]$ thus:

$$\hat{x}[n] = \begin{cases} x[n] & 0 \leq n \leq L-1 \\ 0 & L \leq n \leq N-1 \end{cases} \tag{1.18}$$

$$\hat{h}[n] = \begin{cases} h[n] & 0 \leq n \leq M-1 \\ 0 & M \leq n \leq N-1 \end{cases} \tag{1.19}$$

Then one can show that if $\hat{y}[n] = \hat{x}[n] \circledN \hat{h}[n]$, then $y[n] = \hat{y}[n]$ for $0 \leq n \leq N-1$. Outside of this range for the index n, $y[n]$ is zero. A situation common in practice is the filtering of an indefinitely long input sequence by an FIR filter whose length is relatively short. To avoid performing a long convolution, the indefinitely long input sequence can be broken into pieces of short length. The indefinitely long output stream is obtained by piecing together the results of convolving the small sections of the input with the impulse response of the filter.

There are two well-known methods for performing this sectioning: the *overlap-add* technique and the *overlap-save* technique. In the overlap-add technique the input signal $x[n]$ is segmented into sections of length L points. Each input section is padded with zeroes according to (1.18) and circular convolution (via the FFT) is used to compute a section of the output. The output sections overlap one another, and the final output is the sum of these sections. In the overlap-save method the input $x[n]$ is segmented into sections of length N, but the sections are overlapped by $M - 1$ points. After the individual small convolutions are calculated, the output is obtained by saving L points of each circular convolution and concatenating these L-point segments to form the continuous data stream $y[n]$.

II
multidimensional signals and transforms

Many of the concepts and signal processing techniques just described can be generalized easily to apply to multidimensional signals and systems. The obvious application for such techniques is image processing and other two-dimensional data. A less obvious application, but one which plays an important role in the derivation and development of fast transform algorithms, is the conversion of a one-dimensional (1-D) problem to a multidimensional (n-D) problem and vice versa. For example, by appropriate reindexing of a 1-D sequence it is often possible to compute the DFT either as the 1-D DFT of the sequence or as a multidimensional DFT of the resulting array. The precise conditions under which this rearrangement can be done will be developed later as a corollary to the Chinese Remainder theorem.

An *n-dimensional sequence* is a function, possibly complex-valued, of n independent integer-valued variables $k_1, k_2, \ldots, k_n$. For example, a sampled picture is a two-dimensional signal, since the light intensity varies with the two spatial coordinates of the image. A second example is the n-dimensional complex exponential sequence

$$x[k_1, k_2, \ldots, k_n] = z_1^{k_1} z_2^{k_2} \ldots z_n^{k_n}$$

The latter is an example of a special type of signal known as a *separable* signal because the complex exponential sequence can be written as the product of n one-dimensional sequences

$$x[k_1, k_2, \ldots, k_n] = x_1[k_1]\, x_2[k_2]\, x_3[k_3] \ldots x_n[k_n]$$

where $x_i[k_i] = z_i^{k_i}$.
Finally, the n-dimensional counterpart of the unit sample signal is

$$\delta[k_1, k_2, \ldots k_n] = \begin{cases} 1 & \text{if } k_1 = k_2 = \ldots = k_n = 0 \\ 0 & \text{otherwise} \end{cases} \tag{2.2}$$

Note that $\delta[k_1, k_2, \ldots k_n]$ is a separable signal, being the product of n one-dimensional unit sample signals. Note the distinction between a multidimensional signal as defined here and a vector-valued signal of one independent variable as commonly used in multivariable system theory.

The definition of a multidimensional system is also straightforward. An *n-dimensional system* is a transformation that takes n-D signals as input and yields n-D signals as output. The n-D system is *linear* if

$$T\{a_1 x_1[\mathbf{k}] + a_2 x_2[\mathbf{k}]\} = a_1 T\{x_1[\mathbf{k}]\} + a_2 T\{x_2[\mathbf{k}]\} \tag{2.3}$$

where T is the system transformation and $\mathbf{k}$ is the n-dimensional vector of indices $[k_1, k_2, \ldots, k_n]$. The system is *shift-invariant* if

$$T\{x[k_1 + \epsilon_1, k_2 + \epsilon_2, \ldots, k_n + \epsilon_n]\} = y[k_1 + \epsilon_1, \ldots, k_n + \epsilon_n] \tag{2.4}$$

when $y[\mathbf{k}] = T\{x[\mathbf{k}]\}$. As in the 1-$D$ case, the class of linear shift-invariant (LSI) systems plays a central role and these systems are easily analyzed by transform techniques.

A linear shift-invariant system can be completely characterized by its impulse response, that is, the response to the signal $\delta[k_1, k_2, \ldots, k_n]$. If $h[k_1, k_2, \ldots, k_n]$ is the impulse response of the linear shift-invariant system, then the input-output relation of the system is given by the convolution sum

$$y[k_1, \ldots k_n] = \sum_{l_1} \sum_{l_2} \ldots \sum_{l_n} h[k_1 - l_1, \ldots k_n - l_n]\, x[l_1, \ldots l_n] \tag{2.5}$$

and will be denoted as $y[\mathbf{k}] = h[\mathbf{k}] * x[\mathbf{k}]$.

Paralleling the 1-D development, it is simple to show that the complex exponential sequences are scaled by a constant when processed by n-D LSI system. Thus, the

z-transform is motivated. The z-transform of the sequence $x[k_1, k_2 \ldots k_n]$ is a function of the n complex variables $z_1, z_2, \ldots z_n$

$$X(z_1, z_2 \ldots z_n) = \sum_{k_1} \sum_{k_2} \cdots \sum_{k_n} x[k_1, \ldots k_n] z_1^{-k_1} z_2^{-k_2} \ldots z_n^{-k_n} \tag{2.6}$$

where the sums are taken over all values of k_i. The inverse z-transform involves an n-fold complex contour integration

$$x[k_1, \ldots k_n] = \left(\frac{1}{2\pi j}\right)^n \oint_{c_1} \oint_{c_2} \cdots \oint_{c_n} X(z_1, \ldots, z_n) z_1^{k_1} z_2^{k_2} \ldots z_n^{k_n} \, dz_1 \ldots dz_n \tag{2.7}$$

where the contours of integration must lie within the region of convergence of $X(z_1, \ldots z_n)$. The important property of the z-transform is the convolution property. If $y[\mathbf{k}] = x[\mathbf{k}] * h[\mathbf{k}]$, then $Y(z_1, \ldots z_n) = X(z_1, \ldots z_n) \cdot H(z_1, \ldots z_n)$.

When the z-transform is evaluated on the n-dimensional unit sphere (that is, $|z_1| = 1, |z_2| = 1, \ldots, |z_n| = 1$) we obtain the n-dimensional Fourier transform of the stable sequence $x[\mathbf{k}]$.

$$X(e^{j\omega_1}, \ldots, e^{j\omega_n}) = \sum_{k_1} \cdots \sum_{k_n} x[k_1, \ldots, k_n] e^{-j(\omega_1 k_1 + \ldots + \omega_n k_n)} \tag{2.8}$$

The n-dimensional DFT of the finite length $N_1 \times N_2 \times \ldots \times N_n$ signal is a sampled version of the n-D Fourier transform

$$X[l_1, \ldots l_n] = \sum_{k_1=0}^{N_1-1} \cdots \sum_{k_n=0}^{N_n-1} x[k_1, \ldots k_n] e^{-j2\pi[(k_1 l_1/N_1) + \ldots + (k_n l_n/N_n)]} \tag{2.9}$$

Since the complex exponential is separable, the right-hand side of (2.9) can be rewritten in a much more convenient form for computation.

$$\sum_{k_1=0}^{N_1-1} e^{-j2\pi(k_1 l_1/N_1)} \sum_{k_2=0}^{N_2-1} e^{-j2\pi(k_2 l_2/N_2)} \cdots \sum_{k_n=0}^{N_n-1} x[k_1, \ldots, k_n] e^{-j2\pi(k_n l_n/N_n)} \tag{2.10}$$

The interpretation of (2.10) is that computation of the n-D DFT requires only the evaluation of (many) 1-D DFT's. The inner sum is an N_n-point DFT that must be computed for all possible combinations of $k_1, k_2 \ldots k_{n-1}$. Thus N/N_n N_n-point DFT's will be required. Likewise, N/N_i N_i-point DFT's will be required for evaluation along the ith dimension. Let $C(r)$ be a measure of the computational effort to compute an r-point (1-D) DFT. Then to compute an $N_1 \times N_2 \times \ldots \times N_n$ point (n-D) DFT requires a computational effort

$$C(N_1, N_2, \ldots, N_n) = \frac{N}{N_1} C(N_1) + \frac{N}{N_2} C(N_2) + \ldots + \frac{N}{N_n} C(N_n) \tag{2.11}$$

In the special case where all N_i are equal to N_1, $N = N_1^n$, and the computational cost reduces to

$$C(N_1, N_1, \ldots, N_1) = n N_1^{n-1} C(N_1) \tag{2.12}$$

The derivation of the FFT algorithm can be cast in a multidimensional framework. Consider the first step of a decimation-in-time derivation. In particular, suppose N factors as $L \times M$.

$$X[l_1 + Ll_2] = \sum_{k_2=0}^{M-1} e^{-j(2\pi/M) k_2 l_2} e^{-j(2\pi/N) l_1 k_2} \left(\sum_{k_1=0}^{L-1} x[Mk_1 + k_2] e^{-j(2\pi/L) k_1 l_1} \right) \tag{2.13}$$

If we define two $L \times M$ arrays $\tilde{x}[k_1, k_2]$ and $\tilde{X}[l_1, l_2]$ as

$$\tilde{x}[k_1, k_2] = x[Mk_1 + k_2] \quad \text{and} \quad \tilde{X}[l_1, l_2] = X[l_1 + Ll_2]$$

then the computation of the 1-*D* DFT can be carried out in three steps:

1. Calculate M L-point DFT's along the first dimension of $\tilde{x}[k_1, k_2]$.
2. Multiply the result of step 1 by

$$e^{-j(2\pi/N)l_1k_2},\ l_1 = 0, 1, \ldots L-1 \qquad k_2 = 0, \ldots M-1.$$

   (These are the so-called twiddle factors of the FFT).
3. Calculate L M-point DFT's along the second dimension to obtain $\tilde{X}[l_1, l_2]$.

If it were not for the twiddles of step 2, the computation of the 1-*D* DFT would be identical to the computation of a two-dimensional DFT. The computational effort of an N-point DFT, where N factors as $N_1 \times N_2 \ldots \times N_n$, is thus

$$C(N) = \sum_{i=1}^{n} \left(\frac{N}{N_i}\right) C(N_i) + (n-1)N \qquad (2.14)$$

In the case of a radix-2 FFT, (2.14) reduces to

$$C(2^n) = n2^{n-1}C(2) + (n-1)2^n$$

If the computational effort is measured in terms of complex multiplications $C(2) = 0$ and we find the $N \log_2 N$ behavior of the radix-2 FFT.

The preceding discussion assumes one particular reindexing scheme to form the arrays $\tilde{X}[l_1, l_2]$ and $\tilde{x}[k_1, k_2]$ from the one-dimensional sequences $X[k]$ and $x[l]$. It is interesting to ask whether a different indexing scheme might not eliminate step 2 and make the computation of the 1-*D* DFT exactly equivalent to the computation of the 2-*D* $L \times M$ DFT. The answer to this question is given in Section VI, where it is shown that the twiddles can be eliminated when L and M have no common factors. The necessary reindexing scheme is based on the Chinese Remainder Theorem and was first published by I. J. Good in 1958 [45].

III
basic number theory and the DFT

In this section we present some fundamental properties of the integers, related to factoring. We introduce the idea of congruence and of residue reduction and show how residue reduction may be accomplished as a simple operation. We present the idea of the greatest common divisor, and of mutually prime integers. We introduce Euler's totient function. We show how Euclid's algorithm can be used to solve certain integer equations. We introduce the Chinese Remainder Theorem and show how it is related to a curious number representation which has signal-processing applications. Finally, we show how number theory can be used to uncover a property of the DFT of a permuted sequence.

Number theory for signal processing begins with the elementary idea of division. Let a and b be any two integers with b positive. Then we can divide a by b to get a quotient and a remainder; that is, we can find numbers r and s satisfying both the equation

$$a = br + s \tag{3.1}$$

and the inequality

$$0 \leq s < b \tag{3.2}$$

where r is called the *quotient* and s is called the *remainder*. A great many computers have an instruction, DIVIDE, which solves for r and s, given a and b.†

When the remainder, s, is zero we say b divides a,

$$b \mid a \tag{3.3}$$

or that b is a divisor of a, or a factor of a. Ignoring possible negative factors, every integer has at least the divisors:

$$1 \mid a, a \mid a \tag{3.4}$$

but it may have other divisors as well. If $a > 1$ has no other divisors than 1 and itself, it is a *prime*. The smallest prime numbers are 2, 3, 5, 7, 11, 13, 17, ... The number 1 is not considered a prime, and 2 is the only even number which is prime. Other integers, those with multiple divisors, are called *composite*. The first few composite numbers are 4, 6, 8, 9, 10, ...

Suppose a is composite and b is a divisor. Then

$$a = br \tag{3.5}$$

and r is another divisor of a. There are two possibilities: Either b and r are both prime, or one or both have factors of their own. In the second case we can continue the factorization and ultimately arrive at an expression for a as a product of prime numbers only. For example,

$$24 = 2 \cdot 2 \cdot 2 \cdot 3$$
$$52 = 2 \cdot 2 \cdot 13$$
$$63 = 3 \cdot 3 \cdot 7$$

and, in general,

$$a = \prod_i p_i^{\alpha_i} \tag{3.6}$$

where p_i is the ith prime and α_i is an integer exponent. The *fundamental theorem of arithmetic* states that when any number is completely factored as in (3.6), the expression is unique.

†If a or b is negative, there may be some question as to whether DIVIDE solves (3.1) and (3.2) with differences from one computer to another.

Returning to (3.1) and (3.2), let b be fixed and called the *modulus*. Then, for infinitely many choices of a, we could obtain the same remainder, s. All these choices of a are said to be *congruent* with respect to the modulus, b. The remainder is also called the *residue modulo* b or simply the residue. Thus, if $b = 6$, $a = 16$, $a = 28$, and $a = -2$, each gives the same remainder, 4, so 16, 28, and -2 are said to be *congruent* modulo 6. Any quantity is congruent to its own residue, and we may think of all the congruent quantities as belonging to a class which we tag with the common residue. For a given modulus, b, there are exactly b such classes in all, one for each possible value of the residue, which could be $0, 1, 2, \ldots, b - 1$. For example, the days of the week can be considered as residue classes modulo 7. If October 5 were a Wednesday then the twelveth, nineteenth, and twenty-sixth days of October would belong to the residue class tagged by Wednesday.

Two integers with the same residue modulo b must differ by a multiple of b. It is common to use the symbol $\equiv$ for congruence. Thus the statement

$$a_1 \equiv a_2 \text{ modulo } b \tag{3.7}$$

is the same as

$$b \mid (a_1 - a_2) \tag{3.8}$$

Of course, if

$$a \equiv 0 \bmod b \tag{3.9}$$

then

$$b \mid a \tag{3.10}$$

The congruence relationship is much like equality, but weaker, in the sense that quantities which are equal are congruent but quantities which are congruent need not be equal. If congruent quantities can be shown to be approximately equal, however, they must be exactly equal, since their difference is known to be a multiple of the modulus.

Closely related to congruence is the idea of extracting a residue. If we are given a and asked to find the residue of a modulo b, we are asking for the value of the remainder, s, without asking for the quotient, r, in (3.1). Extracting the residue is a functional transformation of a into s. It occurs often enough in what follows that we introduce a special symbol, $\langle \cdot \rangle$

$$\langle a \rangle_b = s \tag{3.11}$$

The subscript b may be omitted if it is understood from context. In actually performing computations, we can always replace a congruence, which describes a relation among whole classes of numbers with the same residue, by an equality, involving that residue. For example, if

$$x + y \equiv z \bmod b \tag{3.12}$$

we can write instead

$$\langle x + y \rangle_b = \langle z \rangle_b \tag{3.13}$$

Computations involving residues are usually simple because we never need to work with quantities larger than the modulus unless we find it convenient to do so. Consider that

$$\begin{array}{lll} \langle x+y\rangle & \text{is the same as} & \langle\langle x\rangle+\langle y\rangle\rangle \\ \langle x-y\rangle & \text{is the same as} & \langle\langle x\rangle-\langle y\rangle\rangle \\ \langle xy\rangle & \text{is the same as} & \langle\langle x\rangle\langle y\rangle\rangle \end{array} \tag{3.14}$$

so that in any computation involving only $+$, $-$, $\times$ we may, at our option, replace the result of any step by its residue.

PROBLEM	Find the residue of 62^{50} with respect to the modulus $b = 5$.

Note an important point about the computation called for in the preceding problem. If asked to compute 62^{50}, most of us would use logarithms to obtain 4.16×10^{89}. This is an approximate answer and it locates the quantity on the real number line with a certain fuzziness. But there is no fuzziness about the residue. The answer tells us the location of 62^{50} on the real number line with pinpoint precision, except for an ambiguity of some multiple of the modulus. There are instances in which we can combine the fuzzy result of one type of computation with the pinpoint result of the other to obtain exact results in spite of the occurrence of computationally difficult forms. There are also many instances in which only the residue is important. Therefore, we shall now discuss the operation of extracting the residue.

We already know one way to compute $\langle a\rangle_b$, namely, dividing a by b and keeping the remainder. With presently available digital hardware, this can be somewhat burdensome and we can often use tricks to simplify the operation of extracting the residue. We all know the simplest case—the residue of a decimal number modulo 10 is its last decimal digit, since

$$\langle a\rangle_{10} = \langle\sum_i a_i 10^i\rangle_{10} = \langle\sum_i \langle a_i 10^i\rangle\rangle \tag{3.15}$$

and $\langle a_i 10^i\rangle_{10}$ is zero except for the $i = 0$ term. We can generalize this to any radix. If b is a power of two, and a is represented on a binary machine, we have a trivial method of extracting $\langle a\rangle_b$

$$\langle a\rangle_{2^k} = \langle\sum_i a_i\langle 2^i\rangle\rangle = \sum_{i=0}^{k-1} a_i 2^i \tag{3.16}$$

This operation is performed by masking out all but the k least significant bits.

Another simple case occurs when the modulus is of the form $2^k - 1$. Suppose

$$a = a_0 + 2a_1 + 4a_2 + \ldots + 2^k a_k + 2^{k+1}a_{k+1} + \ldots \tag{3.17}$$

Let the first k bits, $a_0, a_1, a_2 \ldots a_{k-1}$, be the binary number A; let the next k bits, a_k, $a_{k+1}, a_{k+2} \ldots a_{2k-1}$, be the binary number $B = a_k + 2a_{k+1} + 4a_{k+2} + \ldots$; etc. Then,

$$a = A + 2^k B + 2^{2k} C + \ldots \tag{3.18}$$

Since $\langle 2^k \rangle_{2^k-1} = 1$

$$\langle 2^{nk} \rangle_{2^k-1} = 1 \tag{3.19}$$

so

$$\langle a \rangle_{2^k-1} = \langle A + B + C + \ldots \rangle_{2^k-1} \tag{3.20}$$

Thus if our modulus is of the form $2^k - 1$, on a binary computer, the residue of a number with very many bits may be found by adding k-bit subwords. If the quantity a arose from an addition of numbers both less than the modulus, as might occur with residues from previous steps, the value of the carry out of the kth bit is worth 2^k and is congruent to 1. Thus if the carry is added into the least significant bit position, the result is the residue desired, except for the unusual case of a result consisting of k 1's, that is $2^k - 1$. This has a residue of 0, but since it produces no carry into the kth bit position, it goes uncorrected. But in a chain of computations, the residue reduction is generally optional except for the last step, so this need produce no difficulty.

Another case in which a residue is easily extracted without division is when the modulus is of the form $2^k + 1$. Let a be as in (3.18). Then since $2^k \equiv -1$, $2^{nk} \equiv -1^n$, and we have

$$\langle a \rangle_{2^k+1} = \langle A - B + C - \ldots \rangle_{2^k+1} \tag{3.21}$$

With this approach, there is the danger that 1 followed by k zeroes, that is, the number 2^k, will be computed as $0 - 1$, giving rise to a recurring cycle of corrections, but this case is easily provided for. Thus we have seen that for moduli of the form 2^k, $2^k \pm 1$, we can extract a residue by methods much simpler than division.

One other case should be mentioned. If our quantity is computed as a sum of residues, this sum, if divided by the modulus, can have a quotient of 0 or 1. If we subtract the modulus from the sum and the result is positive, that result is the residue. Otherwise, the quotient was zero, and we had the residue before we subtracted the modulus. Thus, for an arbitrary modulus we can extract the residue of a sum (or difference) with a simple procedure.

We shall now introduce the concept of the greatest common divisor. Consider the equation

$$6x + 4y = 2 \tag{3.22}$$

where x and y *must be integers*. Equations which require integral solutions are called *diophantine* equations. Equation (3.22) has the obvious solution $x = 1$, $y = -1$ as well as other solutions like $x = 11$, $y = -16$. On the other hand,

$$6x + 4y = 3 \tag{3.23}$$

has no integral solutions at all. The properties of equations like (3.22) and (3.23) are governed by the idea of a greatest common divisor. In (3.22) or (3.23), both the coefficient of x and the coefficient of y are divisible by 2, the largest number which is a factor of both 6 and 4. Thus the left side of either equation could be rewritten as $2(3x + 2y)$ and it must have a factor of 2, no matter what integer values are taken by x and y. Hence, there is hope for solving (3.22) but no hope for solving (3.23).

The largest number which is a divisor of both a and b is called the *greatest common divisor* (GCD) of a and b. It is customarily written (a, b). If a and b have no common divisor, that is, if $(a, b) = 1$, they are called *mutually prime*. Now let a, b, and c be given coefficients in the diophantine equation

$$ax + by = c \tag{3.24}$$

We extract the factor (a, b) from the left side, giving

$$(a, b)[a'x + b'y] = c \tag{3.25}$$

where $a' = a/(a, b)$; $b' = b/(a, b)$. Any factor of the left side must be a factor on the right side. If (a, b) does not divide c, we simply cannot solve (3.25). Therefore suppose

$$(a, b) \mid c \tag{3.26}$$

and let $c' = c/(a, b)$. We are reduced to investigating solutions of equations like

$$ax + by = c \tag{3.27}$$

where $(a, b) = 1$.
We could consider a simpler system

$$ax_1 + by_1 = 1 \tag{3.28}$$

where $(a, b) = 1$.
If we found x_1 and y_1 to solve (3.28), the solution to (3.27) would be

$$x = cx_1; \qquad y = cy_1 \tag{3.29}$$

Therefore we devote some attention to (3.28). We shall discover that we can *always* solve (3.28) and therefore we can always solve (3.27).

In what must at first seem a change of subject, we introduce the concept of a *continued fraction*. Let a and b be positive integers. We can assume $a \geq b$, otherwise the process we describe will quickly interchange their roles. We begin by dividing a by b, giving

$$\frac{a}{b} = r_0 + \frac{s_0}{b} = r_0 + \frac{1}{b/s_0}; \qquad 0 \leq s_0 < b \tag{3.30a}$$

We now divide b by s_0 giving

$$\frac{b}{s_0} = r_1 + \frac{s_1}{s_0} = r_1 + \frac{1}{s_0/s_1}; \qquad 0 \leq s_1 < s_0 \tag{3.30b}$$

We continue the process until some remainder, s_μ, is zero. Then we can write

$$\frac{a}{b} = r_0 + \cfrac{1}{r_1 + \cfrac{1}{r_2 + \cfrac{1}{r_3 + \cdots + \cfrac{1}{r_\mu}}}} \tag{3.30c}$$

Note that each remainder is, by definition, non-negative, but smaller than its predecessor. Therefore the process terminates. In carrying out the expansion we actually generated a sequence of equations:

$$\left.\begin{aligned} a &= b\,r_0 + s_0 \\ b &= s_0 r_1 + s_1 \\ s_0 &= s_1 r_2 + s_2 \\ s_1 &= s_2 r_3 + s_3 \\ &\vdots \\ s_{\mu-2} &= s_{\mu-1} r_\mu + 0. \end{aligned}\right\} \tag{3.30d}$$

The continued fraction provides us with an interesting sequence of approximations to a/b, called *convergents*.

$$\left.\begin{aligned} \frac{a_0}{b_0} &= r_0 \\ \frac{a_1}{b_1} &= r_0 + \frac{1}{r_1} \\ \frac{a_2}{b_2} &= r_0 + \cfrac{1}{r_1 + \cfrac{1}{r_2}} \\ &\vdots \\ \frac{a_\mu}{b_\mu} &= r_0 + \cfrac{1}{r_1 + \cfrac{1}{r_2 + \cfrac{\ddots}{+\cfrac{1}{r_\mu}}}} = \frac{a}{b} \end{aligned}\right\} \tag{3.31}$$

We can expand each convergent into a complicated numerator and a complicated denominator, and equate them separately to a_i and b_i—thus,

$$\left.\begin{aligned} a_0 &= r_0 & b_0 &= 1 \\ a_1 &= r_0 r_1 + 1 & b_1 &= r_1 \\ a_2 &= r_2(r_0 r_1 + 1) + r_0 & b_2 &= r_2 r_1 + 1 \\ a_3 &= r_3(r_2(r_0 r_1 + 1) + r_0) + (r_0 r_1 + 1); & b_3 &= r_3(r_2 r_1 + 1) + r_1 \end{aligned}\right\} \tag{3.32}$$

We immediately see the relationships

$$\left.\begin{aligned} a_n &= r_n a_{n-1} + a_{n-2} \\ b_n &= r_n b_{n-1} + b_{n-2} \end{aligned}\right\} \quad n \geq 2 \tag{3.33}$$

which are easily proved by induction. We can similarly prove that

$$b_n a_{n+1} - a_n b_{n+1} = (-1)^n \qquad n \geq 0 \tag{3.34}$$

Most important, the numerator and denominator of a convergent to any continued fraction are *always in lowest terms*. We can see this immediately from (3.34) because, if α is a divisor of both a_n and b_n, it must be a divisor of $(-1)^n$ which has no positive divisors except unity.

We summarize the three theorems about convergents.

THEOREM Numerator and denominator satisfy the iterations

$$\left.\begin{aligned} a_n &= r_n a_{n-1} + a_{n-2} \\ b_n &= r_n b_{n-1} + b_{n-2} \end{aligned}\right\} \quad n \geq 2$$

THEOREM

$$b_n a_{n+1} - a_n b_{n+1} = (-1)^n \qquad n \geq 0$$

THEOREM

$$(a_n, b_n) = 1$$

PROBLEM Prove these theorems by mathematical induction.

When the process terminates, we have

$$b_{\mu-1} a_\mu - a_{\mu-1} b_\mu = (-1)^{\mu-1} \tag{3.35}$$

But $a_\mu/b_\mu = a/b$. If $(a, b) = 1$ we must have

$$a_\mu = a; \qquad b_\mu = b \tag{3.36}$$

so that (3.35) has provided us with the means to solve (3.28), letting

$$y = (-1)^\mu a_{\mu-1}; \qquad x = (-1)^{\mu+1} b_{\mu-1} \tag{3.37}$$

PROBLEM Equation (3.37) gives one solution to (3.28). How could we find other solutions?

The same process allows us to find (a, b) if it is not known, since

$$\frac{a}{a_\mu} = \frac{b}{b_\mu} = (a, b) \tag{3.38}$$

It can be seen that the last nonzero remainder in (3.30d), $s_{\mu-1}$, is the greatest common divisor. When the continued fraction method is used for finding the greatest common divisor, it is known as *Euclid's algorithm*. It is perhaps the most famous of all algorithms. In Euclid's algorithm, only the remainders matter, and the process takes the form

$$\left.\begin{aligned} s_0 &= \langle a \rangle_b \\ s_1 &= \langle b \rangle_{s_0} \\ s_2 &= \langle s_0 \rangle_{s_1} \\ &\vdots \\ s_\mu &= \langle s_{\mu-2} \rangle_{s_{\mu-1}} = 0 \\ (a, b) &= s_{\mu-1} \end{aligned}\right\} \tag{3.39}$$

To summarize, the continued fraction expansion of a/b allows us to find (a, b), and if $(a, b) = 1$, it allows us to find integers, x and y, which satisfy $ax + by = 1$. We can therefore solve $ax + by = c$ if and only if $(a, b)|c$. As a corollary, we can solve

Part One

$$\langle ax \rangle_b = 1 \tag{3.40a}$$

or

$$\langle ax + by \rangle_{ab} = 1 \tag{3.40b}$$

if and only if $(a, b) = 1$. Furthermore, we will discover that $(y, a) = (x, b) = 1$ in such solutions. (Why?) Clearly, then, there is special interest in the set of integers prime to b. If we are interested in residue classes, modulo b, there is special interest in the residue classes whose residue is prime to b. It is frequently necessary to know how many of the b different residue classes have a residue prime to b, that is, how many positive integers are less than b and prime to b. This can be considered as a function of the integer variable b, and it is called Euler's *totient function* and written $\phi(b)$. We define $\phi(1) = 1$. By simple enumeration, $\phi(2) = 1$, $\phi(3) = 2$, $\phi(4) = 2$, $\phi(5) = 4$, $\phi(6) = 2$, $\phi(7) = 6, \ldots$ There happens to be a fairly easy formula for $\phi(b)$ once we have the canonical prime factorization of b. We can sneak up on this fact by considering, first, the case of b a power of a prime.

$$b = p^\alpha \tag{3.41}$$

The numbers less than b and not prime to b are just the multiples of p, every pth number. Thus there remain $\left(\frac{p-1}{p}\right)p^\alpha$ numbers less than b and prime to it. Thus,

$$\left.\begin{aligned} \phi(p) &= p - 1 \\ \phi(p^\alpha) &= p^{\alpha-1}(p-1) \end{aligned}\right\} \tag{3.42}$$

The general expression for $\phi(N)$, with $N = p_1^{\alpha_1} p_2^{\alpha_2} \ldots p_k^{\alpha_k}$ is a product

$$\phi(p_1^{\alpha_1} p_2^{\alpha_2} \ldots p_k^{\alpha_k}) = \phi(p_1^{\alpha_1})\,\phi(p_2^{\alpha_2}) \ldots \phi(p_k^{\alpha_k}) \tag{3.43}$$

This is equivalent to the following theorem:

If $(P, Q) = 1$, then $\phi(PQ) = \phi(P)\phi(Q)$. **THEOREM**

The theorem is proved by constructing and considering quantities of the form $\langle aQ + bP \rangle_{PQ}$ with a chosen from among the $\phi(P)$ choices with $(a, P) = 1$, $a < P$ and with b chosen from among the $\phi(Q)$ choices with $(b, Q) = 1$, $b < Q$. We readily see that aQ is prime to P because both a and Q are prime to P. Then no divisor of P can divide $aQ + bP$ and $aQ + bP$ must be prime to P. A similar argument shows that it is prime to Q. Therefore $aQ + bP$ is prime to PQ. Thus $\langle aQ + bP \rangle_{PQ}$ is one of the quantities which must be tabulated to enumerate $\phi(PQ)$. Since we can form $\langle aQ + bP \rangle_{PQ}$ in $\phi(P)\phi(Q)$ different ways, we prove our theorem by showing, first, that they are all distinct, and second, that there are no other quantities to enumerate.

To show that they are distinct, let

$$a_1Q + b_1P \equiv a_2Q + b_2P \mod PQ \tag{3.44}$$

Then

$$(a_1 - a_2)Q = (b_2 - b_1)P + nPQ \tag{3.45}$$

We can divide the right side by P but not the left side, unless $(a_1 - a_2)$ is a multiple of P. But both a_1 and a_2 are less than P. Since the only multiple of P less than P is zero, a_1 must equal a_2. Similarly, we show that $b_1 = b_2$. This means that the various constructions of $\langle aQ + bP \rangle_{PQ}$ are all distinct. We have therefore proved that $\phi(PQ) \geq \phi(P)\phi(Q)$. It remains to be shown that any quantity less than PQ and prime to it can be expressed as $\langle aQ + bP \rangle_{PQ}$. But let d be such a quantity, with $(d, P) = (d, Q) = 1$. We have seen that we can solve $Qx + Py = 1$ and therefore, with $a = dx$, $b = dy$, we have

$$Qa + Pb = d = \langle Qa + Pb \rangle_{PQ} \tag{3.46}$$

Then (3.46) is of the form we have been investigating and our theorem is proved.

The proof of the preceding theorem is much more important than the theorem itself. We could, after all, compute $\phi(b)$ in any case of interest by simple enumeration. In proving the theorem, however, we have displayed important principles and methods of argument; hence the reader should study the proof critically.

A more important theorem will now be presented. Because it was known in ancient China, it is called the *Chinese Remainder Theorem*. We illustrate the theorem with a puzzle of the type often found in Sunday supplements:

Puzzle

The soles of four boys' feet measure as follows:

Jim—7 inches
John—8 inches
Joe—9 inches
Bill—11 inches

To measure a certain room, each boy measures with his feet. However, in making the measurements, Jim has 2 inches left over, John has 5 inches left over, Joe has 5 inches left over, and Bill has 6 inches left over. Unfortunately none of the boys can remember how many lengths of his sole filled the room. However, Bill claims he can reproduce the measurement by the amounts left over. What is the length of the room?

The puzzle reduces to four congruences involving the unknown length of the room, L.

$$\begin{aligned} \langle L \rangle_7 &= 2 \\ \langle L \rangle_8 &= 5 \\ \langle L \rangle_9 &= 5 \\ \langle L \rangle_{11} &= 6 \end{aligned} \tag{3.47}$$

The reader may check that 149 inches (12 feet, 5 inches) is a solution. Other solutions may be obtained by adding to 149 any multiple of $7 \times 8 \times 9 \times 11$, that is, of 5544. Since a room over four hundred feet long is unlikely, the puzzle can ask for a unique answer, although the complete solution to the equations (3.47) is a residue class with respect to the modulus 5544.

The Chinese Remainder Theorem tells us that the solution to (3.47) is unique; and also tells us how to construct the solution.

The congruences $\langle x\rangle_{m_i} = r_i$ possess a unique solution among the residue classes modulo $M = \prod m_i$, if the moduli m_i are mutually prime in pairs.

First we construct the composite modulus

$$M = \prod_{i=1}^{k} m_i \tag{3.48}$$

Then we construct the quantities

$$M_1 = \frac{M}{m_1},\ M_2 = \frac{M}{m_2},\ \ldots,\ M_k = \frac{M}{m_k}$$

Each M_i is mutually prime to its corresponding m_i.

$$(M_i, m_i) = 1 \qquad i = 1, 2, \ldots, k \tag{3.49}$$

Therefore we can find numbers N_i solving

$$\langle N_i M_i\rangle_{m_i} = 1 \tag{3.50}$$

as a direct consequence of (3.40a).
Now consider the quantity

$$L = r_1N_1M_1 + r_2N_2M_2 + \ldots + r_kN_kM_k \tag{3.51}$$

Let us take the residue of L modulo m_1. We note that $M_2, M_3, \ldots M_k$ each contains m_1 as a factor, so

$$\langle L\rangle_{m_1} = \langle r_1N_1M_1\rangle_{m_1} \tag{3.52}$$

but since $\langle N_1M_1\rangle_{m_1} = 1$ we have

$$\langle L\rangle_{m_1} = r_1 \tag{3.53}$$

Similarly,

$$\langle L\rangle_{m_i} = r_i \qquad i = 1, 2, \ldots, k \tag{3.54}$$

Therefore, L is a solution to the given set of congruences. It is easy to see that if y is a solution to (3.47), $y - L$ must be a multiple of M, so the solution (3.51) is unique (as a residue class with respect to the modulus M).

Example: For the problem with the boys' shoes, we have

$$m_1 = 7, \quad m_2 = 8, \quad m_3 = 9, \quad m_4 = 11$$
$$M = 7 \cdot 8 \cdot 9 \cdot 11 = 5544$$
$$M_1 = 8 \cdot 9 \cdot 11 = 792, \quad M_2 = 7 \cdot 9 \cdot 11 = 693, \quad M_3 = 7 \cdot 8 \cdot 11 = 616$$
$$M_4 = 7 \cdot 8 \cdot 9 = 504$$

Then N_1 is the solution to $792N_1 \equiv 1 \bmod 7$. This is equivalent to the much simpler congruence $\langle 792\rangle_7 N_1 \equiv 1 \bmod 7$, giving $N_1 = 1$. N_2 is the solution to $\langle 693\rangle_8 N_2 \equiv 1 \bmod 8$; $N_2 = 5$. N_3 is the solution to $\langle 616\rangle_9 N_3 \equiv 1 \bmod 9$; $N_3 = 7$. N_4 is the solution to $\langle 504\rangle_{11} N_4 \equiv 1 \bmod 11$; $N_4 = 5$. (If the moduli m_i were larger, we could not solve for N_i by inspection. We would use the method of continued fractions presented earlier.) We therefore have that

$$L \equiv r_1(792 \times 1) + r_2(693 \times 5) + r_3(616 \times 7) + r_4(504 \times 5) \quad \bmod 5544 \tag{3.55}$$

If our four boys were to measure another room, their remainder measurements, r_1, r_2, r_3, r_4 would let them size the room using (3.55).

Although we may use the Chinese Remainder Theorem to construct solutions to numerous puzzles, its real value to us has to do with its ability to describe a one-to-one mapping from one to many dimensions, different from the type we usually use. To see it as a mapping, let a set of data $f[i]$ be given for $i = 0, 1, \ldots, M - 1$. Now define the k-dimensional array $\hat{f}[r_1, r_2, \ldots, r_k]$ from $f[i]$

$$\hat{f}[r_1, r_2, \ldots, r_k] = f[\langle r_1 N_1 M_1 + r_2 N_2 M_2 + \ldots + r_k N_k M_k \rangle_M] \tag{3.56}$$

where the digits r_i are allowed to take on values $0, 1, \ldots, m_i - 1$. The k-dimensional array $\hat{f}$ has a different number of points in each dimension. The Chinese Remainder Theorem justifies this as a one-to-one mapping. This mapping has both theoretical and practical importance, as we shall see in Section VI.

The Chinese Remainder Theorem is also the basis of a remarkable number system often proposed for high-speed arithmetic (but seldom used). Let us represent each of the integers $0, 1, 2, \ldots, M - 1$ by k digits

$$a = \boxed{a_1 \mid a_2 \mid \ldots \mid a_k} \tag{3.57}$$

where the ith digit is related to the represented number as its residue modulo m_i

$$a_i = \langle a \rangle_{m_i} \tag{3.58}$$

The Chinese Remainder Theorem tells us how to recover the value of a number from its digits. This representation of numbers is here† called the *Sino* correspondence, or Sino representation. To see the high-speed arithmetic property, consider addition. The sum is obtained from the digit representation of $c = a + b$

$$c = \boxed{c_1 \mid c_2 \mid \ldots \mid c_k}$$

where

$$c_i = \langle a_i + b_i \rangle_{m_i} \tag{3.59}$$

that is, by digit-by-digit modular addition of the numbers a and b. Equation (3.59) shows that we can add without any carries propagating from digit to digit. A similar

†In the literature on computer arithmetic [57, 119, 120] this number representation is called *modular*.

argument holds for multiplication, with

$$c_i = \langle a_i b_i \rangle_{m_i} \tag{3.60}$$

We can represent numbers as large as we want in this system by choosing more and more mutually prime moduli, m_i.

PROBLEM The Sino representation is also sometimes proposed as an error-detection scheme. Discuss how to use it to error-check an 8 bit $\times$ 8 bit multiplier?

PROBLEM Let the number a be represented in the Sino representation with $m_1 = p_1^{k_1}$, $m_2 = p_2^{k_2}$, etc. and $M = \prod_i p_i^{k_i}$. How could we recognize whether $(a, M) = 1$?

We shall now note another relationship satisfied by $\phi(N)$ that will be useful later. Suppose d is a divisor of N and suppose $f(n)$ is any function of n. By

$$\sum_{d|N} f(d)$$

we mean the sum of terms $f(d)$ taken over all divisors of N. It is clear that

$$\sum_{d|N} \phi(d) = \sum_{d|N} \phi\left(\frac{N}{d}\right) \tag{3.61}$$

since $\left(\frac{N}{d}\right)$ divides N whenever $d|N$. We can show, surprisingly, that

$$\sum_{d|N} \phi(d) = N \tag{3.62}$$

We will need this relationship when we consider primitive roots. To prove it consider the sets

$$\{x : (x, N) = d \quad \text{and} \quad 1 \le x \le N\}\dagger \tag{3.63}$$

where d is a divisor N. Each integer in the range from 1 to N will belong to one and only one such set. To determine how many elements are contained in each set in (3.63) consider the equivalent set

$$\left\{x : \left(\frac{x}{d}, \frac{N}{d}\right) = 1 \quad \text{and} \quad \frac{x}{d} \le \frac{N}{d}\right\} \tag{3.64}$$

which has exactly $\phi\left(\frac{N}{d}\right)$ elements. Summing up the number of elements in the sets, (3.64), gives N and this proves (3.62).

Example: The divisors of $N = p^2$ are $1, p, p^2$. $\phi(1) = 1$; $\phi(p) = p - 1$; $\phi(p^2) = p(p - 1)$. $\phi(1) + \phi(p) + \phi(p^2) = p^2$.

$\dagger\{x : A\}$ means the set of all x satisfying condition A.

We now turn our attention to sequences of N terms, where the form of the general term of the sequence is $\langle an\rangle_N$, where $n = 0, 1, 2, \ldots, N-1$. The following theorem shows that any such sequence is simply a permutation of the numbers $0, 1, 2, \ldots N-1$, if $(a, N) = 1$.

THEOREM For $n = 0, 1, 2, \ldots N-1$, $\langle an\rangle_N$ takes on all the N possible residues if $(a, N) = 1$.

We prove this by proving that the N members of $\{\langle an\rangle_N\}$ are different and since they all lie in the range 0 to $N-1$, they must be $0, 1, 2, \ldots, N-1$. To see that they are all different, assume two members of the sequence are the same: $\langle an_1\rangle = \langle an_2\rangle$. Then

$$N \mid a(n_1 - n_2) \tag{3.65}$$

But because $(a, N) = 1$, a cannot provide any of the factors of N, so

$$N \mid (n_1 - n_2) \tag{3.66}$$

Clearly, $0 < |n_1 - n_2| < N$, so N cannot be a divisor of $n_1 - n_2$. This proves that no two members of the sequence are the same and the theorem is proved.

As an immediate corollary, any sequence of N terms, $\{f[n]\}$, has the permutations $\{f[\langle an\rangle_N]\}$ for every a prime to N. Let us now examine the DFT of such permuted sequences. If

$$g[n] = f[\langle an\rangle_N] \tag{3.67}$$

we will show that the DFT's, $F[k]$ and $G[k]$ satisfy

$$F[k] = G[\langle ak\rangle_N] \tag{3.68}$$

We write the expression for the DFT† of $\{g[n]\}$

$$G[k] = \sum_{n=0}^{N-1} f[\langle an\rangle] W^{nk} \qquad k = 0, 1, \ldots, N-1 \tag{3.69}$$

hence, with k replaced by $\langle ak\rangle$ (this is only a renumbering of the equations),

$$G[\langle ak\rangle] = \sum_{n=0}^{N-1} f[\langle an\rangle] W^{\langle an\rangle k} \qquad k = 0, 1, \ldots, N-1 \tag{3.70}$$

Now let $\langle an\rangle$ be replaced by n, effectively reordering the terms in a summation, and we have

$$G[\langle ak\rangle] = \sum_{n=0}^{N-1} f[n] W^{nk} = F[k] \tag{3.71}$$

The fact that $W^x = W^{\langle x\rangle}$ is used here and provides the *basic* connection between number theory and the DFT.

We are more familiar with the special case $a = N-1$, relating the DFT of a time-reversed sequence to the DFT of the original sequence.

†W here is the same as W_N in (1.12), namely $W = \exp\left(-j\frac{2\pi}{N}\right)$.

PROBLEM Describe the sequence $\{\langle an \rangle\}$ when $(a, N) = d \neq 1$.

Suppose we had an algorithm which, given a sequence $\{f[n]\}$, computes one DFT point, say $F[1]$. Suppose we wish to compute $F[b]$, where $(b, N) = 1$. First we solve for a such that

$$\langle ab \rangle_N = 1 \tag{3.72}$$

Then we present as input to the algorithm the permuted sequence $\{f[\langle an \rangle]\}$. The algorithm proceeds to compute $F[b]$. For every choice of a, with $(a, N) = 1$, $a < N$, we come up with a different permutation to present to the original algorithm, and we compute a different DFT value, $F[b]$. We can compute a total of $\phi(N)$ different DFT values in this way, all those whose index is prime to N.

Suppose a second algorithm computes $F[c]$ instead of $F[1]$, where $(c, N) = c$, that is, c is some factor of N. As a ranges over the $\phi(N)$ choices prime to N, this second algorithm, with permuted input $\{f[\langle an \rangle]\}$, will compute $F[\langle bc \rangle]$ with $\langle bc \rangle = d$ ranging over all the cases $(d, N) = c$, perhaps with repetitions. This second algorithm thus allows us to compute some points not available to us with the first. How many such algorithms would we need in order to compute all the DFT points? The question is seen to be equivalent to the question "How many factors does N have?" To compute all DFT points using permuted inputs in this fashion we need one algorithm for each factor of N. If

$$N = p_1^{\alpha_1} p_2^{\alpha_2} p_3^{\alpha_3} \dots = \prod_k p_k^{\alpha_k}$$

its divisors are those numbers of the form

$$d = p_1^{\beta_1} p_2^{\beta_2} \dots = \prod_k p_k^{\beta_k} \qquad \beta_k \leq \alpha_k \tag{3.73}$$

and the number of divisors of N is

$$d(N) = (\alpha_1 + 1)(\alpha_2 + 1) \dots = \prod_k (\alpha_k + 1) \tag{3.74}$$

IV

consequences of Euler's theorem for signal processing

In this section, we consider divisibility properties of integers raised to integer powers. We present Euler's theorem and introduce the concept of primitive roots. These are used in expressions which describe permutations of data within a sequence, and allow us to recognize that, for some N, an N-point DFT is essentially a circular convolution. This development is one of the pillars on which Winograd's DFT algorithm stands. The concepts of this section will also be applied in Section V, when we introduce number theoretic transforms.

Having considered sequences $\{\langle an \rangle\}$, we now consider a second kind of sequence, $\{\langle a^n \rangle\}$. Any such sequence could be considered as the output of some finite-state machine whose transition rule is

$$x_{n+1} = \langle ax_n \rangle_N \tag{4.1}$$

The first member of the sequence is always taken to be $x_0 = 1$. Every other member must be divisible by (a, N). A finite-state machine must ultimately have a periodic output. If $(a, N) \neq 1$ (4.1), the initial conditions imply that the machine's periodic output must have an aperiodic preamble. We shall see, however, that if $(a, N) = 1$, the sequence $\{\langle a^n \rangle\}$ is periodic without a preamble. Let us now study the periodicity of the sequence $\{\langle a^n \rangle\}$ with $(a, N) = 1$. Now any member of the sequence must be prime to N.

Consider $N = 7$ and $a = 3$. The sequence $\langle 3^n \rangle_7$ is $\{1, 3, 2, 6, 4, 5, 1, 3, 2, \ldots\}$ and the period is 6. The general rule is that if $(a, N) = 1$, the period of $\langle a^n \rangle$ must be a divisor of $\phi(N)$. This remarkable fact is related to a theorem of Euler.

EULER'S THEOREM If $(a, N) = 1$, then

$$\langle a^{\phi(N)} \rangle_N = 1 \tag{4.2}$$

To prove the theorem, consider the two sequences $\{n\}$ and $\{\langle an \rangle\}$ where only the integers for which $(n, N) = 1$ are included. These two sequences are permutations of one another by an argument much like the one used in Section III on the full sequences $\{\langle n \rangle\}$ and $\{\langle an \rangle\}$. Since the two sequences have the same terms, the product of all the terms in one sequence must equal the product of all the terms in the other.

$$\langle \prod_{(n,N)=1} n \rangle = \langle \prod_{(n,N)=1} an \rangle = \langle a^{\phi(N)} \prod_{(n,N)=1} n \rangle \tag{4.3}$$

The last product is clearly prime to N and can therefore be eliminated from both sides of the equation (why?), giving

$$\langle a^{\phi(n)} \rangle = 1 \tag{4.4}$$

This theorem gives us another method of solving the equation $\langle ax \rangle_N = 1$, namely,

$$x = \langle a^{\phi(N)-1} \rangle_N \tag{4.5}$$

Returning to the sequence $\{\langle a^n \rangle\}$, let the unknown period of the sequence be d. Since $\langle a^0 \rangle = 1$, so does $\langle a^d \rangle, \langle a^{2d} \rangle, \ldots$. It is clear that $\langle a^n \rangle \neq 1$ unless n is a multiple of d, otherwise the period would be less than d. Therefore $\phi(N)$ is a multiple of the period.

$$d \mid \phi(N) \tag{4.6}$$

PROBLEM Find the residue of 62^{50} with respect to the modulus 5.

Another consequence of Euler's theorem is that

$$\langle a^n \rangle_N = \langle a^{\langle n \rangle_{\phi(N)}} \rangle_N \tag{4.7}$$

That is, the exponent may be reduced modulo $\phi(N)$, if $(a, N) = 1$, without changing the residue modulo N.

A special case of Euler's theorem for a prime modulus is

FERMAT'S THEOREM

If p is a prime

$$a^p \equiv a \bmod p \tag{4.8}$$

Having seen that the period of the sequence $\{\langle a^n \rangle_N\}$ is equal to the first integer n for which $\langle a^n \rangle_N = 1$, we define this number as the *order* of a with respect to the modulus N.

Is there a choice of the element a for which the maximum period, $\phi(N)$, is obtained? Such an element would be called a *primitive root* of N. This a fairly complicated question to explore, depending heavily on the form of N. The simplest case is when N is prime. Suppose we take as a starting point the following theorem which is not as obvious as it may sound.

THEOREM

The congruence $\sum_{n=0}^{N} a_n x^n \equiv 0 \bmod p$ has at most N incongruent solutions, if p is prime.

To see the necessity of a prime modulus consider $x^2 - 1 \equiv 0 \bmod 8$, which has the four incongruent solutions 1, 3, 5, 7. We omit the proof of this theorem and proceed directly to another theorem:

THEOREM

If $d \mid p - 1$, there are exactly $\phi(d)$ residue classes of order d modulo the prime p.

Proof: Suppose there is at least one element a of order d. Then $a^0, a^1, a^2, \ldots, \langle a^{d-1} \rangle$ are d distinct residue classes. Since each satisfies the equation $x^d - 1 \equiv 0 \bmod p$, there can be no other elements of order d by the previous theorem. Not all of these solutions are of order d; some are of lower order. But the sequence $\{\langle a^n \rangle\}$ is periodic with period d and can be put in one-to-one correspondence with the integers $\{0, 1, 2, \ldots, d - 1\}$. We see that the sequence $\{\langle (a^r)^n \rangle\}$ is a permutation of the original sequence $\{\langle a^n \rangle\}$ if $(r, d) = 1$; otherwise, it gives a shorter period because of repetitions. Thus if p is prime there are exactly $\phi(d)$ incongruent elements of order d if there are any at all. To complete the proof, we must show that there is some element of order d for each divisor, d, of $p - 1$.

Let $N = p - 1$. We know that $d \mid N$. Let the number of elements of order d be $\alpha(d)$. Then $\alpha(d) = 0$ or $\phi(d)$. The sum, over all d, of the number of elements of order d must be N because each of the N possible elements has some order.

$$\sum_{d \mid N} \alpha(d) = N \tag{4.9}$$

But recall (3.62). If any $\alpha(d)$ were 0, (4.9) would contradict (3.62). Therefore, the theorem is proved.

In particular there are precisely $\phi(p - 1)$ elements of order $p - 1$ and these are primitive roots. Thus we have shown that every prime, p, has primitive roots. If g is a primitive root of the prime p, then $\{g^0, g^1, \langle g^2\rangle, \langle g^3\rangle, \ldots\}$ generates all the incongruent residues modulo p, except 0, and we have the remarkable theorem:

THEOREM If p is prime, there exists an integer g for which $\{\langle g^n\rangle_p\}$ is a permutation of $\{1, 2, \ldots, p-1\}$.

The question of primitive roots is simple enough when N is prime. The next case to consider is $N = p^k$, the power of a prime. It can be shown that if g is a primitive root of p, then g or $(g+p)$ is a primitive root of p^k, except for the case $p = 2$.

Any number, a, suspected of being a primitive root can be quickly tested by raising it to suitable powers. We know that $a^{\phi(N)} \equiv 1$ and that if a is a primitive root we cannot have $a^d \equiv 1$ for $d < \phi(N)$. Also we need to test only those powers d which divide $\phi(N)$. Say we want a primitive root of $25 = 5^2$. We quickly find that 2 is a primitive root of 5. Since $\phi(25) = 20 = 2^2 \times 5$, we need to test only the exponents 1, 2, 4, 5, 10. Those powers of 2, mod 25 are:

$$\begin{aligned} 2^1 &= 2 \\ 2^2 &= 4 \\ 2^4 &= 16 \\ \langle 2^5\rangle &= 7 \\ \langle 2^{10}\rangle &= 24 \end{aligned}$$

Thus 2 is a primitive root of 25. If 2 were not satisfactory we would use $2 + 5 = 7$.

For completeness we mention that there are primitive roots for $N = p^k$, $N = 2p^k$ (p an odd prime) and for $N = 2$, $N = 4$, but not for other choices of N.†

Armed with an understanding of primitive roots, we can prove a remarkable theorem about the DFT.

THEOREM For p a prime, a p-point DFT is essentially a circular convolution.

Let $F[k] = \sum_{n=0}^{p-1} f[n]W^{nk}$ and let $F[0]$ be computed separately. In the other cases we separate $f[0]$ from other terms in the sum

$$F[0] = \sum_{n=0}^{p-1} f[n] \tag{4.10}$$

$$F[k] = \sum_{n=1}^{p-1} f[n]W^{nk} + f[0] \qquad (k, p) = 1 \tag{4.11}$$

The existence of a primitive root g permits us to make the change of variables in (4.11),

$$\begin{aligned} k &\longrightarrow \langle g^k\rangle_p \\ n &\longrightarrow \langle g^n\rangle_p \end{aligned} \tag{4.12}$$

The first is a renumbering of the equations and the other is the reordering of terms in a sum. Thus

$$F[\langle g^k\rangle] = \sum_{n=1}^{p-1} f[\langle g^n\rangle]W^{g^{n+k}} + f[0] \tag{4.13}$$

†This result was first proved by Gauss.

The sum in (4.13) is recognized as a $(p-1)$-point circular correlation since $\{f[\langle g^n \rangle]\}$ is a sequence on n whereas $\{W^{g^{n+k}}\}$ is a sequence on $(n+k)$. The correlation is circular because the exponents of g are taken modulo $\phi(p)$ by (4.7).

To express the DFT as a circular convolution, we define the expression $\langle g^{-n} \rangle$ as $\langle g^{\phi(p)-n} \rangle$. This is consistent with our use of the laws of exponents; $\langle g^{-n} \rangle$ is a sort of reciprocal since

$$\langle g^{-n} g^n \rangle = \langle g^{\phi(p)} \rangle = 1 \tag{4.14}$$

Now if, in place of (4.12), we made the substitutions,

$$\begin{aligned} k &\longrightarrow \langle g^k \rangle \\ n &\longrightarrow \langle g^{-n} \rangle \end{aligned} \tag{4.15}$$

we should find that $F[\langle g^k \rangle]$ is a circular convolution of length $\phi(N)$.

PROBLEM	A 7-point DFT can be written as a matrix vector product. Show the preceding rearrangements in matrix-vector notation.
PROBLEM	Show how to use a 256-point FFT program to compute a 257-point DFT.

For p a prime we have shown that a DFT is essentially a convolution. This was so because all but one of the integers $0, 1, 2, \ldots, p-1$ were prime to p and could be permuted with the relationship $k \to \langle g^k \rangle$. We could have used a similar, but less successful, argument for an N-point DFT if N were not prime but still had a primitive root. We first separate the general expression for $F[k]$ into two cases $(k, N) = 1$ and $(k, N) > 1$.

$$F[k] = \sum_{n=0}^{N-1} f[n] W^{nk} \qquad (k, N) > 1 \tag{4.16}$$

$$F[k] = \sum_{(n,N)>1} f[n] W^{nk} + \sum_{(n,N)=1} f[n] W^{nk} \qquad (k, N) = 1 \tag{4.17}$$

In (4.17) we replace k by $\langle g^k \rangle_N$ and, in the second sum, n by $\langle g^{-n} \rangle_N$, giving

$$F[\langle g^k \rangle] = \sum_{(n,N)>1} f[n] W^{ng^k} + \sum_{n=0}^{\phi(N)-1} f[\langle g^{-n} \rangle] W^{\langle g^{k-n} \rangle} \tag{4.18}$$

Equation (4.18) contains a convolution and a correction term. We regard this as less successful than for the case when p is prime, but it is still amenable to further reduction. Suppose N is a power of a prime.

$$N = p^\alpha \tag{4.19}$$

We first examine (4.16), the case of k not prime to N. This means that (4.16) is for k equal to a multiple of p. Making this restriction, we have

$$F[pk] = \sum_{n=0}^{p^\alpha - 1} f[n] W^{pnk} \tag{4.20}$$

Now we make the substitution $n \to p^{\alpha-1}m + r$, with $m = 0, 1, \ldots, p-1$ and $r = 0, 1, \ldots, p^{\alpha-1} - 1$.

$$F[pk] = \sum_{r=0}^{p^{\alpha-1}-1} \left(\sum_{m=0}^{p-1} f[p^{\alpha-1}m + r] \right) (W^p)^{rk} \tag{4.21}$$

Equation (4.21) tells us that if the p^α-point sequence $\{f[n]\}$ is summed to form a subsequence $\{h[r]\}$

$$h[r] = \sum_{m=0}^{p-1} f[p^{\alpha-1}m + r] \qquad r = 0, 1, \ldots, p^{\alpha-1} - 1 \tag{4.22}$$

then $F[pk]$ is the $p^{\alpha-1}$-point DFT of $\{h[r]\}$. We originally set out to convert a p^α-point DFT to a convolution. We were able to make part of the computation into a convolution plus an added term. We have now shown that the other part of the p^α-point DFT can be computed as a $p^{\alpha-1}$ point DFT. This in turn can be shown to have a convolution part and a DFT part. Eventually the second part will be a DFT of p points and can be converted to a convolution part plus a trivial correction term.

We now turn our attention to the expression

$$C[k] = \sum_{(n,N)>1} f[n]W^{nk} \tag{4.23}$$

which is the correction term in (4.17). We could regard (4.17) as three equations: one is (4.23); the others are

$$B[k] = \sum_{(n,N)=1} f[n]W^{nk} \qquad F[k] = B[k] + C[k]$$

We have seen that $B[k]$ can be computed as a $\phi(p^\alpha)$-point convolution. We shall see that $C[k]$ can be recognized as a DFT of $p^{\alpha-1}$ points. Note that $(n, p^\alpha) > 1$ implies n is a multiple of p. Thus,

$$C[k] = \sum_{n=0}^{p^{\alpha-1}-1} f[pn]W^{pnk} \tag{4.24}$$

But this is the DFT of the $p^{\alpha-1}$-point sequence $\{f[pn]\}$. Of course, it may be further reduced to a convolution of $\phi(p^{\alpha-1})$ points and some $p^{\alpha-2}$-point DFT's, etc.

The reduction of a DFT of p^α points to a collection of circular convolutions is complicated and cumbersome. Repeated separations, permutations of sequences, and additions of partial results to one another are required. We present a block diagram of a 9-point DFT reduced in this manner (Figure 4-1).

We have seen that not every N has a primitive root. For the important case of $N = 2^\alpha$ there is no primitive root except if $N = 2$ or 4. However, we can always find some sort of exponential representation for the set of residues prime to N. It might be of the form $\langle g_1^{n_1} g_2^{n_2} \rangle$, or more complicated still. In the case of a power of 2, the residues prime to N (that is, the odd numbers less than N) are of the form $\langle (-1)^{n_1} 3^{n_2} \rangle_N$ with $n_1 = 0, 1$; $n_2 = 0, 1, \ldots, \dfrac{\phi(N)}{2} - 1$. Confirm that this is the case for $N = 8$, 16, 32, and 64, minimizing effort by noting that the order of 3 must divide $\phi(N)$.

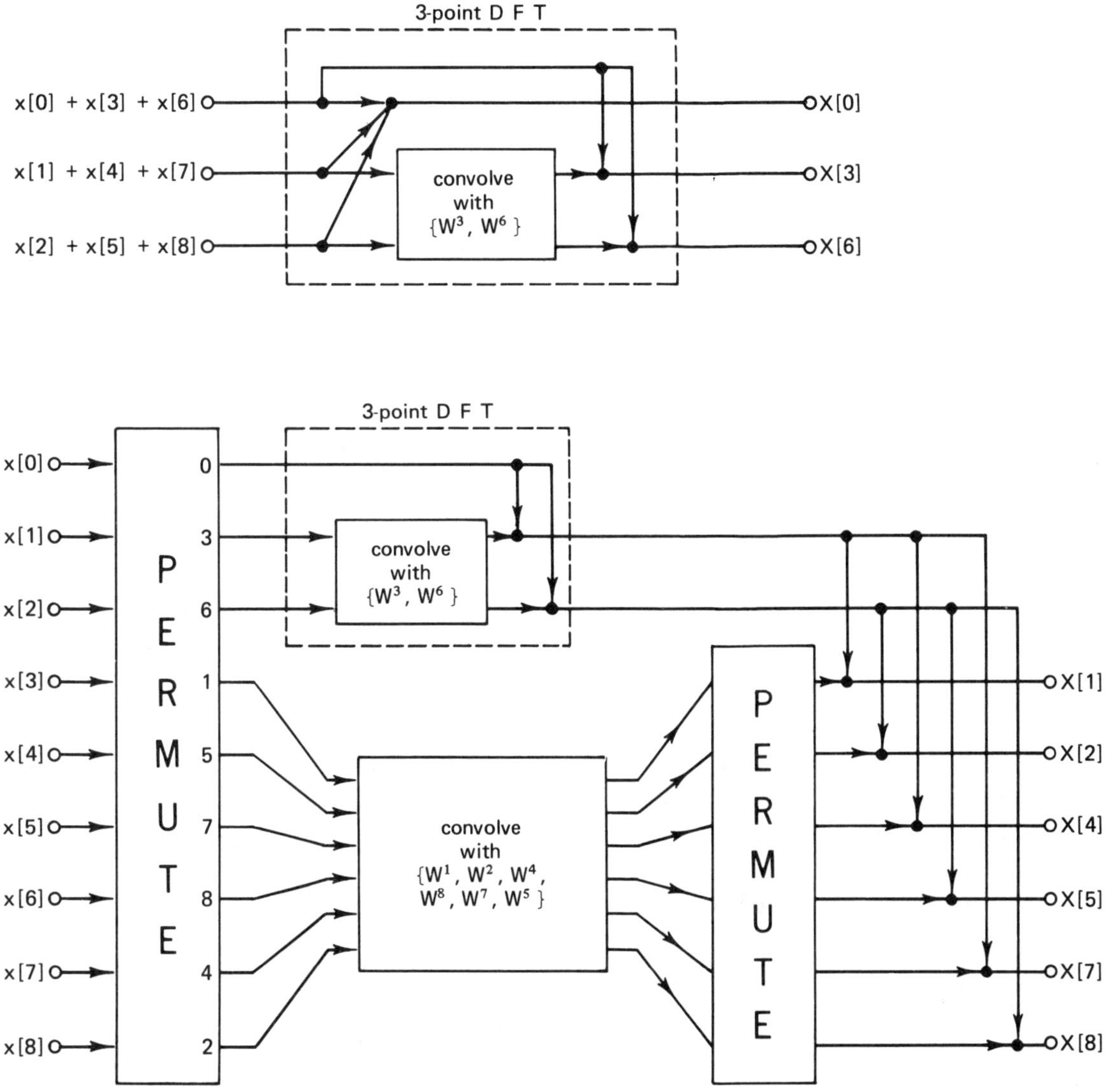

Figure 4-1 Nine-point DFT reduced to convolutions, permutations, and additive of correction terms.

If the elements prime to N possess a representation $\langle g_1^{n_1} g_2^{n_2} \rangle$, then in reducing a DFT to convolution a two-dimensional convolution will result. Let us take the case of $N = 2^\alpha$ in detail. Equation (4.16) is modified to a $2^{\alpha-1}$-point DFT as before. In (4.17) we treat the $C[k]$ term (4.23) as before and it also becomes a $2^{\alpha-1}$-point DFT. The remaining term, $B[k]$, is

$$B[k] = \sum_{(n,\,2^\alpha)=1} f[n] W^{nk} = \sum_{\text{n odd}} f[n] W^{nk} \tag{4.25}$$

We let $k \rightarrow \langle (-1)^{k_1} 3^{k_2} \rangle$, $n \rightarrow \langle (-1)^{n_1} 3^{n_2} \rangle$ giving

$$\hat{B}[k_1, k_2] = B[\langle -1^{k_1} 3^{k_2} \rangle] = \sum_{n_1=0}^{1} \sum_{n_2=0}^{2^{\alpha-2}-1} \hat{f}[n_1, n_2] W^{-1^{n_1+k_1} 3^{n_2+k_2}} \tag{4.26}$$

V

groups, rings, and fields in digital signal processing

We have given some attention to the set of integers $\{0, 1, 2, \ldots, N-1\}$ and the operations $+$ and $\times$ with respect to a modulus, N. A modern mathematical point of view sees these as special cases of a large class of more general relationships, and perhaps to save the trouble of proving analogous theorems in each case, these general relationships are considered abstractly. We shall consider three of these abstract relationships, known as *groups*, *rings*, and *fields*.

Many of the properties of the set $\{0, 1, \ldots, N-1\}$ when subject to mod N addition, can be generalized in the following way. Let $a, b, c, \ldots$ be all the legitimate elements of a set which may contain an infinite number of elements. Let $\cdot$ be an operation. In one case of interest, the elements of the set are the non-negative integers less than N, and $\cdot$ is the operation of addition followed by residue reduction modulo N. The following statements are all true for mod N addition:

1. Closure—$a \cdot b$ is a legitimate element of the set.
2. Commutative law—$a \cdot b = b \cdot a$
3. Associative law—$(a \cdot b) \cdot c = a \cdot (b \cdot c)$
4. Identity—there is some element, denoted 0, such that for any element a, $0 \cdot a = a$
5. Inverse—corresponding to every element, a, there is another element, $\tilde{a}$, such that $\tilde{a} \cdot a = 0$.

One can prove that the inverse of an element is unique. The fifth law allows us to define another operation, *subtraction*, which is the inverse of $\cdot$.

Any set together with an operation which satisfies all these properties is called a *group*†. The variety of sets and relationships which are groups is very large.

PROBLEM Let the elements of a set be all possible resistance values, including negative resistances. Let the operation $\cdot$ be series connection. Is this a group? What is the identity? Suppose the operation were connection in parallel. What would be the identity in this case?

†Strictly speaking, it is called a *commutative* (or Abelian) group. Groups which do not satisfy the second, *commutative* law are also sometimes considered. We justify the abbreviated terminology *group* because we consider only commutative groups.

If the set is finite we can describe the group operation by a table giving the result for every possible combination. Thus for {0, 1, 2, 3} and addition mod 4, we have the following addition table

·	0	1	2	3
0	0	1	2	3
1	1	2	3	0
2	2	3	0	1
3	3	0	1	2

PROBLEM What do properties 1, 2, 4, 5 say about such a table? What about property 3? The latter is a more difficult question.

Now consider a different set, the integers {1, 2, 3, 4} and the operation of multiplication modulo 5. We shall see that these also form a group. Consider the properties one by one.

1. Closure: $(a \times b)$ modulo 5 could be 0, 1, 2, 3, or 4, by definition, but if a and b are both limited to 1, 2, 3, 4, they are both prime to 5 and their product cannot be divisible by 5. Thus $(a \times b)$ modulo 5 cannot be zero, and we have closure.
2. Commutative law: $(a \times b)$ modulo 5 $= (b \times a)$ modulo 5 because ordinary multiplication commutes.
3. Associative law: See (3.14)
4. Identity: The element 1 is the identity. This may be confusing since "1" in this set plays the role assumed by "0" in the definition of a group, but this is only a change of symbol.
5. Inverse: "1" is its own inverse since $\langle 1 \times 1 \rangle_5 = 1$
   "4" is its own inverse since $\langle 4 \times 4 \rangle_5 = 1$
   "2" and "3" are inverses of one another since $\langle 2 \times 3 \rangle_5 = 1$

The operation table for this group is written in the following multiplication table:

×	1	2	3	4
1	1	2	3	4
2	2	4	1	3
3	3	1	4	2
4	4	3	2	1

A surprise comes when we permute the rows and columns of the multiplication table:

×	1	2	4	3
1	1	2	4	3
2	2	4	3	1
4	4	3	1	2
3	3	1	2	4

The surprise is that the table for the multiplication group modulo 5 has the *same structure* as the table for the addition group modulo 4. There is a one-to-one correspondence of elements between the two groups as follows:

addition group mod 4		*multiplication group mod 5*
0	$\longleftrightarrow$	1
1	$\longleftrightarrow$	2
2	$\longleftrightarrow$	4
3	$\longleftrightarrow$	3

In fact, the correspondence can be generalized to hold for any prime, p. The elements $1, 2, 3, \ldots, p-1$ are all powers of some primitive root, g, and the law of exponents modulo $p-1$, see (p. 27), tells us that

$$\langle g^i g^j \rangle_p = \langle g^{\langle i+j \rangle_{p-1}} \rangle_p$$

so that the correspondence is nothing more than

$$\begin{array}{ccc} 0 & \longleftrightarrow & \langle 2^0 \rangle_5 \\ 1 & \longleftrightarrow & \langle 2^1 \rangle_5 \\ 2 & \longleftrightarrow & \langle 2^2 \rangle_5 \\ 3 & \longleftrightarrow & \langle 2^3 \rangle_5 \end{array}$$

In summary, we have

THEOREM The complete set of residues modulo N, with mod N addition form a group.

THEOREM For any prime p, the non zero residues mod p, with mod p multiplication, form a group. This multiplicative group is *isomorphic* to the additive group modulo $\phi(p)$, that is, it has the same structure.

PROBLEM Prove that the integers less than M and relatively prime to M form a group with the operation multiplication modulo M. This group has $\phi(M)$ elements. When is this group isomorphic to the additive group modulo $\phi(M)$?

A group is called *cyclic* when all the elements of the group can be generated from one element of the group in the form $a, a \cdot a, a \cdot a \cdot a, a \cdot a \cdot a \cdot a, \ldots$. For example, the additive group mod N is always cyclic because it is generated by the integer "1." In a finite cyclic group the rows of the table for the group operation must, after suitable permutation, be cyclic rotations of one another, as the table for the additive group mod 4 illustrates. Not all groups of N elements are isomorphic to one another, but all cyclic groups of N elements are isomorphic to one another. When we showed that DFT of p points was essentially a circular convolution, we were exploiting this isomorphism.

We can take the idea of group isomorphism and the relation of DFT's to convolution somewhat further. Consider a group which is *not* cyclic, for example, the integers less than 15 and relatively prime to 15, with modulo 15 multiplication. This set is $\{1, 2, 4, 7, 8, 11, 13, 14\}$. We can construct each element in the form $\langle g_3^i g_5^j \rangle_{15}$, where $g_3 = 11$ and $g_5 = 7$.† Here i takes on only values 0 and 1 and j takes on only values 0, 1, 2, 3 because

$$\langle g_3^i \rangle_{15} = \langle g_3^{i \bmod 2} \rangle_{15} \quad \text{and} \quad \langle g_5^j \rangle_{15} = \langle g_5^{j \bmod 4} \rangle_{15}$$

We therefore have the correspondence:

$$\begin{aligned}
|0, 0| &\longleftrightarrow 1 \\
|0, 1| &\longleftrightarrow 7 \\
|0, 2| &\longleftrightarrow 4 \\
|0, 3| &\longleftrightarrow 13 \\
|1, 0| &\longleftrightarrow 11 \\
|1, 1| &\longleftrightarrow 2 \\
|1, 2| &\longleftrightarrow 14 \\
|1, 3| &\longleftrightarrow 8
\end{aligned}$$

We have constructed a group isomorphic to the original multiplicative group. The elements of the new group are pairs of integers $|i, j|$ where i is from the set of residues mod 2 and j is from the set of residues mod 4. The combining operation of the new group is addition mod 2 on the first member of each pair and addition mod 4 on the second member of each pair.

$$|i_1, j_1| \cdot |i_2, j_2| = |\langle i_1 + i_2 \rangle_2, \langle j_1 + j_2 \rangle_4|$$

The combining rule is seen to be one group operation (mod 2 addition) on one member of the pair and another group operation (mod 4 addition) on the other member of the pair. The resulting group is said to be the *product* of the two simpler groups, that is, the product of the additive group mod 2 and the additive group mod 4. In a similar manner, we can show that every group composed of the integers less than M and prime to M, with multiplication modulo M as the group operation, is isomorphic to a cyclic group or to a product of several cyclic groups.

PROBLEM In a certain sequence of 35 points, $\{f[n]\}$, all those values for which $(n, 35) \neq 1$ are identically zero. Show that part of the DFT of $\{f[n]\}$ is exactly a two-dimensional circular convolution of dimension 6 by 4.

In summary, the recognition that one group is isomorphic to another may sometimes be used to show that one type of computation is the same as another. In the case of the DFT, the data of one sequence, $\{f[n]\}$, are multiplied by the members of another sequence, $\{h[n]\} = \{W^n\}$, and the member to be multiplied by $f[n]$ in the mth sum is

†The reader may wonder how g_3 and g_5 were selected. g_3 is the sum of a primitive root of 3 and a multiple of 3 such that $g_3 \equiv 1 \bmod 5$. g_5 is a primitive root of 5 plus some multiple of 5 such that $g_5 \equiv 1 \bmod 3$.

$h[\langle mn \rangle_N]$. We recognize that we may treat the operation $\langle mn \rangle_N$ as a group operation, and the *addresses* of some $f[n]$ and $h[n]$ as elements of the group. Such a group is isomorphic to either a cyclic group or a product of cyclic groups. By relabeling the indices of $f[n]$ and $h[n]$ according to the isomorphism, we obtain a computation in which the element to be multiplied by $f[n]$ is $h[n \cdot m]$, where $\cdot$ is either mod-$\phi(N)$ addition or a multidimensional modular addition. Such a computation becomes a circular convolution in one or several dimensions.

We now consider a set of elements which can be combined by two different operations, $\cdot$ and $\blacksquare$. In the examples to follow we will often use cases where $\cdot$ is some kind of addition and $\blacksquare$ is some kind of multiplication. Other possible pairs of operations are the logical *or* and the logical *and*, or the set operations of *union* and *intersection*. Now suppose that our elements and operations satisfy the following laws:

1. Closure: $\left.\begin{matrix} a \cdot b \\ a \blacksquare b \end{matrix}\right\}$ are valid elements of the set.
2. Commutative law: $a \cdot b = b \cdot a$
   $a \blacksquare b = b \blacksquare a$
3. Associative law: $(a \cdot b) \cdot c = a \cdot (b \cdot c)$
   $(a \blacksquare b) \blacksquare c = a \blacksquare (b \blacksquare c)$
4. Identity: There is some element, denoted 0, such that $0 \cdot a = a$.
   There is some element, denoted 1, such that $1 \blacksquare a = a$.
5. Inverse: For any element, a, there is an element $\tilde{a} = -a$, such that $a \cdot \tilde{a} = 0$.
   ***Note that for only some elements, a, are there elements a^{-1}, such that $a \blacksquare a^{-1} = 1$.
6. Distributive law: $a \blacksquare (b \cdot c) = (a \blacksquare b) \cdot (a \blacksquare c)$

Apart from the distributive law, the other laws amount to saying that the elements form a group with respect to the operation $\cdot$ and come close to forming a group with respect to the operation $\blacksquare$, except that the inverse law is too weak.

PROBLEM For resistances, with parallel connections as $\cdot$ and series connections as $\blacksquare$, we do not satisfy one of these laws. Which law fails?

Elements and operations satisfying these laws form what is called a *commutative ring with identity*. We call them simply a *ring*.

Consider the set $\{0, 1, \ldots, N-1\}$ with the operations of addition mod N for $\cdot$ and multiplication mod N for $\blacksquare$. This structure forms a ring.

PROBLEM Show that the following are rings, with $\cdot$, $\blacksquare$ as $+$, $\times$.

1. The integers, positive, negative and zero
2. The rational numbers
3. Real numbers
4. Complex numbers
5. Polynomials
6. Ratios of polynomials

Most, but not all, of the familiar laws of algebra are valid for any ring. We make use of this fact when we use the concepts developed by manipulating integers to manipulate polynomials, complex numbers, residues, etc.

PROBLEM Justify each step in the following proof that for any element a, in a ring, $a \blacksquare 0 = 0$:

1. $a \blacksquare 0 = 0 \cdot a \blacksquare 0$
2. $a \blacksquare 0 = -(a \blacksquare a) \cdot (a \blacksquare a) \cdot (a \blacksquare 0)$
3. $a \blacksquare 0 = -(a \blacksquare a) \cdot (a \blacksquare (a \cdot 0))$
4. $a \blacksquare 0 = -(a \blacksquare a) \cdot (a \blacksquare a)$
5. $a \blacksquare 0 = 0$

The other familiar laws of algebra can be similarly proved for a ring, except that in a ring it is possible for $a \blacksquare b$ to be zero when neither a nor b is zero. Also, in a ring, the operation of division is usually (but not always) meaningless, because the usual proscription against dividing by zero must be extended to other divisors as well.

If we strengthen the law of inverse for $\blacksquare$ to read "for every element, a, except 0, there is some unique element, a^{-1}, such that $a \blacksquare a^{-1} = 1$", then the ring is called a *field*. In the preceding problem, four of the six rings are also fields. (Which are not? Why?) The strengthened inverse law gives meaning to division as the inverse of $\blacksquare$. In a field, $a \blacksquare b$ can be zero only if a or b is 0.

In our further discussion of rings and fields, we drop the generality of $\cdot$ and $\blacksquare$ and use the common symbols of addition and multiplication. Although there can be rings and fields in which the operations are not multiplication and addition, they do not concern us here. Multiplication and addition can, however, mean different things in different rings, as can be seen by comparing multiplication of polynomials and of complex numbers, or addition of complex numbers or ratios of polynomials.

In any ring or field, an element repeatedly combined with itself by multiplication is justifiably represented in an exponential notation, for example $aa = a^2$, $aaa = a^3$, etc. Note that while a^n is an element of the field or ring (by closure) the exponent, n, need not be an element. Now consider the equation

$$f(x) = x^5 + ax^4 + bx^3 + cx^2 + dx + e = 0 \tag{5.1}$$

where a, b, c, d, e are given elements of a certain field and x is a variable. We do not know, at this time, that $f(x) = 0$ has any solutions in the field of interest. Suppose, however, that it has the solution $x = u$. Then the factorization

$$f(x) = (x - u)(x^4 + (u + a)x^3 + (u^2 + au + b)x^2 + (u^3 + au^2 + bu + c)x + (u^4 + au^3 + bu^2 + cu + d)) \tag{5.2}$$

can be verified. The reader will recognize that (5.2) can be extended to any root of any polynomial over any field or ring. Now suppose there is another solution, $x = v$. Since it is a different solution, $v - u \neq 0$. In a ring, this buys us nothing, but in a field, we know that $(x - u)^{-1}$ exists (except for $x = u$) and we can multiply (5.1) by $(x - u)^{-1}$ to deduce that the other solution, $x = v$ must satisfy

$$x^4 + a'x^3 + b'x^2 + c'x + d' = 0 \tag{5.3}$$

where a', b', c', d' are the coefficients in (5.2). Therefore in a field, but not in a ring, we can deduce by iteration that:

THEOREM In any field, an nth degree polynomial equation can have no more than n solutions.

The reader may have thought this theorem obvious and perhaps this is so. Recall that we assumed a special case of this theorem, where the field in question was the residue classes modulo a prime, p, in the discussion leading to (4.9). That equation was then used to demonstrate the existence of a primitive root in that field. A closely related argument can be used to demonstrate that *in any finite field*† there is an element whose powers generate, in succession, all the elements of the field except 0. (Note that the definition of a primitive root for a modulus, N, was not as demanding as this—the primitive root was required to generate only the residues prime to N. A primitive root is defined more restrictively in a field—it must generate all the elements except 0. The only finite fields among the integers modulo N are when N is a prime, so the two definitions coalesce.) We shall not attempt to discuss the structure of finite fields in detail here. The reader is encouraged to pursue the subject in one of the references listed. Several important results are called out for special attention, however.

THEOREM If p is a prime, the elements $\{0, 1, 2, \ldots, p-1\}$ form a field with addition and multiplication modulo p. This field is denoted GF(p).

THEOREM In any finite field, the number of elements must be a power of a prime, p^k. This field is the Galois field GF(p^k).

THEOREM Every finite field has a primitive root, an element whose powers generate all the field elements except 0.

THEOREM Every field with N elements is isomorphic to every other field with N elements.

Let us see how a finite field of p^k elements would be constructed and what form the operations "multiplication" and "addition" would take. First, we need to list the p^k elements. Let these elements take the form of polynomials†† of degree $k-1$, with coefficients from the field GF(p), that is, coefficients restricted to $0, 1, \ldots, p-1$. For example, the elements of GF(3^2) would be the set $\{0, 1, 2, x, x+1, x+2, 2x, 2x+1, 2x+2\}$.

It is evident that there are exactly p^k field elements. The field elements could have been written just as well as $k$tuples of digits, each restricted to GF(p); the advantage of the polynomial form for field elements comes when we consider the operations of multiplication and addition.

It would be natural to define addition and multiplication in our field as ordinary polynomial addition and ordinary polynomial multiplication. But such definitions would not satisfy the field axioms because closure would be violated. Instead, we define both addition and multiplication in GF(p^k) as polynomial addition and multiplication

†A field with a finite number of elements is also called a *Galois* field.

††It is important to avoid confusion between these polynomials and expressions involving elements of the field. Thus if y is a variable field element, we might have an expression in y that contains constant terms which look superficially like functions of x.

followed by a correction process. For addition, the correction process must restore each coefficient to the set $\{0, 1, \ldots, p - 1\}$.

Rule for Addition

1. Add the elements as polynomials.
2. Replace each coefficient in the sum polynomial by its residue modulo p. Note that the polynomial 0 is the additive identity element.

PROBLEM Show that the elements of $GF(p^k)$ and the foregoing addition rule form a group.

The correction process for multiplication is more complicated because it must correct not only for the range of the coefficients but for the fact that multiplication of polynomials can produce kth and higher powers of x, which cannot be present in the elements of $GF(p^k)$. Suppose that we are given a special polynomial, $P(x)$, of degree k. We discuss the choice of $P(x)$ later, but for now let us say that it is *monic* (its leading coefficient is one) and its coefficients are integers in $GF(p)$. Then, to correct a polynomial multiplication to eliminate x^k and higher powers of x, we can divide the product polynomial by $P(x)$ and keep only the remainder.†

Rule for Multiplication

1. Multiply the elements as polynomials; call the product $Q(x)$.
2. Divide $Q(x)$ by $P(x)$ giving a remainder, $R(x)$. The degree of $R(x)$ will be less than k.
3. Replace the coefficients of $R(x)$ by their residues modulo p. Note that the polynomial 1 is the multiplicative identity element.

Example: The elements of $GF(5^2)$ are represented by $ax + b$, where a and b are allowed to be 0, 1, 2, 3, 4. The sum of $ax + b$ and $cx + d$ is an element $ex + f$. According to our addition rule, e and f are given by

$$e = \langle a + c\rangle_5; \qquad f = \langle b + d\rangle_5 \tag{5.4}$$

To compute products we need a second-degree polynomial $P(x)$. We use $P(x) = x^2 + 2$. For now, this is an arbitrary choice but its properties will be discussed. To multiply $ax + b$ by $cx + d$, giving $gx + h$, we follow the product rule, first performing the natural polynomial multiplication:

$$Q(x) = (ax + b)(cx + d) = acx^2 + (ad + bc)x + bd$$

†This is the polynomial counterpart of residue reduction and is discussed further in Section VII.

Then we divide $Q(x)$ by $P(x)$ and keep the remainder, $R(x)$:

$$\begin{array}{r|l} & ac \\ x^2+2 & acx^2 + (ad+bc)x + bd \\ & acx^2 \qquad\qquad\quad + 2ac \\ \hline & \qquad (ad+bc)x + (bd - 2ac) \Longrightarrow R(x) \end{array}$$

Finally we replace each coefficient by its residue modulo p:

$$g = \langle ad + bc \rangle_5; \qquad h = \langle bd - 2ac \rangle_5 \tag{5.5}$$

PROBLEM Show that the rules for multiplication and addition in GF(p^k) satisfy the ring axioms.

After having worked the preceding problem, we need only show that every nonzero element of GF(p^k) has a multiplicative inverse to show that we have constructed a field. The existence of the multiplicative inverse property will depend on the choice of $P(x)$. In the foregoing example, if we had used $P(x) = x^2 + 1$ we would not have constructed a field, only a ring. (Why?) Without proof we state that $P(x)$ must be an irreducible polynomial (that is, one which cannot be factored as a product of field elements) in order for the multiplication rule to satisfy the field axioms, particularly the existence of a multiplicative inverse for every element except 0.

Although we shall not prove the field properties of GF(p^k) in general, it is not difficult to prove that $P(x) = x^2 + 2$ leads to a multiplication rule satisfying field axioms in GF(5^2). The nontrivial part of the proof is the existence of an element, $cx + d$, inverse to $ax + b$, such that $(cx + d) \blacksquare (ax + b) = 1$.

Proof: An inverse to $ax + b$ exists if we can find c and d such that

$$\text{(1)} \quad \langle ad + bc \rangle_5 = 0 \tag{5.6a}$$

$$\text{(2)} \quad \langle bd - 2ac \rangle_5 = 1 \tag{5.6b}$$

We can satisfy (5.6a) by choosing

$$\text{(3)} \quad d = \langle -qb \rangle_5 \quad \text{and} \quad c = \langle qa \rangle_5 \tag{5.7}$$

for any choice of q. Then substituting these choices for c and d into (5.6b) we have

$$\text{(4)} \quad \langle q(2a^2 + b^2) \rangle_5 = 1 \tag{5.8}$$

Clearly an inverse to $ax + b$ exists in GF(5^2) if an inverse exists to $\langle 2a^2 + b^2 \rangle$ in GF(5). The possible residues of b^2 in GF(5) are 0, 1, 4, with 0 occurring only when $b = 0$. The possible residues of $2a^2$ are 0, 2, 3, with 0 occurring only when $a = 0$. Therefore, unless a and b are simultaneously 0, $\langle 2a^2 + b^2 \rangle$ cannot be 0, and therefore we can solve (5.8) for q and then compute c and d using (5.7).

PROBLEM Show that $x + 1$ is a primitive root in GF(5^2). HINT: it is only necessary to compute $(x + 1)^d$ for $d = 2, 3, 4, 6, 8, 12$, and to verify that none of these is 1.

The rule for multiplication in GF(p^k) is admittedly not very natural or familiar. Consider GF(p^2), however, and suppose that we are lucky enough that $P(x) = x^2 + 1$ is an irreducible polynomial. Then the rule for multiplication becomes

$$(ax + b)(cx + d) = \langle (bc + ad) \rangle_p x + \langle (bd - ac) \rangle_p \tag{5.9}$$

This rule is very familiar—it reminds us of the rule for multiplication of complex numbers. For this reason, GF(p^2), for certain choices of p, can be used as an analogy to the complex numbers, with x playing the role of the imaginary operator, $j = \sqrt{-1}$.

In all the fields and rings we have been considering so far, the elements of the field or ring have been either numbers or $k$tuples of numbers. The combining rules have been close enough to ordinary multiplication and addition that we could at least find special cases where they coincided with ordinary multiplication or addition. For example, if the field were the integers modulo p, and if any sum or product were less than p in ordinary arithmetic, then it would be identical in mod p arithmetic. This would even be true if we were to carry out a long chain of computations with some confidence that the result would be in the range of 0 to $p - 1$. Therefore it is possible to carry out certain computations in a finite field or ring even though the result is desired in the usual numerical domain of integers.

Now let us carefully consider carrying out computations in the familiar number system of integers by using the elements and operations of some finite computational structure (FCS), a finite ring or field. Though this may seem like a strange idea, we will give some examples in which it clearly works, as well as some in which it does not. Suppose we do computations in the ring of integers modulo 64 and we assume that the familiar integers 0, 1, . . . , 63 correspond to the same-named ring elements. We proceed according to Figure 5-1:

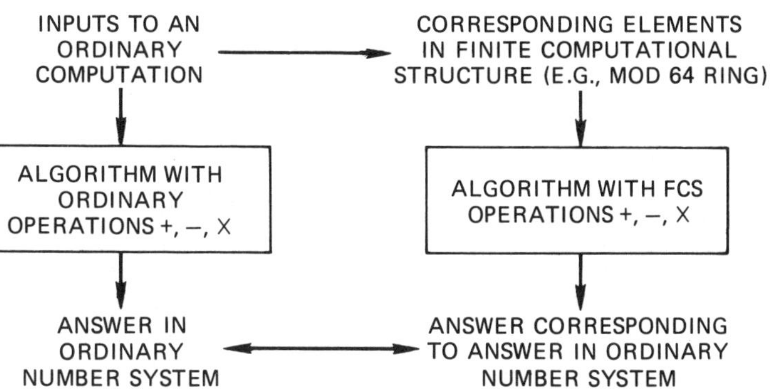

Figure 5-1 Ordinary computations carried out in a finite computational structure.

Example 1: We want to compute $9^2 - 8^2$. We compute it, instead, mod 64:

$$\langle 9^2 \rangle_{64} = 17; \qquad \langle 8^2 \rangle_{64} = 0; \qquad \langle 17 - 0 \rangle_{64} = 17$$

The answer is correct, even though both intermediate quantities were incorrect.

Example 2: We want to compute $9^2 - 3^2$. We compute it, instead, mod 64:

$$\langle 9^2 \rangle_{64} = 17; \qquad \langle 3^2 \rangle_{64} = 9; \qquad \langle 17 - 9 \rangle_{64} = 8$$

The answer, modulo 64, is incorrect since the correct answer is not in the range of numbers 0 to 63.

Example 3: We want to compute $9^2/3$. We compute it, instead, mod 64, noting that 3 has an inverse in the ring ($3 \times 43 \equiv 1 \bmod 64$).

$$\langle 9^2 \rangle_{64} = 17; \qquad \langle 3^{-1} \rangle_{64} = 43; \qquad \langle 17 \times 43 \rangle_{64} = 27$$

The answer is correct.

Example 4: We want to compute $8^2/4$. We cannot compute it mod 64 because division by 4 has no meaning in the ring.

The correspondence between ordinary numbers and FCS elements is many to one. We can exploit this to handle negative numbers.

Example: We want to compute $6 \times 15 + 9 \times (-11)$. We compute it instead modulo 64. $\langle -11 \rangle_{64}$ is 53.

$$\langle 6 \times 15 \rangle_{64} = 26; \qquad \langle 9 \times 53 \rangle_{64} = 29; \qquad \langle 26 + 29 \rangle_{64} = 55 \equiv -9 \bmod 64.$$

In some finite computational structures there are many elements with no correspondence to ordinary numbers. For example, in the field of 7^2 elements, the elements are

0	1	2	3	4	5	6
x	$x + 1$	$x + 2$	$x + 3$	$x + 4$	$x + 5$	$x + 6$
$2x$	$2x + 1$	$2x + 2$	$2x + 3$	$2x + 4$	$2x + 5$	$2x + 6$
$3x$	$3x + 1$	$3x + 2$	$3x + 3$	$3x + 4$	$3x + 5$	$3x + 6$
$4x$	$4x + 1$	$4x + 2$	$4x + 3$	$4x + 4$	$4x + 5$	$4x + 6$
$5x$	$5x + 1$	$5x + 2$	$5x + 3$	$5x + 4$	$5x + 5$	$5x + 6$
$6x$	$6x + 1$	$6x + 2$	$6x + 3$	$6x + 4$	$6x + 5$	$6x + 6$

Only the elements in the first row correspond to ordinary numbers. The other elements cannot arise as answers to a computation whose answer is known to be an ordinary number. However, they may be intermediate results.

One sort of computation we can profitably carry out in a FCS is a discrete convolution. We shall see that there is a complete analogy to the discrete Fourier transform, the fast Fourier transform, and the convolution theorem, and this permits us to compute discrete convolutions in suitable FCS's chosen so that the arithmetic is particularly simple. The first step in this chain of demonstrations is the existence of an analogy to the discrete Fourier transform. The finite computational structures we might want to examine are: (1) the ring of integers R_M with respect to modulus M; (2) the field of integers GF(p) with respect to prime modulus p—a special case of (1)—; and (3) the Galois field GF(p^k) of p^k elements. Since the demonstrations are simpler in the latter two cases,† we shall deal only with the ring. As a running example, let $M = 637 = 7^2 \times 13$.

If there were a DFT analogy in the ring R_M we would expect the transform of a set of N elements, $\{a[0], a[1], \ldots, a[N-1]\}$ to be a set of elements, $\{A[0], A[1], \ldots, A[N-1]\}$ given by the formula:

$$A[k] = \left\langle \sum_{n=0}^{N-1} a[n] w^{nk} \right\rangle_M \tag{5.10}$$

†See, for example, Pollard [6].

where w is an element of the ring chosen to have a property like that of $\exp(j2\pi/N)$ for the complex numbers, namely,

$$\begin{aligned} \langle w^N \rangle_M &= 1 \\ \langle w^n \rangle_M &\neq 1 \qquad 0 < n < N \end{aligned} \tag{5.11}$$

That is, w is an element of order N. We see at once that this prescription may be difficult to fill—there may very well be no element in the ring whose order is N (in which case, there could not be a transform of length N). First of all, we are restricted to w prime to M, and thus we must have $N|\phi(M)$ because of (4.2). In our example of $M = 637$, $\phi(M) = 7 \times 6 \times 12$ so the only conceivable transform lengths are the divisors of $7 \times 6 \times 12$. Even that is not a sufficient guarantee; for instance, there is no element of order 8 in mod-16 arithmetic.

Suppose now that we have found a suitable w and we compute the transform. We would also like to have an inverse transform of the form

$$a[m] = \left\langle N^{-1} \sum_{k=0}^{N-1} A[k]w^{-mk} \right\rangle_M \tag{5.12}$$

Even this modest equation makes some demands on ring arithmetic. We have written w^{-mk} and N^{-1}, meaning that the quantities N and w^{mk} must have inverses in the ring. With w^{mk}, we have no difficulty in demonstrating an inverse, namely w^r, where $r = \langle -mk \rangle_N$, but the requirement that N have an inverse is new. Let the factorization of M be given:

$$M = p_1^{k_1} p_2^{k_2} p_3^{k_3} \dots \tag{5.13}$$

Clearly N must contain none of the factors $p_1, p_2, p_3, \dots$. But N must be a divisor of

$$\phi(M) = (p_1 - 1)p_1^{k_1 - 1}(p_2 - 1)p_2^{k_2 - 1}(p_3 - 1)p_3^{k_3 - 1} \dots$$

Thus N must be a divisor of the product $(p_1 - 1)(p_2 - 1)(p_3 - 1) \dots$. In our example of $M = 637$, we have now shown that $N|6 \times 12$.

PROBLEM Show that there are no transform pairs of the form of (5.10) and (5.12) in mod-2^k arithmetic.

Note, however, that we have not shown that (5.12) is indeed an inverse of (5.10), but only that the operations it calls for may be carried out in the ring. To show that it is actually an inverse, we need to demonstrate that, when we substitute (5.12) into (5.10), we get an identity. This, in turn, hinges on the validity of the identity

$$(1 + w^k + w^{2k} + \dots + w^{(N-1)k}) = 0 \qquad \text{for } k \neq 0 \bmod N \tag{5.14†}$$

which is valid in certain rings for certain w and N, but not in general. If we were working in a field, even a finite field, we would multiply the right side of the equation by $(w^k - 1)$

†Compare this with (1.13).

and obtain, by cancellation of terms, that the result is zero. Since $(w^k - 1)$ is not zero because w is an element of order N, the other factor would have to be zero. In a ring this argument is valid if and only if $(w^k - 1)$ is invertible (that is, mutually prime to M) for every $k \neq 0 \bmod N$. However, we can use the Chinese Remainder Theorem to clarify the conditions for a transform in a ring. Recall the Sino representation defined in (3.57) based on the Chinese Remainder Theorem, which works nicely and naturally for modular arithmetic. For $M = p_1^{k_1} p_2^{k_2} \ldots p_\mu^{k_\mu}$ we would represent a quantity x by μ digits, the ith of which was

$$x_i = \langle x \rangle_{p_i^{k_i}} \tag{5.15}$$

For example, with $M = 637$, x would be represented by

$\langle x \rangle_{49}$	$\langle x \rangle_{13}$

Then w would be represented by μ digits

w_1	w_2	$\ldots$	w_μ

and $w^k - 1$ would be represented by digits like $\langle w_i^k - 1 \rangle_{p_i^{k_i}}$. In our example, $w^k - 1$ would be represented by

$\langle w_1^k - 1 \rangle_{49}$	$\langle w_2^k - 1 \rangle_{13}$

In the Sino number system, if the ith digit is 0 or divisible by p_i, the number is not prime to M, and conversely, if none of the digits is zero or divisible by its own prime, the number is prime to M. (Why?) Thus, if $w^k - 1$ is to be prime to M, we require that

$$\langle w_i^k - 1 \rangle_{p_i^{k_i}}$$

have no factor p_i for any digit position and for $0 < k < N$. Therefore,

$$\langle w_i^k - 1 \rangle_{p_i} \neq 0 \qquad 0 < k < N \tag{5.16}$$

or, equivalently,

$$\langle w_i^k \rangle \neq 1 \qquad 0 < k < N \tag{5.17}$$

Also $\langle w_i^N - 1 \rangle_{p_i} = 0$ since by (5.11) w is order N. But by Fermat's theorem, $w_i^{p_i - 1} \equiv 1 \bmod p_i$. The only tenable conclusion is that N must be a divisor of $p_i - 1$ for each separate prime, p_i, dividing M.

Applying this to our example of $M = 637$, we see that N must divide both 6 and 12. Therefore, the transform lengths possible are $N = 2, 3, 6$. We can also use the Sino system to construct w. We first find a primitive root g_{49} of 49, and a primitive root g_{13} of 13. From a table such as in *Handbook of Mathematical Functions* by M. Abramowitz and I. Stegun, we find

$$g_{49} = 3 \qquad g_{13} = 2$$

Now g_{49} is an element of GF(7^2) of order 42, so g_{49}^7 is an element of order 6. Similarly g_{13} is an element of GF(13) whose order is 12, so g_{13}^2 is an element of order 6. We compute

$$w_1 = \langle 3^7 \rangle_{49} = 31; \qquad w_2 = \langle 2^2 \rangle_{13} = 4$$

Therefore w is the element of R_{637} whose Sino representation is | 31 | 4 |. Note that the answer is not unique.

PROBLEM Find w in ordinary notation.

An interesting possibility is to represent both w and our data in Sino notation and use the Sino representation arithmetic described in (3.59) and (3.60) to carry out the transform. This reduces to two completely independent computations.

To avoid confusion, we emphasize that we have not described a means for computing values of the DFT of a sequence by computing the FCS analogy to a DFT. We can use the FCS analogy to the DFT for other purposes, but not as a way directly to compute a spectrum.†

Having established the existence of an invertible transform for certain data sequence lengths in certain rings, we leave it to the reader to establish the appropriate analogy to the DFT in finite fields. Once we have established the transform and, using (5.14), the inverse transform, a circular convolution theorem follows easily in any finite computational structure.

PROBLEM State and prove the circular convolution theorem in R_M.

PROBLEM What is the analogy in R_M to the fast Fourier transform?

We established the analogy between convolution computations in ordinary arithmetic and in FCS arithmetic to permit us to perform a convolution by taking the right-hand path in Figure 5-1. To perform a useful convolution, the FCS should have the following properties

1. The FCS should support a "DFT-like" transform of length N where N is appropriate to our problem.
2. N should be highly composite so that a high-speed convolution algorithm exists.
3. All possible numerical answers should be representable in the FCS.

†It is possible to calculate samples of the DFT with the FCS analogy by computing a convolution, as in the chirp z-transform or by the methods of Section IV.

4. Multiplication by any power of w (as an FCS multiplication) should be an easy operation. One of the more attractive cases is convolution carried out in R_M where M is a Fermat Number $(2^{(2^k)} + 1)$. For this case, the transform lengths are powers of two, the powers of w used in the transform are almost always powers of two, and the arithmetic in the ring is of the simple type discussed in (3.21).

Several of the papers included discuss computation of convolution using transforms in a finite field or ring. The methods are superficially attractive but less so after examination because there are fundamental conflicts hidden in the four preceding properties. In particular, property 4 places a limitation on the achievable sequence length since there are just so many elements for which multiplication is simple. Also, since the methods give *exact* answers, property 3 forces us to carry a high-precision representation of each number, since we cannot allow approximate answers. For example, to convolve a sequence of 8-bit numbers with a sequence of 12-bit numbers means to sum 20-bit products, and if any term in the convolution involves the sum of 64 such products, we might need 26 bits for the answer. This means that we must choose a FCS with at least 2^{26} elements, and therefore all operations will involve at least a 26-bit word. Since few signal-processing problems require such an accurate answer, the constraint imposed by the method seems, and is, a disadvantage.

VI
rearranging one-dimensional data into multidimensional arrays

We shall now use the number theory we have been developing to explore connections between one-dimensional data and multidimensional data. Specifically, we shall see that both circular convolution and the DFT, for one-dimensional sequences whose lengths factor appropriately, can be computed as multidimensional convolutions and DFT's of the same data, rearranged into arrays.

To set the stage, consider a sequence of thirty data points, $\{a[0], a[1]\, a[2], \ldots a[29]\}$. Let us form a two-dimensional array, $\{\hat{a}[n_1, n_2]\}$, by rearranging $\{a[n]\}$, such as

$$\begin{bmatrix} \hat{a}[0,0] & \hat{a}[0,1] & \hat{a}[0,2] & \hat{a}[0,3] & \hat{a}[0,4] & \hat{a}[0,5] \\ \hat{a}[1,0] & \hat{a}[1,1] & \hat{a}[1,2] & \hat{a}[1,3] & \hat{a}[1,4] & \hat{a}[1,5] \\ \hat{a}[2,0] & \hat{a}[2,1] & \hat{a}[2,2] & \hat{a}[2,3] & \hat{a}[2,4] & \hat{a}[2,5] \\ \hat{a}[3,0] & \hat{a}[3,1] & \hat{a}[3,2] & \hat{a}[3,3] & \hat{a}[3,4] & \hat{a}[3,5] \\ \hat{a}[4,0] & \hat{a}[4,1] & \hat{a}[4,2] & \hat{a}[4,3] & \hat{a}[4,4] & \hat{a}[4,5] \end{bmatrix} = \begin{bmatrix} a[0] & a[1] & a[2] & a[3] & a[4] & a[5] \\ a[6] & a[7] & a[8] & a[9] & a[10] & a[11] \\ a[12] & a[13] & a[14] & a[15] & a[16] & a[17] \\ a[18] & a[19] & a[20] & a[21] & a[22] & a[23] \\ a[24] & a[25] & a[26] & a[27] & a[28] & a[29] \end{bmatrix}$$

We call this a rearrangement in *lexicographic order*. The idea is to write the subscript, n, of the original sequence in radix notation, in this case $n = 6n_1 + n_2$. This gives the subscripts n_1 and n_2 of the $\hat{a}$ sequence, locating where $a[n]$ goes in the two-dimensional array.

We could similarly form a three-dimensional array, $\{\bar{a}[n_1, n_2, n_3]\}$, by lexicographic rearrangement of $\{a[n]\}$, using

$$n = 15n_1 + 5n_2 + n_3 = 3 \times 5n_1 + 5n_2 + n_3$$
$$a[15n_1 + 5n_2 + n_3] = \bar{a}[n_1, n_2, n_3]$$

We recognize lexicographic rearrangement of the data from a one-dimensional sequence into a multidimensional sequence as closely related to number representation. If we find that an index, n, can be represented by digits n_1, n_2, n_3 in some number system, we use these three digits as indices in the three-dimensional array.

Both the DFT and circular convolution, as introduced in Section II, have meaning for multidimensional sequences. We shall now show that in certain important cases one-dimensional DFT's and convolutions can be computed as multidimensional DFT's and convolutions, and vice versa. This, however, is not the case when lexicographic rearrangement is used.

Let us use the rearrangement first discussed in (3.56), which corresponds to expressing the one-dimensional index, n, in the Sino number system (3.57). To use this rearrangement, it is necessary that N, the number of data points, be factorable into μ mutually prime factors m_i, $i = 1, 2, \ldots, \mu$

$$N = \prod_{i=1}^{\mu} m_i; \qquad (m_i, m_j) = 1 \qquad \text{for } i \neq j \tag{6.1}$$

The requirement (6.1) would be impossible to fulfill if N were a power of a prime so all discussion in this section is not applicable to such important cases as N a power of two. Moreover, the number of data samples in each dimension of the rearranged data must be unequal.

Recalling (3.48) to (3.51), we construct quantities M_i

$$M_i = \frac{N}{m_i} \qquad i = 1, 2, \ldots, \mu \tag{6.2}$$

and then we construct quantities N_i by solving

$$\langle N_i M_i \rangle_{m_i} = 1 \qquad i = 1, 2, \ldots, \mu \tag{6.3}$$

Note that (4.5) allows us to express N_i as $\langle M_i^{\phi(m_i)-1} \rangle_{m_i}$. Then the representation

$$n = \boxed{n_1 \mid n_2 \mid \cdots \mid n_\mu} \tag{6.4}$$

is obtained by computing each n_i from n

$$n_i = \langle n \rangle_{m_i} \tag{6.5}$$

and n can be recovered from the collection of n_i using

$$n = \left\langle \sum_{i=1}^{\mu} n_i N_i M_i \right\rangle_N \tag{6.6}$$

Let us take our 30-point sequence $\{a[n]\}$ and form a three-dimensional rearrangement $\{\tilde{a}[n_1, n_2, n_3]\}$ with $m_1 = 5$, $m_2 = 3$, and $m_3 = 2$. From (6.2) we obtain

$$M_1 = 6 \qquad M_2 = 10 \qquad M_3 = 15$$

$$N_1 = 1 \qquad N_2 = 1 \qquad N_3 = 1$$

from (6.5)

$$n = \begin{array}{|c|c|c|}\hline \langle n\rangle_5 & \langle n\rangle_3 & \langle n\rangle_2 \\ \hline\end{array} = \begin{array}{|c|c|c|}\hline n_1 & n_2 & n_3 \\ \hline\end{array}$$

and from (6.6)

$$n = \langle 6n_1 + 10n_2 + 15n_3\rangle_{30}$$

Using these equations as data rearrangement rules, we have the mapping

$$\tilde{a}[n_1, n_2, n_3] = a[n]$$

Now suppose we have two N-point sequences, $\{a[n]\}$ and $\{b[n]\}$, and we define their circular convolution

$$c[n] = \sum_{l=0}^{N-1} a[l]b[\langle n - l\rangle_N] \tag{6.7}$$

Now let us form three-dimensional sequences $\{\tilde{a}[n_1, n_2, n_3]\}$ and $\{\tilde{b}[n_1, n_2, n_3]\}$ as above, and define their three-dimensional circular convolution

$$\tilde{c}[n_1, n_2, n_3] = \sum_{l_1=0}^{m_1-1} \sum_{l_2=0}^{m_2-1} \sum_{l_3=0}^{m_3-1} \tilde{a}[l_1, l_2, l_3]\tilde{b}[\langle n_1 - l_1\rangle_{m_1}, \langle n_2 - l_2\rangle_{m_2}, \langle n_3 - l_3\rangle_{m_3}] \tag{6.8}$$

We now make the startling observation that $\{\tilde{c}[n_1, n_2, n_3]\}$ is exactly the sequence we would obtain by the three-dimensional rearrangement of $\{c[n]\}$. To see this, write (6.7) with the indices in Sino notation, remembering that in that system of notation, mod N addition is implemented by digit-by-digit mod m_i addition

$$c\left[\begin{array}{|c|c|c|}\hline n_1 & n_2 & n_3 \\ \hline\end{array}\right] = \sum_{\begin{array}{|c|c|c|}\hline l_1 & l_2 & l_3 \\ \hline\end{array}} a\left[\begin{array}{|c|c|c|}\hline l_1 & l_2 & l_3 \\ \hline\end{array}\right] b\left[\begin{array}{|c|c|c|}\hline \langle n_1 - l_1\rangle_{m_1} & \langle n_2 - l_2\rangle_{m_2} & \langle n_3 - l_3\rangle_{m_3} \\ \hline\end{array}\right] \tag{6.9}$$

This means that if we have good algorithms for computing multidimensional convolutions we can use those algorithms to compute long one-dimensional convolutions. On the other hand, an algorithm for computing a one-dimensional convolution can be used to compute a multidimensional convolution.

Let us now see how close we can come to finding a similar relationship for the DFT. For an N point sequence, $\{a[n]\}$, we have the DFT, $\{A[k]\}$, given by the familiar equation

$$A[k] = \sum_{n=0}^{N-1} a[n]W^{nk} \tag{6.10}$$

and if we form $\{\tilde{a}[n_1, n_2, n_3]\}$ by

$$\tilde{a}[n_1, n_2, n_3] = a\left[\begin{array}{|c|c|c|}\hline n_1 & n_2 & n_3 \\ \hline\end{array}\right]$$

its multidimensional DFT, $\{\tilde{A}[k_1, k_2, k_3]\}$ would be given by

$$\tilde{A}[k_1, k_2, k_3] = \sum_{n_1=0}^{m_1-1} \sum_{n_2=0}^{m_2-1} \sum_{n_3=0}^{m_3-1} \tilde{a}[n_1, n_2, n_3](W^{M_1})^{n_1k_1}(W^{M_2})^{n_2k_2}(W^{M_3})^{n_3k_3} \qquad (6.11)$$

One hopes that the reader can see how this generalizes to μ different dimensions.

How is $\{\tilde{A}[k_1, k_2, k_3]\}$ related to $\{A[k]\}$? If we express n and k in their Sino representations (6.10) becomes

$$A\left[\begin{array}{|c|c|c|}\hline k_1 & k_2 & k_3 \\ \hline\end{array}\right] = \sum_{n_1=0}^{m_1-1} \sum_{n_2=0}^{m_2-1} \sum_{n_3=0}^{m_3-1} a\left[\begin{array}{|c|c|c|}\hline n_1 & n_2 & n_3 \\ \hline\end{array}\right] W^{\begin{array}{|c|c|c|}\hline n_1 & n_2 & n_3 \\ \hline\end{array} \times \begin{array}{|c|c|c|}\hline k_1 & k_2 & k_3 \\ \hline\end{array}} \qquad (6.12)$$

The multiplication

$$\begin{array}{|c|c|c|}\hline n_1 & n_2 & n_3 \\ \hline\end{array} \times \begin{array}{|c|c|c|}\hline k_1 & k_2 & k_3 \\ \hline\end{array}$$

called for in the exponent in (6.12) is modulo N and can therefore be done digit by digit modulo m_i. Thus the exponent becomes

$$\begin{array}{|c|c|c|}\hline \langle n_1k_1\rangle_{m_1} & \langle n_2k_2\rangle_{m_2} & \langle n_3k_3\rangle_{m_3} \\ \hline\end{array}$$

and reconstructing its value from (6.6) the exponent is

$$\sum_{i=1}^{\mu} \langle n_ik_i\rangle_{m_i} N_i M_i$$

Since this is an exponent of W, the term

$$W^{\begin{array}{|c|c|c|}\hline n_1 & n_2 & n_3 \\ \hline\end{array} \times \begin{array}{|c|c|c|}\hline k_1 & k_2 & k_3 \\ \hline\end{array}}$$

becomes

$$\prod_{i=1}^{\mu} (W^{M_i})^{N_i\langle n_ik_i\rangle_{m_i}} \qquad (6.13)$$

But (6.13) can be simplified still further because the residue reduction mod m_i in the exponent of (W^{M_i}) is superfluous. (Why?) We see, then, that

$$A\left[\begin{array}{|c|c|c|}\hline k_1 & k_2 & k_3 \\ \hline\end{array}\right] = \sum_{n_1=0}^{m_1-1} \sum_{n_2=0}^{m_2-1} \sum_{n_3=0}^{m_3-1} \tilde{a}[n_1, n_2, n_3](W^{M_1})^{n_1k_1N_1}(W^{M_2})^{n_2k_2N_2}(W^{M_3})^{n_3k_3N_3} \qquad (6.14)$$

We can now see how $\{\tilde{A}[k_1, k_2, k_3]\}$ is related to $\{A[k]\}$. Specifically let $k_1' = \langle N_1k_1\rangle_{m_1}$, let $k_2' = \langle N_2k_2\rangle_{m_2}$, and let $k_3' = \langle N_3k_3\rangle_{m_3}$.

$$k_i' = \langle N_ik_i\rangle_{m_i} \qquad (6.15)$$

Then

$$\tilde{A}[k_1', k_2', k_3'] = A\left[\begin{array}{|c|c|c|}\hline k_1 & k_2 & k_3 \\ \hline\end{array}\right] \qquad (6.16)$$

Because N_i is mutually prime to m_i (6.15) merely permutes indices and therefore the correspondence in (6.16) is one to one. The multidimensional DFT (6.11) can be computed using (6.10), and the one-dimensional DFT (6.10) can be computed using (6.11), with only the need to rearrange the inputs and outputs to make the two equations completely equivalent.

PROBLEM We have seen that a number-theoretic transform in R_{637} cannot compute convolutions with more than 6 points. Suppose we express a convolution of more than 6 points as a multidimensional convolution. Suppose also that we generate a multidimensional number theoretic DFT analogy to the foregoing. Our goal is to compute long convolutions in the arithmetic of R_{637}. Unfortunately it proves to be impossible. (Why?)

Before the FFT paper by Cooley and Tukey in 1965, I. J. Good, in 1958, showed how the equivalence of (6.10) and (6.11) could be used to compute one-dimensional DFT's. Before the advent of the FFT it took much less time to compute an m_1 by m_2 by m_3 point multidimensional DFT than to compute an $m_1m_2m_3$ point one-dimensional DFT. Good's algorithm showed that the two were essentially equivalent for m_1, m_2, and m_3 mutually prime.

VII
polynomial theory

One basic operation in linear system theory is that of convolving two signals. The direct method of calculating a convolution involves a number of multiplications proportional to the product of the lengths of the two signals, as is obvious from the convolution sum

$$y[n] = \sum_{k=-\infty}^{\infty} h[k]x[n-k] = h[n] * x[n] \tag{7.1}$$

Furthermore, it is well known that the problem of convolving $h[n]$ with $x[n]$ is the same as that of determining the coefficients of the product of two polynomials whose coefficients are given. This is, in fact, the meaning of the convolution theorem for z-transforms. That is, $Y(z) = H(z)\,X(z)$ where $Y(z)$, $X(z)$, and $H(z)$ are the z-transforms of $y[n]$, $x[n]$, and $h[n]$, respectively. Recall that $Y(z) = \sum_{n=-\infty}^{\infty} y[n]z^n$, etc.†

We are concerned in this section with the problem of computing $y[n]$ in (7.1) with

†We use z-transforms in positive powers of z to simplify our presentation with polynomial arithmetic.

an algorithm that has the least number of multiplications among all algorithms that compute $y[n]$. Our immediate goal is not to present a rigorous mathematical derivation of these minimal algorithms, but rather to explain in simple terms the basic approach employed and thus to motivate the study of elementary notions from the theory of residue classes of polynomials. The development of residue classes of polynomials will parallel the discussion of residue classes of integers presented earlier. In fact, both are specific examples of the abstract ring and field theory of Section V.

Direct computation of the convolution sum requires LM multiplications to compute the $L + M - 1$ points of the output sequence $\{y[n]\}$, assuming that the length of $\{x[n]\}$ is L points and the length of $\{h[n]\}$ is M points. The direct method is clearly not the algorithm with the minimum number of multiplications. The FFT technique of high-speed convolution is quite often more efficient. But how close is the method of high-speed convolution to the optimal algorithm with the minimum number of multiplications? More importantly, what is the minimum number of multiplications required and how does one synthesize a minimal algorithm in a systematic fashion? Before providing a partial answer to these questions let's consider an example.

Example: (3-point convolution)

$$X(z) = x[0] + x[1]z^1 + x[2]z^2$$
$$H(z) = h[0] + h[1]z^1 + h[2]z^2$$
$$Y(z) = H(z)X(z) = \sum_{n=0}^{4} y[n]z^n$$

The five output points are

$$y[0] = x[0]h[0]$$
$$y[1] = x[1]h[0] + x[0]h[1]$$
$$y[2] = x[2]h[0] + x[1]h[1] + x[0]h[2]$$
$$y[3] = x[2]h[1] + x[1]h[2]$$
$$y[4] = x[2]h[2]$$

Nine multiplications are needed by the direct algorithm. This number can be reduced to seven using the following trick to compute $y[2]$:

$$y[2] = (x[0] + x[1] + x[2]) \cdot (h[0] + h[1] + h[2]) - y[0] - y[1] - y[3] - y[4]$$

The direct method is still used to compute $y[0]$, $y[1]$, $y[3]$, and $y[4]$, and we have reduced the number of multiplications from nine to seven. Further reduction is possible by inspection but we prefer a systematic method of algorithm synthesis.

As we have seen, the problem of convolving two sequences is equivalent to the problem of multiplying two polynomials. This latter problem is easier to work with and we shall henceforth concentrate our efforts on polynomial multiplication. In the process we shall obtain results that apply to a wider class of problems than convolution. Note, however, that the actual answers to the polynomial multiplication problems are the *coefficients* of the answer polynomial.

Recall that an Nth order polynomial can be interpolated exactly through $N + 1$ points using Lagrange's interpolation formula. This suggests a method for determining $Y(z) = H(z)X(z)$ by first evaluating the constants $Y(z_0)$, $Y(z_1)$, . . . $Y(z_{L+M-2})$, and then constructing $Y(z)$ using Lagrange's identity

$$Y(z) = \sum_{i=0}^{L+M-2} Y(z_i)L_i(z) \tag{7.2}$$

where the interpolation polynomials are

$$L_i(z) = \prod_{j \neq i} \frac{(z - z_j)}{(z_i - z_j)}$$

The evaluation of $Y(z_i)$ requires one multiplication; $Y(z_i) = H(z_i) \cdot X(z_i)$. If the evaluations of $H(z_i)$ and $X(z_i)$ require no multiplications (or very few multiplications) and if the Lagrange interpolation (7.2) requires no multiplications, then we can obtain an algorithm for convolution that involves only $L + M - 1$ multiplications. In fact, the evaluation of $X(z_i)$ will require multiplications by the fixed constants z_i^n, but these multiplications are, in some sense, easier than general multiplication of two arbitrary numbers. A similar disclaimer might be made for $H(z_i)$ and for the multiplication of $Y(z_i)$ by the fixed coefficients of $L_i(z)$. Consider the application of this approach to the example of 3-point convolution

Example: Five evaluations are needed to employ the Lagrange interpolation formula. Choose the points z_i to be:

$$\begin{aligned} z_1 &= 0 \\ z_2 &= 1 \\ z_3 &= -1 \\ z_4 &= 2 \\ z_5 &= -2 \end{aligned}$$

Then

$$\begin{aligned} Y(0) &= X(0) \cdot H(0) = x[0] \cdot h[0] \\ Y(1) &= (x[0] + x[1] + x[2]) \cdot (h[0] + h[1] + h[2]) \\ Y(-1) &= (x[0] - x[1] + x[2]) \cdot (h[0] - h[1] + h[2]) \\ Y(2) &= (x[0] + 2x[1] + 4x[2]) \cdot (h[0] + 2h[1] + 4h[2]) \\ Y(-2) &= (x[0] - 2x[1] + 4x[2]) \cdot (h[0] - 2h[1] + 4h[2]) \end{aligned}$$

The interpolating polynomials $L_i(z)$ are a function of the z_i's but not of the coefficients $x[n]$ and $h[n]$.

$$\begin{aligned} L_1(z) &= \frac{(z-1)(z+1)(z-2)(z+2)}{(0-1)(0+1)(0-2)(0+2)} = \frac{1}{4}(z^4 - 5z^2 + 4) \\ L_2(z) &= \frac{(z-0)(z+1)(z-2)(z+2)}{(1-0)(1+1)(1-2)(1+2)} = -\frac{1}{6}(z^4 + z^3 - 4z^2 - 4z) \\ L_3(z) &= -\frac{1}{6}(z^4 - z^3 - 4z^2 + 4z) \\ L_4(z) &= \frac{1}{24}(z^4 + 2z^3 - z^2 - 2z) \\ L_5(z) &= \frac{1}{24}(z^4 - 2z^3 - z^2 + 2z) \end{aligned}$$

Collecting terms in like powers of z in (7.2), we obtain:

$$y[0] = Y(0)$$

$$y[1] = \frac{1}{24}\{16Y(1) - 16Y(-1) - 2Y(2) + 2Y(-2)\}$$

$$y[2] = -\frac{5}{4}Y(0) + \frac{1}{24}\{16Y(1) + 16Y(-1) - Y(2) - Y(-2)\}$$

$$y[3] = \frac{1}{24}\{-4Y(1) + 4Y(-1) + 2Y(2) - 2Y(-2)\}$$

$$y[4] = \frac{1}{4}Y(0) + \frac{1}{24}\{-4Y(1) - 4Y(-1) + Y(2) + Y(-2)\}$$

Although it may appear that the algorithm just derived has more than five multiplications, all except five of the multiplications are by fixed constants. The algorithm has been arranged so that most of these constants are powers of two. The factor of $\frac{1}{24}$ may present a problem, but can be taken care of in one of two ways. If $h[n]$ were the impulse response of a filter that was to be used to filter a long sequence of input $x[n]$, then $H(z_i)$ would be precomputed and the scale factor of $\frac{1}{24}$ could be incorporated here. Alternatively, one might be willing to drop the factor of $\frac{1}{24}$ and accept a gain of 24 in the convolution.

One multiplication by the constant 5 must be performed but this can be done using one shift and one add because $5 = 2^2 + 1$. Finally, the five-multiply algorithm is given by the flow diagram of Figure 7.1.

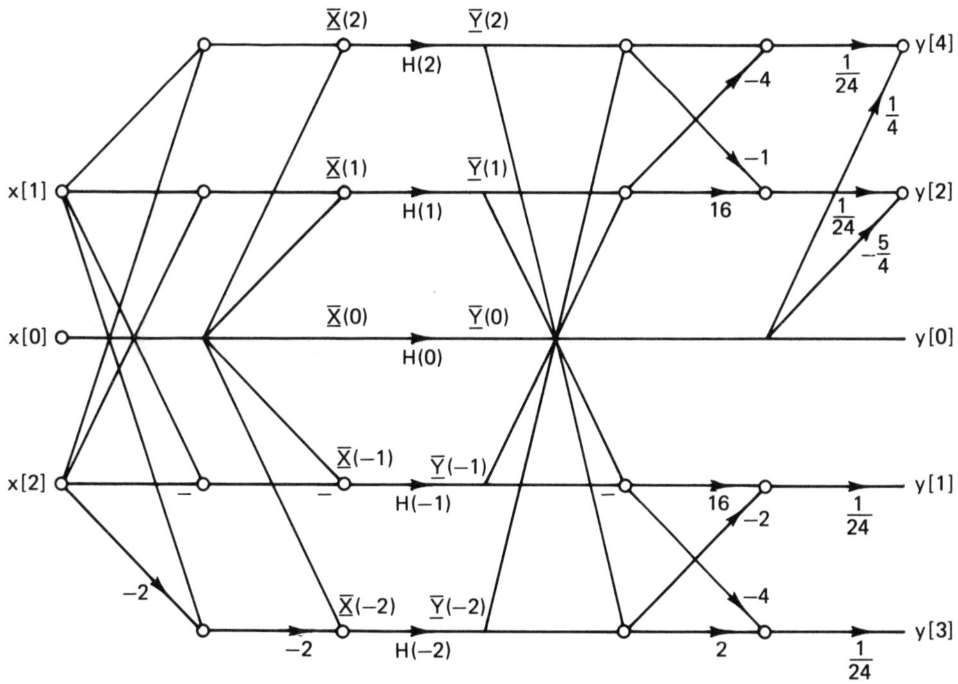

Figure 7-1 Algorithm for 3-point linear convolution using five multiplications. Synthesis is based on Toom-Cook procedure.

This general method of algorithm synthesis is usually referred to as the *Toom-Cook algorithm* [57]. The same method of derivation will be followed as we generalize to other situations, such as circular convolution. Thus far we have not answered the question of whether the Toom-Cook algorithm has the minimum number of multiplications. Before answering this question we shall introduce the concept of residue polynomials.

Residue Polynomials

Our discussion of residue polynomials will follow quite closely the earlier discussion of residue classes of integers. The objective here is to develop those results that are necessary for an understanding of Winograd's recent work on the synthesis of minimum complexity algorithms. We begin with the definition of an nth degree polynomial over a field.

Definition

An n*th degree polynomial* p(z) *over a field* F is an expression of the form

$$p(z) = \sum_{i=0}^{n} a_i z^i \qquad n \geq 0 \tag{7.3}$$

where the coefficients of the polynomial, a_i, are elements of some field F, and $a_n \neq 0$. If $a_n = 1$ the polynomial is termed *monic*. The degree of $p(z)$ is denoted by deg $[p(z)]$.

It is important to emphasize that the field of coefficients can make a difference in some of the properties of polynomials. The most obvious case in point is that $(z^2 + 1)$ can be expressed as the product of two first-degree polynomials if the field of coefficients is the set of complex numbers but not over the real or rational numbers. When speaking of several polynomials, we shall assume that all polynomials under discussion are defined over the same field unless specifically noted otherwise. The rules for combining polynomials via addition and multiplication are assumed to be familiar to the reader and we first consider the concept of divisibility.

Definition

A polynomial $d(z)$ *divides* a second polynomial $p(z)$ if there exists a polynomial $q(z)$ such that the equality $p(z) = q(z)d(z)$ holds. $d(z)$ is called a *divisor* of $p(z)$ and we write $d(z) \mid p(z)$. A polynomial $p(z)$ whose only divisors have degree equal to zero or deg $[p(z)]$ is said to be *irreducible* (over the field F). Note that irreducibility depends on the field of coefficients as the preceding discussion indicates for $z^2 + 1$.

Irreducible polynomials play an important role in a canonical factorization for polynomials. Every polynomial can be written uniquely in the form

$$p(z) = K \prod_{i=1}^{s} [p_i(z)]^{r_i} \tag{7.4}$$

where K is a constant, the $p_i(z)$ are irreducible monic polynomials, and $\sum_{i=1}^{s} r_i \deg [p_i(z)] = \deg [p(z)]$. The Fundamental Theorem of Algebra, which states that an nth degree polynomial can have no more than n roots, is equivalent to the statement that its unique factorization can contain no more than n first-degree factors.

For two polynomials $p(z)$ and $d(z)$, it is always possible to write the following equality:

$$p(z) = q(z)d(z) + r(z) \tag{7.5}$$

where $r(z)$ satisfies deg $[r(z)] <$ deg $[d(z)]$. By analogy with integer division, $r(z)$ is called the *remainder polynomial*. The representation of (7.5) is unique. Indeed, assume a second representation for $p(z)$,

$$p(z) = \hat{q}(z)d(z) + \hat{r}(z) \tag{7.6}$$

It follows that

$$r(z) - \hat{r}(z) = [\hat{q}(z) - q(z)]d(z) \tag{7.7}$$

The degree of the left side of (7.7) is less than deg $[d(z)]$ while the degree of the right side of (7.7) must be at least equal to deg $[d(z)]$ unless $q(z) - q(z)$ is identically zero. Therefore we must have $r(z) = \hat{r}(z)$ and $q(z) = \hat{q}(z)$. Note the role of the degree of polynomials in providing enough order to make division unique.

If the remainder polynomial $r(z)$ is zero, then $d(z)$ divides $p(z)$. The operation of obtaining the remainder $r(z)$ from $p(z)$ and $d(z)$ is called *polynomial residue reduction* and will be denoted in either of two ways:

$$r(z) = [p(z)] \bmod d(z)$$

or

$$r(z) = \langle p(z) \rangle_{d(z)}$$

Two polynomials $p_1(z)$ and $p_2(z)$ are said to be *congruent* modulo $d(z)$ if they yield the same result when reduced modulo $d(z)$ (that is, $\langle p_1(z) \rangle_{d(z)} = \langle p_2(z) \rangle_{d(z)}$). Equivalently, $p_1(z)$ and $p_2(z)$ are congruent modulo $d(z)$ if their difference is divisible by $d(z)$. We denote congruence by $p_1(z) \equiv p_2(z) \bmod d(z)$.

PROBLEM Find $\langle z^4 + z^3 + z^2 + 2z + 1 \rangle_{z^2 - 4z + 3}$.

Although residue reduction seems to require long division of polynomials, some shortcuts can simplify the operation. For instance, the reduction of $[p(z)] \bmod (z - a)$ yields the constant $p(a)$ because $[z^i] \bmod (z - a) = a^i$ holds for all i. In general, the reduction of $p(z)$ modulo the monic polynomial

$$d(z) = z^n + \sum_{i=0}^{n-1} d_i z^i$$

can be simplified by noting that

$$z^n \equiv -\sum_{i=0}^{n-1} d_i z^i \bmod d(z)$$

and thus z^n can be replaced with the polynomial

$$-\sum_{i=0}^{n-1} d_i z^i.$$

PROBLEM	Rework the last problem using replacements.

If $d(z)$ is not a monic polynomial, that is, if d_n is not unity, $\tilde{d}(z) = (1/d_n)d(z)$ is monic, and it is simple to see that $[p(z)] \bmod \tilde{d}(z) = [p(z)] \bmod d(z)$. This follows from a rearrangement of (7.5); $p(z) = (d_n q(z))\tilde{d}(z) + r(z)$. In other words, residue reduction and congruence modulo $d(z)$ are not affected by scalar multiples of $d(z)$. Therefore, without loss of generality we shall always take $d(z)$ to be monic in the following discussions.

We can now state the idea of the Toom-Cook algorithm in the following more general form:

$$\text{(1) Compute } Y_i(z) \equiv Y(z) \bmod P_i(z) \qquad i = 0(1)L + M - 2 \tag{7.8}$$

using

$$Y_i(z) = \{X(z) \bmod P_i(z)\} \cdot \{H(z) \bmod P_i(z)\} \bmod P_i(z)$$

(2) Construct the unique $Y(z)$ satisfying the congruences of (7.8), and the condition $\deg[Y(z)] \leq L + M - 2$.

In the Toom-Cook algorithm, $P_i(z) = z - z_i$, and the Lagrange interpolation formula provides one method of constructing $Y(z)$. In general, we need to prove a polynomial version of the Chinese Remainder Theorem in order to give the conditions on $P_i(z)$ such that (7.8) has a unique solution.

Chinese Remainder Theorem for Polynomials

There exists a unique polynomial $Y(z)$ satisfying

$$\text{(1) } Y(z) \equiv Y_i(z) \bmod P_i(z) \qquad i = 1(1)\nu$$

$$\text{(2) } 0 \leq \deg[Y(z)] < \sum_{i=1}^{\nu} \deg[P_i(z)]$$

if the monic polynomials $P_i(z)$ have no common factors (when taken in pairs).

Proof: The method of proof we shall employ is constructive in that it gives an explicit procedure for computing $Y(z)$. The problem is first reduced to a simpler problem that involves finding ν auxiliary polynomials $R_i(z)$, each satisfying

$$R_i(z) \equiv \delta_{ij} \bmod P_j(z) \qquad j = 1(1)\nu \tag{7.9}$$

The solution $Y(z)$ can then be given by

$$Y(z) = \sum_{i=1}^{\nu} R_i(z) Y_i(z) \bmod P(z) \tag{7.10}$$

where $P(z) = \prod_{i=1}^{\nu} P_i(z)$.

The solution of (7.9) must be of the form

$$R_i(z) = S_i(z) \prod_{\substack{j=1 \\ j \neq i}}^{\nu} P_j(z) \tag{7.11}$$

where $S_i(z)$ is determined from the single congruence

$$S_i(z) \prod_{\substack{j=1 \\ j \neq i}}^{\nu} P_j(z) \equiv 1 \bmod P_i(z) \tag{7.12}$$

The other $(\nu - 1)$ congruences in (7.9) are automatically satisfied by the form of $R_i(z)$ taken in (7.11). For notational convenience we let

$$Q_i(z) = \prod_{\substack{j=1 \\ j \neq i}}^{\nu} P_j(z)$$

Equation (7.12) can be written as the equivalent problem of finding two polynomials $S_i(z)$ and $T_i(z)$ satisfying

$$S_i(z)Q_i(z) + T_i(z)P_i(z) = 1 \tag{7.13}$$

This problem is the polynomial equivalent of the diophantine equation solved by Euclid's algorithm. A polynomial version of Euclid's algorithm would seem to be in order, and in fact, we can construct the solutions to (7.13) under the condition that the monic polynomials $Q_i(z)$ and $P_i(z)$ have no common factors. The derivation parallels our earlier discussion for integers given in Section III. We start with the continued fraction expansion of $P(z)/Q(z)$. Note that the subscript i will be suppressed in this discussion; also we assume that $\deg [P(z)] \geq \deg [Q(z)]$.

$$\frac{P(z)}{Q(z)} = C_0(z) + \cfrac{1}{C_1(z) + \cfrac{1}{C_2(z) + \cfrac{1}{\ddots + \cfrac{1}{C_\mu(z)}}}} \tag{7.14}$$

where the $C_i(z)$ are determined successively from the following long-division operations

$$\begin{aligned}
\frac{P(z)}{Q(z)} &= C_0(z) + \frac{r_0(z)}{Q(z)}, && \deg [r_0(z)] < \deg [Q(z)] \\
\frac{Q(z)}{r_0(z)} &= C_1(z) + \frac{r_1(z)}{r_0(z)}, && \deg [r_1(z)] < \deg [r_0(z)] \\
\frac{r_0(z)}{r_1(z)} &= C_2(z) + \frac{r_2(z)}{r_1(z)}, && \deg [r_2(z)] < \deg [r_1(z)] \\
&\vdots \\
\frac{r_{\mu-2}(z)}{r_{\mu-1}(z)} &= C_\mu(z) + 0
\end{aligned} \tag{7.15}$$

The polynomials $r_0(z)$, $r_1(z)$, . . . $r_\mu(z)$ are the result of the following sequences of residue

Part One reductions and form a sequence of polynomials for which the degree is strictly decreasing.

$$\begin{aligned} r_0(z) &= \langle P(z) \rangle_{\text{mod } Q(z)} \\ r_1(z) &= \langle Q(z) \rangle_{\text{mod } r_0(z)} \\ r_2(z) &= \langle r_0(z) \rangle_{\text{mod } r_1(z)} \\ &\vdots \\ r_\mu(z) &= 0 \end{aligned} \tag{7.16}$$

PROBLEM Express $\dfrac{z^3 + z^2 + z - 1}{z^2 - z - 1}$ as a continued fraction.

If we consider the approximation to $P(z)/Q(z)$ obtained by truncating the continued fraction expansion after a finite number of terms, we obtain the polynomial convergents

$$\begin{aligned} \frac{\psi_0(z)}{\theta_0(z)} &= C_0(z) \\ \frac{\psi_1(z)}{\theta_1(z)} &= C_0(z) + \frac{1}{C_1(z)} \\ \frac{\psi_2(z)}{\theta_2(z)} &= C_0(z) + \cfrac{1}{C_1(z) + \cfrac{1}{C_2(z)}} \\ &\vdots \\ \frac{\psi_\mu(z)}{\theta_\mu(z)} &= \frac{P(z)}{Q(z)} \end{aligned} \tag{7.17}$$

The numerators $\psi_i(z)$ and denominators $\theta_i(z)$ can be generated recursively from

$$\psi_i(z) = C_i(z)\psi_{i-1}(z) + \psi_{i-2}(z) \tag{7.18a}$$

$$\theta_i(z) = C_i(z)\theta_{i-1}(z) + \theta_{i-2}(z) \qquad i \geq 2 \tag{7.18b}$$

One can also verify the following identity

$$\theta_i(z)\psi_{i-1}(z) - \psi_i(z)\theta_{i-1}(z) = (-1)^i \qquad i = 1(1)\mu \tag{7.19}$$

Using the special case when $i = \mu$

$$\theta_\mu(z)\psi_{\mu-1}(z) - \psi_\mu(z)\theta_{\mu-1}(z) = (-1)^\mu \tag{7.20}$$

we conclude that $\psi_\mu(z)$ and $\theta_\mu(z)$ have no common factors (of degree greater than zero). Since the same is true of $P(z)$ and $Q(z)$ we obtain from (7.17)

$$\begin{aligned} \psi_\mu(z) &= KP(z) \\ \theta_\mu(z) &= KQ(z) \end{aligned} \tag{7.21}$$

The constant K is obviously the leading coefficient of $\psi_\mu(z)$ and $\theta_\mu(z)$ since $P(z)$ and $Q(z)$ are monic. Plugging (7.21) into (7.20) and regrouping terms we obtain

$$Q(z)\,[(-1)^\mu K\psi_{\mu-1}(z)] - P(z)[(-1)^\mu K\theta_{\mu-1}(z)] = 1 \qquad (7.22)$$

and thus identify the solution for (7.13) as

$$\begin{aligned} S(z) &= (-1)^\mu K\psi_{\mu-1}(z) \\ T(z) &= (-1)^\mu K\theta_{\mu-1}(z) \end{aligned} \qquad (7.23)$$

Recall that subscripts i in (7.13) were suppressed for clarity. Thus our proof of the Chinese Remainder Theorem is complete.

Example: (applicable to 3-point circular convolution)

Solve

$$R(z) \equiv \begin{cases} 0 \bmod (z-1) \\ 1 \bmod (z^2+z+1) \end{cases}$$

From above we must solve

$$S(z)(z-1) - (z^2+z+1)T(z) = 1$$

for $S(z)$ and $T(z)$. We thus take the continued fraction expansion

$$\frac{z^2+z+1}{z-1} = (z+2) + \frac{1}{\frac{1}{3}(z-1)}$$

Hence, $\mu = 1$ and $C_0(z) = z+2$ and $C_1(z) = \frac{1}{3}(z-1)$. The approximants are

$$\frac{\psi_0(z)}{\theta_0(z)} = z+2$$
$$\frac{\psi_1(z)}{\theta_1(z)} = \frac{\frac{1}{3}(z^2+z+1)}{\frac{1}{3}(z-1)}$$

From (7.20) we obtain

$$\tfrac{1}{3}(z-1)(z+2) - \tfrac{1}{3}(z^2+z+1) = -1$$

and thus $S(z) = -\frac{1}{3}(z+2)$ and $T(z) = -\frac{1}{3}$

Finally, we obtain the solution for $R(z)$ from (7.11)

$$R(z) = \frac{-(z-1)(z+2)}{3} = -\frac{1}{3}(z^2+z-2)$$

We now return to the general problem of computing

$$Y(z) = \{H(z)X(z)\} \bmod P(z) \qquad (7.24)$$

in the minimum number of multiplications. Two special cases of (7.24) are explored in some depth in the following: linear convolution and circular convolution. *Linear convolution* results when the degree of $P(z)$ is chosen to be greater than deg $[X(z)] +$ deg $[H(z)]$. In this case the residue reduction will have no effect and $Y(z) = X(z)H(z)$. One possible choice for $P(z)$ is

$$P(z) = \prod_{i=0}^{N} (z - z_i)$$

where the z_i are arbitrary but distinct constants from some field (usually the field of coefficients) and

$$N = \deg [X(z)] + \deg [H(z)].$$

The resulting algorithm will be identical to the method of Lagrange interpolation described in (7.2).

On the other hand, N-*point circular convolution results when* P(z) *is chosen to be* $z^N - 1$. The residue reduction mod $z^N - 1$ of $Y(z)$ will effectively make all sequences periodic with period N. Thus, if two $(N - 1)$st degree polynomials, $X(z)$ and $H(z)$, are multiplied, the result is a polynomial $Y(z)$ of degree $2N - 2$. The mod $z^N - 1$ residue reduction will cause the coefficients of z^i and z^{i+N} to be added together to get the final result. For example, the constant coefficient of $Y(z)$ mod $(z^N - 1)$ will be $x[0]h[0] + (x[N-1]h[1] + x[N-2]h[2] + \ldots + x[1]h[N-1])$, which matches the result of circular convolution.

The problem posed in (7.24) can be simplified somewhat if the polynomial $P(z)$ can be factored as

$$P(z) = \prod_{i=1}^{k} P_i(z)$$

The simplification results from noting that we can first calculate $Y_i(z) = \langle X(z)H(z)\rangle_{P_i(z)}$ and then use the Chinese Remainder Theorem to construct $Y(z)$ from the $Y_i(z)$, if the $P_i(z)$ have no common factors. Since the factorization of a polynomial depends on the field from which the coefficients of the polynomial factors are taken, it follows that the "minimal" algorithms will depend upon the field over which $P(z)$ is factored. For example, $P(z) = z^5 - 1$ factors as $\prod_{k=0}^{4} (z - W_5^k)$ over the complex numbers; as $(z - 1)\left(z^2 + 2\cos\frac{2\pi}{5}z + 1\right)\left(z^2 + 2\cos\frac{4\pi}{5}z + 1\right)$ over the real numbers; and as $(z - 1)(z^4 + z^3 + z^2 + z + 1)$ over the rationals. Thus over the complex numbers the problem of (7.24) would split into five first-order problems; over the reals there would be three smaller problems; and over the rationals two problems. For the time being, the choice of the field of constants will be left unspecified. After developing the structure of the minimal algorithms we shall return to a discussion of the impact of the field of constants on the minimal algorithms.

If we assume that

$$P(z) = \prod_{i=1}^{k} P_i(z)$$

then (7.24) is broken into k subproblems

$$X(z)H(z) \bmod P_i(z) \qquad i = 1(1)k \tag{7.25}$$

where $P_i(z)$ is a monic irreducible polynomial. The evaluation of (7.25) can certainly be accomplished in three steps by first reducing $X(z)$ and $H(z)$ modulo $P_i(z)$ to obtain $X_i(z)$ and $H_i(z)$, then calculating $X_i(z)H_i(z)$, and finally performing another residue reduction modulo $P_i(z)$. The calculation of $X_i(z)H_i(z)$ can be done using the Toom-Cook algorithm as discussed earlier. Since the degree of the polynomials $X_i(z)$ and $H_i(z)$ is, in general, $\deg[P_i(z)] - 1$, the Toom-Cook algorithm will involve $2\deg[P_i(z)] - 1$ multiplications in which both the multiplier and the multiplicand depend on the coefficients of the polynomials $X_i(z)$ and $H_i(z)$. All other multiplications in the Toom-Cook algorithm are multiplications by fixed constants. These constants depend on the particular choices that are made for the z_i in the Lagrange interpolation formula of (7.2). If the z_i are chosen to be rational numbers, then all the constant multipliers will be rational. In an actual computation these constant multipliers must be counted as actual multiplications, but the theoretical development of minimal algorithms cannot account for such multiplications in a systematic fashion and hence they will NOT be counted.

The operation of residue reduction modulo $P_i(z)$ will require additions, subtractions, and multiplications by fixed constants that belong to the field that contains the coefficients of $P_i(z)$. As before these multiplications are not counted under the theory developed here. Thus each subproblem (7.25) can be computed with $2\deg[P_i(z)] - 1$ multiplications. In order to finish the computation of $Y(z)$ in (7.24) we need only to observe that (7.25) gives $Y_i(z) \equiv Y(z) \bmod P_i(z)$; hence $Y(z)$ can be reconstructed using the Chinese Remainder Theorem. The form for $Y(z)$ is

$$Y(z) = \left[\sum_{i=1}^{k} S_i(z)Y_i(z)\right] \bmod P(z) \tag{7.26}$$

From the construction of the $S_i(z)$ just given, it is easy to see that the coefficients of each polynomial $S_i(z)$ lie in the same field as the coefficients of the $P_i(z)$. Thus the implementation of (7.26) requires only additions, subtractions, and multiplications from the field of constants for the coefficients of $P_i(z)$.

Putting all this information together shows that an algorithm exists for the computation of $X(z)H(z) \bmod P(z)$ in which the number of multiplications is

$$\sum_{i=1}^{k} \{2\deg[P_i(z)] - 1\} = 2\deg[P(z)] - k \tag{7.27}$$

The number of multiplies in (7.27) does not include multiplications by constants. We have in fact proved half of the following important theorem due to Winograd.

THEOREM (WINOGRAD)

The minimum number of multiplications required to compute the expression $X(z)H(z) \bmod P(z)$ is

$$\boxed{2N - k}$$

where $N = \deg[P(z)]$ and k is the number of irreducible factors $P_i(z)$ of the polynomial $P(z)$ over the field F. Multiplications by *fixed* elements of F are *not* counted in this theorem.

We have demonstrated how to construct a "minimal" algorithm with $2N - k$ multiplications. To show that at least $2N - k$ multiplies are necessary and that the minimal algorithm must use the Chinese Remainder Theorem decomposition is much more difficult. The paper by Winograd [13] includes the proof of these facts.

A further component of Winograd's complexity theorem is that the structure of the minimal algorithms can be presented in a canonical form. If we represent the coefficients of the polynomials $X(z)$, $H(z)$, and $Y(z)$ as N-dimensional vectors $\mathbf{x} = [x_0\, x_1 \ldots x_{N-1}]^T$, $\mathbf{h} = [h_0\, h_1 \ldots h_{N-1}]^T$, and $\mathbf{y} = [y_0\, y_1 \ldots y_{N-1}]^T$, then the algorithm can be written in matrix form

$$\mathbf{y} = C(B\mathbf{h} \otimes A\mathbf{x}) \tag{7.28}$$

where A, B, and C are rectangular matrix transformations whose entries are fixed constants in the field F. The matrices A and B are of dimension $(2N - k) \times N$ and the C matrix is $N \times (2N - k)$. All the multiplications in this structure are contained in the pointwise multiplication $A\mathbf{x} \otimes B\mathbf{h}$. The matrix transformations may require multiplications but these will be by fixed constants from the field F. The A and B transformations implement the residue reductions of $X(z)$ and $H(z)$ modulo $P_i(z)$ and the preconditioning steps of the Toom-Cook algorithm. The transformation C implements the reconstruction phase of the Toom-Cook algorithm as well as the reconstruction steps of the Chinese Remainder Theorem decomposition. Figure 7-2 shows a block

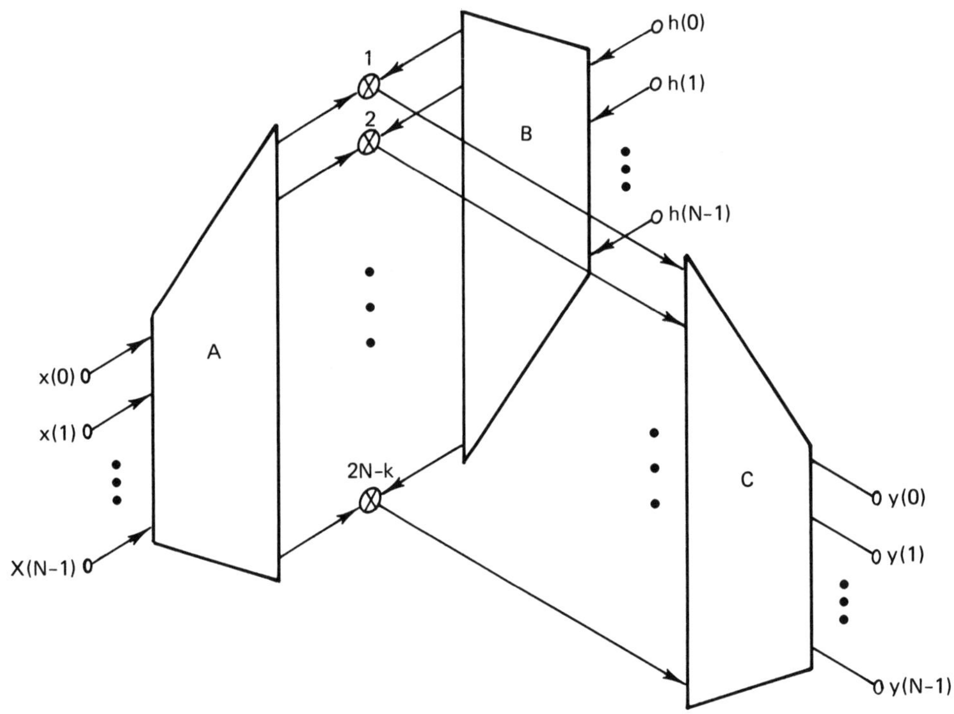

Figure 7-2 General form of convolution algorithm.

diagram of the canonical structure, stressing the rectangular nature of the matrix transformations A, B, and C. As a final note, observe that the DFT implementation of circular convolution is a special case of the structure given in Figure 7-2 in which the A and B transformations are DFT's and C is the inverse DFT. We shall show next that if the field F is chosen to be the set of complex numbers, then the result of the preceding derivation is in fact the DFT.

In order to investigate the implications of Winograd's complexity theorem more closely we shall consider its application to circular convolution (that is, the special case

$P(z) = z^N - 1$). We first take the field of constants F to be the complex numbers. In this case† $P(z) = \prod_{i=0}^{N-1} (z - W_N^i)$, and $P_i(z) = z - W_N^i$. Therefore, $k = N$ and a minimal algorithm with $2N - N = N$ multiplications must exist. We first calculate $X(z)$ mod $P_i(z) = X(W_N^i)$ and $H(z)$ mod $P_i(z) = H(W_N^i)$. Then

$$Y_i(z) = X(z)H(z) \bmod P_i(z) = X(W_N^i) \cdot H(W_N^i)$$

The calculation of $Y_i(z)$ requires one multiplication because both $X(W_N^i)$ and $H(W_N^i)$ are constants. Expanding $X(W_N^i)$ as $\sum_{n=0}^{N-1} x[n]W_N^{ni}$ shows quite clearly that the A matrix is the $N \times N$ DFT matrix; likewise for $H(W_N^i)$. It is important to note that since the numbers W_N^i are constants in the field of complex numbers, none of the multiplications in the DFT are counted!

Now we need to show that the reconstruction phase of the Chinese Remainder Theorem yields the inverse DFT. Our knowledge of the circular convolution property for DFT's tells us that this is true but we shall show the mechanics of this approach. We must produce the reconstruction polynomials $S_i(z)$. Since $S_i(z)$ satisfies $S_i(z) \equiv \delta_{ij}$ mod $P_j(z)$, $S_i(z)$ must be of the form $S_i(z) = C_i \prod_{r \neq i} (z - W_N^r)$. The constant C_i can be evaluated, using $S_i(z) \equiv 1$ mod $P_i(z)$ to obtain

$$C_i = \frac{1}{\prod_{\substack{r=1 \\ r \neq i}}^{N} (W_N^i - W_N^r)} = \frac{1}{W_N^{i(N-1)} \prod_{r=1}^{N-1} (1 - W_N^r)} = W_N^i C_0$$

Note that $S_i(z)$ can be written as

$$S_i = C_i \frac{(z^N - 1)}{(z - W_N^i)} = C_i(z^{N-1} + W_N^i z^{N-2} + W_N^{2i} z^{N-3} + \ldots + W_N^{(N-1)i})$$

Using $S_0(z) \equiv 1$ mod $(z - 1)$, we obtain $C_0 = 1/N$, and $C_i = W_N^i/N$. Following (7.10), we can write $Y(z)$ as

$$Y(z) = \left(\sum_{i=0}^{N-1} Y_i(z) \left(\frac{W_N^i}{N} \right) \sum_{l=0}^{N-1} z^l W_N^{i(N-1-l)} \right) \bmod (z^N - 1)$$

Interchanging the order of summation we obtain

$$Y(z) = \left(\sum_{l=0}^{N-1} z^l \left(\frac{1}{N} \right) \sum_{i=0}^{N-1} Y_i W_N^{-il} \right) \bmod (z^N - 1)$$

and we can identify the coefficients of $Y(z)$ as

$$y[l] = \frac{1}{N} \sum_{i=0}^{N-1} Y_i W_N^{-il}$$

which is precisely the inverse DFT. The multiplications by W_N^{-i} in the IDFT are not counted within Winograd's theory because they are fixed constants in the field of complex numbers.

The previous example demonstrates that the choice for the field of constants can be critical in the application of Winograd's methods and the derivation of "minimal"

†Recall that W_N is defined as $e^{-j(2\pi/N)}$.

algorithms. Since we are not counting multiplications by fixed constants from the field F we would like to force all such multiplications to be simple so that, in fact, they need not be counted. The first step in this direction is choosing the field of constants to be the rational numbers. On the one hand rational numbers, such as 0, 1, −1, 2, or $\frac{1}{2}$, would make simple multipliers; on the other hand, $\frac{9}{47}$ and $\frac{181}{257}$ could hardly be considered simple. There is no way to force the constant multipliers to be simple, but it turns out that use of the rational numbers as the field of constants results in a number of algorithms (for small values of N) that contain only simple multipliers.

Therefore, we now consider the development of minimum-complexity circular convolution algorithms over the field of rational numbers. First, we must show how to factor the polynomial $z^N - 1$ as a product of polynomials whose coefficients are rational numbers. Such a factorization can be given, in general, in terms of *cyclotomic polynomials*. The cyclotomic polynomial $C_k(z)$ is a polynomial whose roots, z_i, are a subset of the roots of $z^N - 1$, that is, a subset of the set $\{W_N^i\}$. The subset is chosen so that z_i is *not* a root of $z^d - 1$ for $0 < d < k$. Formally we can write

$$C_k(z) = \prod_{\substack{(i,k)=1 \\ i \le k}} (z - W_k^i) \tag{7.29}$$

Consider the example of $C_6(z)$. The roots of $z^6 - 1$ are the set $\{e^{-j(2\pi i/6)};\ i = 0, 1, 2, 3, 4, 5\}$. Of this set, e^{-j0} is rejected because it is a root of $z - 1$; $e^{-j(2\pi 2/6)}$ and $e^{-j(2\pi 4/6)}$ are rejected because they are roots of $z^3 - 1$; and $e^{-j(2\pi 3/6)}$ is rejected because it is a root of $z^2 - 1$. Therefore,

$$C_6(z) = (z - e^{-j(2\pi/6)})(z - e^{-j(2\pi 5/6)}) = z^2 - z + 1$$

PROBLEM Compute $C_4(z)$ from the definition.

PROBLEM Show that the degree of $C_k(z)$ is $\phi(k)$.

PROBLEM Show that for p a prime, $C_p(z) = z^{p-1} + z^{p-2} + \dots + z + 1$.

We tabulate below some of the cyclotomic polynomials.

$$\begin{aligned}
C_1(z) &= z - 1 \\
C_2(z) &= z + 1 \\
C_3(z) &= z^2 + z + 1 \\
C_4(z) &= z^2 + 1 \\
C_5(z) &= z^4 + z^3 + z^2 + z + 1 \\
C_6(z) &= z^2 - z + 1 \\
C_7(z) &= z^6 + z^5 + z^4 + z^3 + z^2 + z + 1 \\
C_8(z) &= z^4 + 1 \\
C_9(z) &= z^6 + z^3 + 1 \\
C_{10}(z) &= z^4 - z^3 + z^2 - z + 1
\end{aligned}$$

From the definition of $C_k(z)$ it is clear that $C_k(z)$ divides $z^k - 1$. It should also be clear that $C_k(z)$ divides $z^{kr} - 1$. In fact, *Part One*

$$z^N - 1 = \prod_{d|N} C_d(z) \tag{7.30}$$

that is, the product of terms $(z - W_N^i)$ collected into subgroups such that the W_N^i of each subgroup all have order d, and all possible orders d are included.

From the definition just given, it is not immediately evident either that $C_k(z)$ has rational coefficients or that $C_k(z)$ cannot be factored further into smaller factors with rational coefficients. Indeed, the reader might be tempted to conjecture that the coefficients of $C_k(z)$ are limited to the set $\{0, 1, -1\}$, although that conjecture is false. We shall defer rigorous proof that the factorization (7.30) is complete, and we shall make use of the property, observed for all $C_k(z)$ encountered in practice ($k < 105$), that the coefficients are indeed in the set $\{0, 1, -1\}$. For now we state as unproven theorems, the following:

The coefficients of $C_k(z)$ are integers. **THEOREM**

For $k < 105$, the coefficients of $C_k(z)$ can only be 0, 1, or -1. **THEOREM**

$C_k(z)$ cannot be factored further over the field of rational numbers. **THEOREM**

The first two theorems are proved in Appendix VIIa; the third is beyond the scope of this book.

We now return to the development of a minimum complexity circular convolution algorithm over the rational numbers. From Winograd's Theorem we know that an algorithm can be synthesized with $2N - k$ multiplications where k is the number of irreducible factors of $(z^N - 1)$ over the rationals. This factorization is given in terms of cyclotomic polynomials

$$z^N - 1 = \prod_{l|N} C_l(z)$$

Thus k equals the number of integers that divide N, including 1 and N. For example, 6-point circular convolution requires $2(6) - 4 = 8$ multiplications because 1, 2, 3, and 6 divide 6; hence

$$z^6 - 1 = C_1(z)C_2(z)C_3(z)C_6(z)$$
$$z^6 - 1 = (z - 1)(z + 1)(z^2 + z + 1)(z^2 - z + 1)$$

It is interesting to observe that 5-point circular convolution also requires 8 multiplications.

PROBLEM How many multiplications are required according to Winograd's complexity theorem to perform 4-point circular convolution? 81-point circular convolution?

EXERCISE Can you give an example where N_1 is *less* than N_2 but the N_1-point circular convolution requires *more* multiplications than the N_2-point circular convolution?

The first step of Winograd's synthesis procedure requires the evaluation of $X(z) \bmod P_i(z)$ and $H(z) \bmod P_i(z)$ which, in this case, are polynomial residue reductions modulo cyclotomic polynomials. The coefficients of the cyclotomic polynomials are simple; hence these residue reductions can be performed using only additions and subtractions! (For very high-order (≥ 105) cyclotomic polynomials, some multiplications by integers may be necessary. In theory, these integer multiplications could be implemented with successive additions; in practice, we shall be concerned only with low-order cyclotomic polynomials.)

In order to facilitate understanding of the synthesis procedure we use 3-point circular convolution as an example. The factorization of $z^3 - 1$ is $C_1(z)C_3(z)$. Four multiplications are required according to Winograd's theorem. The first step of the synthesis results in

$$\begin{aligned} X_1(z) &= X(z) \bmod C_1(z) \\ &= X(z) \bmod (z - 1) = x_{10} \\ &= X(1) \\ &= x[0] + x[1] + x[2] \end{aligned}$$

and

$$\begin{aligned} X_2(z) &= X(z) \bmod C_3(z) \\ &= X(z) \bmod (z^2 + z + 1) = x_{20} + x_{21}z \\ &= (x[0] + x[1]z + x[2]z^2) \bmod (z^2 + z + 1) \\ &= (x[0] - x[2]) + (x[1] - x[2])z \end{aligned}$$

This means our algorithm will require the following computations:

$$\begin{aligned} x_{10} &= x[0] + x[1] + x[2] \\ x_{20} &= x[0] - x[2] \\ x_{21} &= x[1] - x[2] \end{aligned}$$

Similarly,

$$H_1(z) = h[0] + h[1] + h[2] = h_{10}$$

and

$$H_2(z) = (h[0] - h[2]) + (h[1] - h[2])z = h_{20} + h_{21}z$$

Thus our algorithm will also require

$$\begin{aligned} h_{10} &= h[0] + h[1] + h[2] \\ h_{20} &= h[0] - h[2] \\ h_{21} &= h[1] - h[2] \end{aligned}$$

Since $X_1(z)$ and $H_1(z)$ are constants, the calculation of $Y_1(z) = Y(z) \bmod C_1(z) = X_1(z) H_1(z) \bmod (z - 1)$ requires one multiplication; $x_{10} \cdot h_{10}$. The calculation of $Y_2(z) = X(z)H(z) \bmod C_2(z)$ can easily be done using four multiplications, but requires additional considerations in order to be done in three multiplications. One approach to this problem is the brute force approach (which will work for low-order examples) in which we first write

$$\begin{aligned} Y_2(z) &= (x_{20} + x_{21}z)(h_{20} + h_{21}z) \bmod (z^2 + z + 1) \\ &= (x_{20}h_{20} - x_{21}h_{21}) + (x_{21}h_{20} + x_{20}h_{21} - x_{21}h_{21})z \end{aligned}$$

It is easy to see that the coefficients of $Y_2(z)$ can be obtained using the following three multiplications:

$$\mu_1 = x_{21} \cdot h_{21}$$
$$\mu_2 = x_{20} \cdot h_{20}$$
$$\mu_3 = (x_{20} - x_{21}) \cdot (h_{20} - h_{21})$$

Then $Y_2(z)$ is

$$\begin{aligned} Y_2(z) &= (\mu_2 - \mu_1) + (\mu_2 - \mu_3)z \\ &= y_{20} + y_{21}z \end{aligned}$$

It is worth recalling the systematic approach for computing expressions like $Y_2(z)$. This method is based on the observation that $Y_2(z)$ can be computed in two steps. First the product $X_2(z)H_2(z)$, corresponding to a linear convolution, is computed using the Toom-Cook algorithm presented in Section VII. Then the residue reduction modulo $C_2(z)$ is performed. The Toom-Cook algorithm requires $\deg\{X_2(z)\} + \deg\{H_2(z)\} + 1$ multiplications in general, and the residue reduction modulo $C_2(z)$ operation can usually be performed without multiplications. Thus in the present example the Toom-Cook algorithm would require three multiplications for $Y_2(z)$ which is precisely the number we are trying to realize. The resulting procedure, however, although requiring the same number of multiplications as the *ad hoc* method, does not necessarily give the same algorithm as was found by brute force search. The resulting set of computations is often different. Unfortunately, the *ad hoc* approach usually yields a superior algorithm because there may be fewer additions, and because there may be no uncounted multiplications by annoying rationals. Thus, although the Toom-Cook algorithm gives a systematic solution to the problem of calculating expressions such as $Y_2(z)$, it is possible to generate better algorithms (in the sense of practical realizations) by using other techniques.

Now we return to the synthesis problem for the 3-point circular convolution algorithm. Up to this point we have calculated $Y(z) \bmod C_1(z)$ and $Y(z) \bmod C_2(z)$. The final answer $Y(z)$ can be constructed using the Chinese Remainder Theorem for polynomials. We must find $S_1(z)$ and $S_2(z)$ satisfying

$$S_i(z) \equiv \delta_{ij} \bmod C_j(z) \qquad i, j = 1, 2$$

The result from a previous example is

$$S_1(z) = \tfrac{1}{3}(z^2 + z + 1)$$

and

$$S_2(z) = -\tfrac{1}{3}(z^2 + z - 2)$$

If we let $Y_1(z) = y_{10}$ and $Y_2(z) = y_{20} + y_{21}z$, we obtain

$$\begin{aligned} Y(z) &= [\tfrac{1}{3}(z^2 + z + 1)y_{10} - \tfrac{1}{3}(z^2 + z - 2)(y_{20} + y_{21}z)] \bmod (z^3 - 1) \\ &= \tfrac{1}{3}[(y_{10} + 2y_{20} - y_{21}) + (y_{10} - y_{20} + 2y_{21})z + (y_{10} - y_{20} - y_{21})z^2] \end{aligned}$$

Collecting together all equations, the following algorithm will calculate a 3-point circular convolution:

$$x_{10} = x[0] + x[1] + x[2]$$
$$x_{20} = x[0] - x[2]$$
$$x_{21} = x[1] - x[2]$$
$$h_{10} = \tfrac{1}{3}(h[0] + h[1] + h[2])$$
$$h_{20} = \tfrac{1}{3}(h[0] - h[2])$$
$$h_{21} = \tfrac{1}{3}(h[1] - h[2])$$
$$y_{10} = x_{10} \cdot h_{10}$$
$$\mu_1 = x_{21} \cdot h_{21}$$
$$\mu_2 = x_{20} \cdot h_{20}$$
$$\mu_3 = (x_{20} - x_{21}) \cdot (h_{20} - h_{21})$$
$$y_{20} = \mu_2 - \mu_1$$
$$y_{21} = \mu_2 - \mu_3$$
$$y[0] = y_{10} + 2y_{20} - y_{21}$$
$$y[1] = y_{10} - y_{20} + 2y_{21}$$
$$y[2] = y_{10} - y_{20} - y_{21}$$

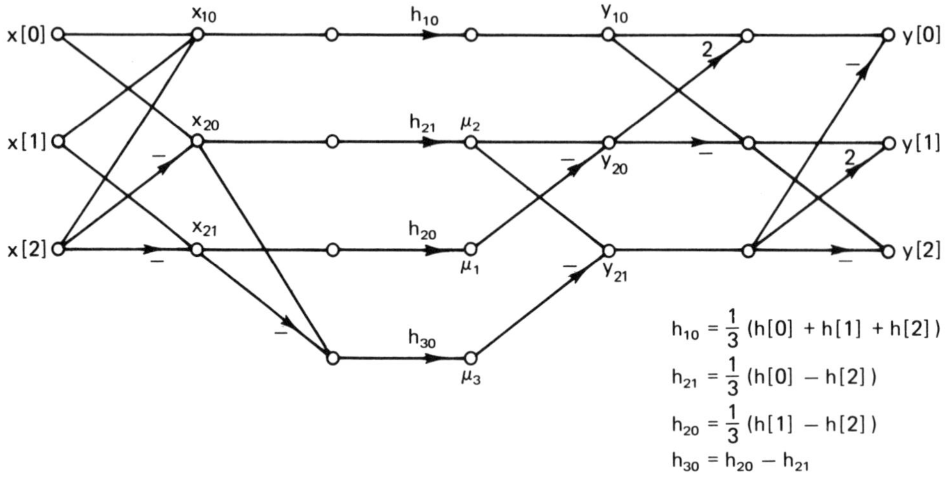

Figure 7-3 Three-point circular convolution implemented using four multiplications.

The 3-point circular convolution can be expressed in two other forms: the matrix form of (7.28) and the signal flowgraph form of Figure 7.3. The A, B, and C matrices would be

$$A = \begin{bmatrix} 1 & 1 & 1 \\ 1 & 0 & -1 \\ 0 & 1 & -1 \\ 1 & -1 & 0 \end{bmatrix} = \begin{bmatrix} 1 & 0 & 0 \\ 0 & 1 & 0 \\ 0 & 0 & 1 \\ 0 & 1 & -1 \end{bmatrix} \begin{bmatrix} 1 & 1 & 1 \\ 1 & 0 & -1 \\ 0 & 1 & -1 \end{bmatrix}$$

$$B = \tfrac{1}{3}A$$

$$C = \begin{bmatrix} 1 & 1 & -2 & 1 \\ 1 & 1 & 1 & -2 \\ 1 & -2 & 1 & 1 \end{bmatrix} = \begin{bmatrix} 1 & 0 & -1 \\ 0 & 1 & 2 \\ 0 & 1 & -1 \end{bmatrix} \begin{bmatrix} 1 & 2 & 0 \\ 1 & -1 & 0 \\ 0 & 0 & 1 \end{bmatrix} \begin{bmatrix} 1 & 0 & 0 & 0 \\ 0 & 1 & -1 & 0 \\ 0 & 1 & 0 & -1 \end{bmatrix}$$

The factored form corresponds to the flowgraph of Figure 7.3.

PROBLEM Synthesize the algorithm for 4-point circular convolution using five multiplications.

Answer

$$A = \begin{bmatrix} 1 & 1 & 1 & 1 \\ 1 & -1 & 1 & -1 \\ 1 & 1 & -1 & -1 \\ 1 & 0 & -1 & 0 \\ 0 & 1 & 0 & -1 \end{bmatrix}$$

$$B = \tfrac{1}{4}\begin{bmatrix} 1 & 1 & 1 & 1 \\ 1 & -1 & 1 & -1 \\ 2 & 0 & -2 & 0 \\ 2 & -2 & -2 & 2 \\ 2 & 2 & -2 & -2 \end{bmatrix}$$

$$C = \begin{bmatrix} 1 & 1 & 1 & 0 & -1 \\ 1 & -1 & 1 & -1 & 0 \\ 1 & 1 & -1 & 0 & 1 \\ 1 & -1 & -1 & 1 & 0 \end{bmatrix}$$

We have seen that we can find an algorithm for $y = x * h$, of the form

$$\mathbf{y} = C(A\mathbf{x} \otimes B\mathbf{h}) \tag{7.31}$$

where A and B are matrices with only ± 1, 0 entries, but where the entries of C are guaranteed only to be rational. In the cases we have worked out, although we had little control over the entries of C, these entries have turned out to be acceptable numbers. We are justly concerned that this might not always be so. On the other hand, in many cases we form BH in advance and therefore we derive no great benefit from control over the entries of B. It would be desirable to derive an algorithm in the form of (7.31) in which we had control over entries in A and C, taking whatever rationals we get in B. Such an algorithm can be found by using a "duality" principle. Winograd [4] explains this principle essentially as follows:

Suppose the sequence $\{y_k : k = 0(1)t - 1\}$ is defined in terms of $\{x_i : i = 0(1)r - 1\}$ and $\{h_j : j = 0(1)s - 1\}$ by the system of sums

$$y_k = \sum_{j=0}^{s-1} \sum_{i=0}^{r-1} \alpha_{ijk} x_i h_j; \qquad k = 0, 1, \ldots, t-1 \tag{7.32}$$

 Winograd defines a "dual" or *transpose* system as the set of sums

$$\hat{y}_k = \sum_{j=0}^{s-1} \sum_{i=0}^{t-1} \alpha_{kji} \hat{x}_i \hat{h}_j; \qquad k = 0, 1, \ldots, r-1 \tag{7.33}$$

System (7.32) is also the transpose of system (7.33). Then if there is an algorithm

$$\mathbf{y} = C(A\mathbf{x} \otimes B\mathbf{h}) \tag{7.34}$$

for the sums of (7.32) there is a dual algorithm for (7.33), namely

$$\hat{\mathbf{y}} = A^t(C^t\hat{\mathbf{x}} \otimes B\hat{\mathbf{h}}) \tag{7.35}$$

Winograd gives an elegant proof of this result.

Now suppose that (7.33) defines our desired result, for example, circular convolution. We discover an algorithm (7.34) for its transpose system (which will turn out to be circular correlation), using methods presented earlier in this section, guaranteeing that the elements of A and B are $\pm 1, 0$. Then (7.35) is an algorithm for circular convolution in which the output matrix, A^t, has the desirable elements.

Although C^t, containing uncontrolled rational entries, is still needed to operate on $\hat{\mathbf{x}}$, we know that for circular convolution we can interchange the roles of $\hat{\mathbf{x}}$ and $\hat{\mathbf{h}}$, giving

$$\hat{\mathbf{y}} = A^t(B\hat{\mathbf{x}} \otimes C^t\hat{\mathbf{h}}) \tag{7.36}$$

an algorithm in which the costly operations are in C^t only, and can be done in advance.

For (7.33) to be circular convolution, we must have

$$r = s = t = N \tag{7.37}$$

and

$$\alpha_{kji} = \delta[\langle i + j - k \rangle_N] \tag{7.38}$$

Thus α_{ijk} needed in (7.32) is

$$\alpha_{ijk} = \delta[\langle k + j - i \rangle_N] \tag{7.39}$$

and (7.32) must be

$$y[k] = \sum_{j=0}^{N-1} h[j]x[\langle j + k \rangle_N] \tag{7.40}$$

We recognize this as circular correlation. We now need to find an algorithm for circular correlation with the properties of (7.34). But this is easy since we can obtain correlation from convolution by time reversing input and output. Equivalently, beginning with an already derived convolution algorithm

$$\mathbf{y} = C_0(A_0\mathbf{x} \otimes B_0\mathbf{h})$$

we find the circular correlation algorithm by permuting the columns of the A_0 matrix and the rows of the C_0 matrix. The permutation leaves the first column (or row) in place and reverses the order of the remaining columns (or rows). Let this operation be expressed symbolically by superscripts R and r

$$A = A_0^R; \qquad C = C_0^r$$

The required correlation (7.34) then becomes

$$\mathbf{y} = C_0^r(A_0^R\mathbf{x} \otimes B_0\mathbf{h})$$

and an algorithm for circular convolution, using (7.36) becomes

$$\hat{\mathbf{y}} = (A_0^R)^t(B_0\hat{\mathbf{x}} \otimes (C_0^r)^t\hat{\mathbf{h}}) \tag{7.41}$$

As an example of the foregoing development, recall that we had a 3-point circular convolution algorithm where

$$\mathbf{x} = \begin{bmatrix} x_0 \\ x_1 \\ x_2 \end{bmatrix}; \qquad \mathbf{h} = \begin{bmatrix} h_0 \\ h_1 \\ h_2 \end{bmatrix}; \qquad A_0 = B_0 = \begin{bmatrix} 1 & 1 & 1 \\ 1 & 0 & -1 \\ 0 & 1 & -1 \\ 1 & -1 & 0 \end{bmatrix}$$

$$C_0 = \begin{bmatrix} \frac{1}{3} & \frac{1}{3} & -\frac{2}{3} & \frac{1}{3} \\ \frac{1}{3} & \frac{1}{3} & \frac{1}{3} & -\frac{2}{3} \\ \frac{1}{3} & -\frac{2}{3} & \frac{1}{3} & \frac{1}{3} \end{bmatrix} = \frac{1}{3}\begin{bmatrix} 1 & 1 & -2 & 1 \\ 1 & 1 & 1 & -2 \\ 1 & -2 & 1 & 1 \end{bmatrix}$$

$$\mathbf{y} = C_0(A_0\mathbf{x} \otimes B_0\mathbf{h})$$

Using (7.41) we find that another 3-point convolution algorithm is

$$A = B_0 = \begin{bmatrix} 1 & 1 & 1 \\ 1 & 0 & -1 \\ 0 & 1 & -1 \\ 1 & -1 & 0 \end{bmatrix}; \qquad B = (C_0^r)^t = \frac{1}{3}\begin{bmatrix} 1 & 1 & 1 \\ 1 & -2 & 1 \\ -2 & 1 & 1 \\ 1 & 1 & -2 \end{bmatrix};$$

$$C = (A_0^R)^t = \begin{bmatrix} 1 & 1 & 0 & 1 \\ 1 & -1 & -1 & 0 \\ 1 & 0 & 1 & -1 \end{bmatrix}$$

The signal flowgraph form is shown in Figure 7.4.

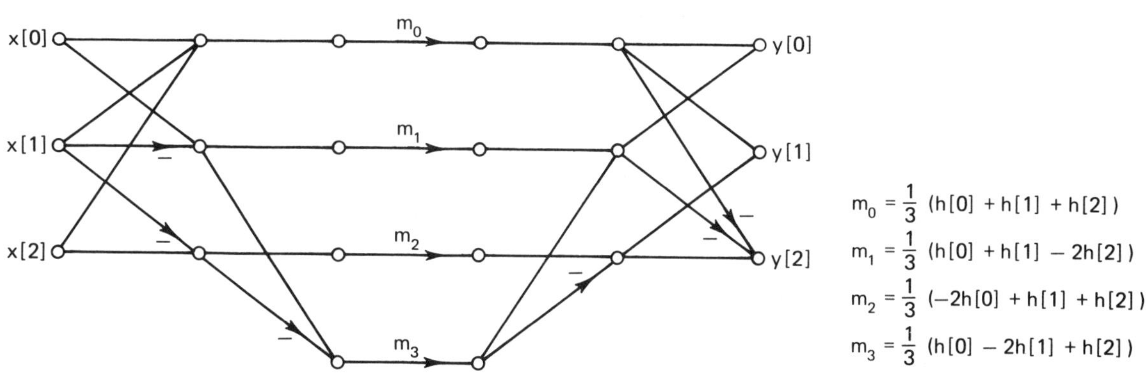

Figure 7-4 Transposed algorithm for 3-point circular convolution.

appendix VIIA *properties of the cyclotomic polynomials*

In this appendix we present some properties of the cyclotomic polynomials referred to in Section VII. We prove that the coefficients are integers and that for $l < 105$ the coefficients of $C_l(z)$ come from the set $\{0, 1, -1\}$. Proof of the irreducibility of $C_l(z)$ over the rationals may be found in [73, pp. 160–64]. The lth cyclotomic polynomial $C_l(z)$ is defined to be

$$C_l(z) = \prod_{\substack{(r,l)=1 \\ 0<r<l}} (z - W_l^r) \tag{7A.1}$$

Recall that the factorization for $z^N - 1$ in terms of cyclotomic polynomials $C_l(z)$ is

$$z^N - 1 = \prod_{l|N} C_l(z) \tag{7A.2}$$

We would like to demonstrate that the coefficients of $C_l(z)$ are rational; in fact they are integers. First, we must prove some simple properties of the cyclotomic polynomials.

Property 1:

If p is a prime,

$$C_p(z) = z^{p-1} + z^{p-2} + z^{p-3} + \ldots + z + 1$$

Proof:

From the definition (7A.1)

$$C_p(z) = \prod_{(r,p)=1} (z - W_p^r) = \prod_{r=1}^{p-1} (z - W_p^r)$$

Thus,

$$C_p(z) = \frac{(z^p - 1)}{(z - 1)} = z^{p-1} + z^{p-2} + \ldots + z + 1$$

Property 2:

For any choice of the integers m and p

$$C_{mp^k}(z) = C_{mp}(z^{p^{k-1}})$$

Proof:

$$\begin{aligned} C_{mp^k}(z) &= \prod_{(r,mp^k)=1} (z - W_{mp^k}^r) \\ &= \prod_{\substack{(r,mp)=1 \\ r<mp}} \prod_{s=0}^{p^{k-1}-1} (z - W_{mp^k}^{r+mps}) \\ &= \prod_{\substack{(r,mp)=1 \\ 0<r<mp}} \prod_{s=0}^{p^{k-1}-1} (z - W_{mp^k}^r W_{p^{k-1}}^s) \end{aligned}$$

Using the identity

$$z^N - \alpha^N = \prod_{s=0}^{N-1} (z - \alpha W_N^s)$$

we obtain

$$\begin{aligned} C_{mp^k}(z) &= \prod_{\substack{(r,mp)=1 \\ 0<r<mp}} (z^{p^{k-1}} - W_{mp^k}^{rp^{k-1}}) \\ &= \prod_{(r,mp)=1} (z^{p^{k-1}} - W_{mp}^r) \\ &= C_{mp}(z^{p^{k-1}}) \end{aligned}$$

Property 3:

If p is prime, and p does not divide m (equivalently, $(p, m) = 1$) then

$$C_{pm}(z) = \frac{C_m(z^p)}{C_m(z)}$$

Proof:

$$\begin{aligned} C_{pm}(z) &= \prod_{(r,pm)=1} (z - W_{pm}^r) \\ &= \prod_{(r,pm)=1} (z - W_{pm}^r) \frac{\prod_{(s,m)=1} (z - W_m^s)}{\prod_{(s,m)=1} (z - W_m^s)} \end{aligned}$$

Recall from our earlier discussions in Section III that the set of integers relatively prime to mp can be generated in the form $\langle am + bp \rangle_{mp}$ where a and b satisfy $(a, p) = 1$ and $(b, m) = 1$. Thus, since p is prime

$$= \prod_{(b,m)=1} \prod_{a=1}^{p-1} (z - W_{mp}^{am+bp}) \frac{\prod_{(s,m)=1} (z - W_m^s)}{C_m(z)}$$

Note the product on s becomes the $a = 0$ term of the product on a.

$$\begin{aligned} &= \prod_{(b,m)=1} \prod_{a=0}^{p-1} \frac{(z - W_p^a W_m^b)}{C_m(z)} \\ &= \prod_{(b,m)=1} \frac{(z^p - W_m^b)}{C_m(z)} \\ &= \frac{C_m(z^p)}{C_m(p)} \end{aligned}$$

Using properties 1, 2, and 3 it is straightforward to show that the coefficients of $C_l(z)$ are always rational. Property 1 shows that $C_p(z)$ has integer coefficients for p a prime number. The operations of properties 2 and 3 involve only additions, subtractions, and multiplications of the integer coefficients; hence the coefficients of $C_l(z)$ must be integers. Note that no divisions are required (even in property 3) because $C_l(z)$ is always a monic polynomial. As a final note, we remark that cyclotomic polynomials can be defined over any field that possesses primitive roots of unity [73].

PROBLEM Prove the following properties of cyclotomic polynomials:

(a) $C_{2n}(z) = C_n(-z)$ for n odd and $n \geq 3$.

(b) $C_n(0) = \begin{cases} 0 & \text{if } n = 1 \\ p & \text{if } n \text{ is a power of a prime, } p. \\ 1 & \text{if } n \text{ has two or more prime factors} \end{cases}$

From our simple properties of cyclotomic polynomials we can calculate some of the $C_l(z)$:

$$C_3(z) = z^2 + z + 1$$
$$C_4(z) = C_2(z^2) = z^2 + 1$$
$$C_5(z) = z^4 + z^3 + z^2 + z + 1$$
$$C_6(z) = \frac{C_2(z^3)}{C_2(z)} = \frac{z^3 + 1}{z + 1} = z^2 - z + 1$$
$$C_7(z) = z^6 + z^5 + z^4 + z^3 + z^2 + 1$$
$$C_8(z) = C_2(z^4) = z^4 + 1$$
$$\vdots$$
$$C_{15}(z) = \frac{C_3(z^5)}{C_3(z)} = \frac{C_5(z^3)}{C_5(z)} = z^8 - z^7 + z^5 - z^4 + z^3 - z + 1$$

$$C_{105}(z) = \frac{C_{15}(z^7)}{C_{15}(z)} = \frac{C_3(z^{35})C_3(z)}{C_3(z^7)C_3(z^5)}.$$

$$\begin{aligned} C_{105} = {} & 1 + z + z^2 - z^5 - z^6 - 2z^7 - z^8 - z^9 + z^{12} + z^{13} + z^{14} + z^{15} \\ & + z^{16} + z^{17} - z^{20} - z^{22} - z^{24} - z^{26} - z^{28} + z^{31} + z^{32} + z^{33} \\ & + z^{34} + z^{35} + z^{36} - z^{39} - z^{40} - 2z^{41} - z^{42} - z^{43} + z^{46} + z^{47} + z^{48} \end{aligned}$$

PROBLEM Calculate $C_{24}(z)$, $C_{41}(z)$, and $C_{81}(z)$.

Notice that the coefficients of $C_l(z)$ are very simple (that is., $0, \pm 1$), at least for $l = 1, 2, \ldots 8$. As a matter of fact, the smallest value of l for which $C_l(z)$ has a coefficient of value other than $0, \pm 1$ is $l = 105$. $C_{105}(z)$ has two coefficients equal to -2.

We can show that $C_l(z)$ has coefficients from the set $\{0, 1, -1\}$ for $l < 105$ by proving the following more general result:

THEOREM $C_l(z)$ has coefficients from the set $\{0, 1, -1\}$ when l has at most two different odd prime factors. Since $105 = 3 \cdot 5 \cdot 7$ is the smallest integer divisible by three odd primes we have the desired property for $C_l(z)$, $l < 105$.

The proof of the theorem when l is the power of an odd prime follows easily from

$$C_{p^k}(z) = C_p(z^{p^{k-1}})$$

Powers of two are also permissible in l because

$$C_{m2^k} = C_{2m}(z^{2^{k-1}}) = C_m(-z^{2^{k-1}})$$

Thus we shall concentrate on the case $l = pq$, where p and q are distinct odd primes. From property 3 we have

$$C_{pq}(z) = \frac{C_p(z^q)}{C_p(z)}$$

Expanding the inverse of $C_p(z)$ as an infinite sum we obtain

$$\begin{aligned}C_{pq}(z) &= C_p(z^q)\sum_{i=0}^{\infty} z^{ip}(1-z)\\ &= \sum_{l=0}^{p-1} z^{lq}\sum_{i=0}^{\infty} z^{ip}(1-z)\\ &= \sum_{i=0}^{\infty}\sum_{l=0}^{p-1} z^{lq+ip} - z\sum_{i=0}^{\infty}\sum_{l=0}^{p-1} z^{lq+ip}\end{aligned}$$

Since the degree of $C_{pq}(z)$ is known to be $\phi(pq) = (p-1)(q-1)$

$$C_{pq}(z) = \sum_{l=0}^{p-1}\sum_{ip\leq\phi(pq)-lq} z^{lq+ip} - z\sum_{l=0}^{p-1}\sum_{ip\leq\phi(pq)-lq-1} z^{lq+ip} \tag{7A.3}$$

The exponent $lq + ip$ will take on distinct values for $0 \leq i \leq q-1$ and $0 \leq l \leq p-1$. The argument is similar to that used for Euler's totient function in Section III. Suppose $l_1 q + i_1 p = l_2 q + i_2 p$ for $l_1 \neq l_2$ and $i_1 \neq i_2$. Then $(l_1 - l_2)q = (i_2 - i_1)p$ and it follows that p divides $(l_1 - l_2)$ and q divides $(i_2 - i_1)$. But this is possible only if $l_1 - l_2 = 0$ and $i_1 = i_2$ because

$$|l_1 - l_2| < p \quad \text{and} \quad |i_2 - i_1| < q$$

Thus the first sum in (V11A.3) is a polynomial whose coefficients are 0 or +1 as is the second sum. Hence, the difference of these two polynomials, $C_{pq}(z)$, has coefficients from the set $\{0, 1, -1\}$.

Part Two

VIII
introduction to reprinted papers

Thirteen papers on number theory in digital signal processing are reprinted in the following collection. These papers are a significant sample of the various approaches and applications of number-theoretic and polynomial theoretic ideas to two important digital signal processing algorithms: circular convolution and the discrete Fourier transform (DFT).

Loosely, the papers are grouped into two broad categories: those pertaining to the Winograd Fourier Transform algorithm (WFTA) and those on the subject of number theoretic transforms (NTT's). Papers 1–5 and Paper 13 deal with topics involved in the development of the WFTA. Papers 6–12 discuss the theory of NTT's, together with some particular transforms and implementations, such as the Fermat number transform. The ordering of papers within these two broad categories is not significant, but the following introduction tries to describe the interrelationships among the papers as well as the logical development of the WFTA and the NTT.

Before discussing the details of each paper, some general comments about the entire set of papers are appropriate. Three basic themes are evident in the algorithms presented. The first type of algorithm treats the addresses of the data (the indices) in a number-theoretic way; the second type acts on the signal amplitudes in a number-theoretic (or algebraic) manner; the third type acts on the data with "rectangular" linear transformations that may be considered a generalization of the DFT which is itself a square matrix. Algorithms of the first type are distinguished by the operations of data reindexing and permutation and might be referred to as *permutation* techniques. The primary examples of such algorithms are the methods presented by Rader in Paper 2 and Good in Paper 5. The obvious examples of the second class are the number-theoretic transforms presented in Papers 6, 7, 11, and 12. The data points are treated as elements in a finite computational structure (for example, a ring or field) rather than members of the field of complex numbers as in the DFT. Since these algorithms involve linear transformations of N-point data vectors, one might refer to them as *transform* techniques. A theoretical development of the third class of algorithms, the "rectangular" transforms, is presented in Paper 13 and in the opening sections of Papers 1, 3, and 4. These rectangular transforms are a generalization of the DFT in that the transform space may be of larger dimension than the sequence being transformed, but the cyclic convolution property of the DFT is maintained. The algorithms are an outgrowth of the theory of computational complexity and can be shown to minimize the number of multiplications necessary for polynomial multiplication. In their application so far, the rectangular transforms have been synthesized only for short sequences and must be combined with algorithms of the permutation type in order to calculate DFTs and circular convolution for long sequences.

In order to interrelate the papers and to tie them to our previous presentation of basic principles, each paper is discussed in some detail. First we consider the five papers that contain both transform- and permutation-type algorithms. These are the papers by Kolba and Parks, by Rader, by Agarwal and Cooley, by Winograd, and by Good. The second set of papers deals with number-theoretic transform techniques. The papers on NTT's divide naturally into three groups. The first two papers, by Pollard, and Agarwal and Burrus, introduce the essential theoretical background; the second pair by McClellan and Leibowitz deal with hardware implementations for Fermat number transforms; the third pair by Reed and Truong, and by Nussbaumer, consider extensions to complex-valued NTT's. Also included is a paper by Agarwal and Burrus on multidimensional mappings applicable to Fermat number transforms. The final paper, by Winograd, deals with the complexity of bilinear forms.

As a final note, before launching into details, the reader should remember that this treatment is not intended to be a critique of these thirteen papers. Each paper makes a unique contribution to the development of number-theoretic algorithms in signal processing. The impact of a given paper changes with time; if one hastens to pass judgment on a certain technique, it often happens that new theoretical or technological developments alter the evaluation. A most dramatic case in point is Rader's 1968 paper, "The DFT when the Number of Data Points is Prime." Regarded as a curiosity for many years, the paper has taken on new significance in the light of Winograd's new DFT algorithm. Thus, this discussion is offered to encourage investigations both by engineers who seek to use new techniques in implementing efficient signal processors and by researchers who try to extend the theoretical work begun in these papers.

Paper 1—A Prime Factor FFT Algorithm Using High-speed Convolution

The first paper in the collection is Kolba and Parks' "A Prime Factor FFT Algorithm Using High-Speed Convolution." This paper contains the theoretical background for two new DFT algorithms and a comparison of those techniques with the FFT in terms of the number of operations and running times on two computers, the IBM 370 and the 8080 microprocessor. The two DFT algorithms considered are the Winograd Fourier Transform algorithm (WFTA) and a prime factor FFT algorithm originally due to Good but improved by the use of efficient short length DFT algorithms developed by Winograd.

The authors point out that the development of these new DFT algorithms rests on two key ideas. The first is the conversion of a DFT to circular convolution as introduced by Rader (see Paper 2). This point is illustrated by specific examples for 5-point and 9-point DFT matrices. The second idea is the minimization of the number of multiplications necessary for circular convolution as shown by Winograd (see Papers 4 and 13). Synthesis of the minimal algorithm for the specific case of 6-point cyclic convolution is worked out in considerable detail. The net result of these two ideas is the ability to derive efficient algorithms for short DFTs. Table I, Paper 1, gives an operation count for DFT's of length 2, 3, 4, 5, 7, 8, and 9. Specific algorithms are given in the appendix for lengths 3, 5, 7, and 9. These algorithms differ slightly from those presented by Winograd in Paper 4.

The short transforms can be combined in several ways to obtain a long transform of length N. Paper 1 describes two techniques based on the Chinese Remainder Theorem. When N has relatively prime factors, Good has shown that the one-dimensional transform is equivalent to a multidimensional DFT (Paper 5). The authors refer to this technique as the *prime factor algorithm*. The second technique is a reordering of operations in the prime factor algorithm to merge all the multiplications and thus reduce the total number of multiplications relative to the prime factor algorithm. The authors have chosen to call this algorithm the *nested* algorithm. The reader should turn to Figures 2, 3, and 4, Paper 1 for a graphic explanation of the differences between the two algorithms.

The rest of Paper 1 compares the FFT, the prime factor algorithm, and the nested algorithm. Table II summarizes the comparison on the basis of number of multiplications and additions. Table IV contains the results of a timing comparison performed by running various programs on an IBM 370/155. The prime factor algorithm is compared with a radix-2 FFT and the mixed radix FFT due to Singleton [117]. Finally, some timing estimates for the 8080 microprocessor are given.

Paper 2—Discrete Fourier Transforms When the Number of Data Samples Is Prime

Paper 1 discusses algorithms that are the outgrowth of research results presented by several other workers in the field. One key result, due to Rader, is presented in Paper 2, "Discrete Fourier Transforms When the Number of Data Samples Is Prime." Rader showed that when the length of the transform is a prime number, the N-point DFT contains an $(N-1)$-point circular convolution. In order to isolate this convolution, a permutation must be applied to the $(N-1)$ signal points $x[1]\,x[2], \ldots, x[N-1]$. A second permutation is applied to the set $\left\{\exp\left(-j\frac{2\pi}{N}k\right),\ k=1,2,\ldots,N-1\right\}$. Both permutations are generated by using a primitive root which exists because the length N is prime. To obtain the DFT from the circular convolution, the point $x[0]$ and the DC frequency value $X[0]$ must be handled separately.

One application suggested in Paper 2 is performing the circular convolution by high-speed convolution techniques, thus yielding an efficient algorithm for prime length DFT's. The example of $N = 563$ is considered in the paper. As a side benefit, this paper shows how to compute a circular convolution of length $N-1$ with an FFT method using transforms of length greater than $2N-3$.

Paper 3—New Algorithm for Digital Convolution

Paper 3 concentrates on circular convolution rather than on the DFT, although it is probably now apparent that the two operations are essentially the same. Paper 3 by Agarwal and Cooley is called "New Algorithms for Digital Convolution." This paper shows how the Chinese Remainder Theorem can be used to convert a one-dimensional circular convolution to a multidimensional convolution that is cyclic in all dimensions. The length of the 1-D convolution, of course, must factor into the product of relatively prime integers. Then special algorithms are presented for computing the short convolutions along each of the dimensions. This idea can also be used in conjunction with number-theoretic transforms to extend the maximum convolution length possible using these transforms. A very interesting development reported in this paper is the use of a formula manipulation program, SCRATCHPAD, to perform many of the tedious manipulations needed in the derivation of the convolution algorithms.

The first topic considered in Paper 3 is the development of algorithms for short convolutions. The authors present the Toom-Cook algorithm as a method applicable to noncyclic convolution. A matrix formulation of the Toom-Cook algorithm is also given to illustrate the form of the short convolution algorithms in terms of rectangular transforms. Next, the work of Winograd on optimal short convolution algorithms is presented. The resulting synthesis technique is aided greatly by use of the SCRATCH-PAD program. Table I, Paper 3, summarizes the operations necessary for the optimal algorithms of length $N = 2, 3, 4, \ldots, 12$. The derivation of the 4-point algorithm is considered in detail.

Since the direct synthesis technique applies only to short convolutions, composite algorithms must be used to obtain longer convolutions of various lengths. The extension of the length is based on the Chinese Remainder Theorem. By using the same mapping Good introduced for the DFT, a one-dimensional circular convolution can be converted to a multidimensional circular convolution. Notice, however, that the multidimensional convolution is not separable, hence use of the rectangular transforms results automatically in a nested algorithm for circular convolution. The authors consider the two-factor case first and then generalize to the multifactor case. An operation count is given in both instances. Their Table III summarizes the number of additions and multiplications for lengths up to 2520; Table IV gives similar information for FFT (high-speed) convolution. The multidimensional mapping may be used in conjunction with the Fermat number transform to give another convolution algorithm. (A comparison is presented in Table V.)

The paper considers various implementation issues. Among these are dynamic range constraints which are important for the FNT, roundoff error, and the optimal block length for noncyclic convolution. Finally, the authors give a tabulation of short $(2 \leq N \leq 9)$ convolution algorithms in an appendix.

Paper 4—On Computing the Discrete Fourier Transform

The fourth paper is "On Computing the Discrete Fourier Transform" by Shmuel Winograd. This paper presents a derivation of the Winograd Fourier transform algorithm, along with some interesting new ideas that are applicable to multidimensional DFT's.

The paper first gives the theoretical background necessary for constructing minimal algorithms for cyclic convolution. The construction is based on polynomial manipulations modulo a fixed polynomial $P(z)$ and the polynomial version of the Chinese Remainder Theorem. One difficult step in the construction occurs when $P(z)$ is irreducible over the field of interest. Two approaches are presented to solve this problem. The first, given in equation (3), Paper 4, is identical to the Toom-Cook algorithm discussed in Paper 3. The second algorithm is given in equation (4). Note that every minimal algorithm must use one of these two forms.

Another interesting theoretical concept is that of the dual or transpose system. The paper's equations (7) and (9) express the concept that a system of bilinear forms can always be computed in one of two ways.

Next, the minimal algorithm for 3-point cyclic convolution is constructed. Notice the use of the dual system in equations (22a) and (22b) to simplify certain constant multipliers to ± 1 only. The case of 6-point cyclic convolution is used to demonstrate the use of multidimensional convolution to obtain longer 1-D convolutions. The mapping from one to two dimensions is based on the Chinese Remainder theorem for integers. Cyclic convolution algorithms for 2, 3, 4, 5, and 6 points are necessary for the WFTA and are given in appendix A.

Application of the cyclic convolution algorithms to the DFT is presented by using the fact that the difficult part of the DFT calculation is a cyclic convolution. Three cases are considered: the length of the DFT, N, is a prime (considered by Rader in Paper 2); N is the power of an odd prime, p^r, in which case we obtain a cyclic convolution of length $\varphi(p^r) = (p - 1)p^{r-1}$ points; and the length, N, is a power of two, 2^r, which results in a two-dimensional cyclic convolution of size $2 \times 2^{r-2}$. Details of the 7-point and 9-point DFT algorithms are given. Appendix B lists all the short DFT algorithms constructed ($N = 2, 3, 4, 5, 7, 8, 9$, and 16). Their multiplicative complexity is summarized in Table 2 (Paper 4). A general proof that this construction will always give an algorithm with purely real or purely imaginary coefficients is also presented.

The final step in the construction of the WFTA is using the index mapping of indices based on the Chinese Remainder Theorem to convert a long 1-*D* DFT into a multidimensional DFT whose dimensions are relatively prime. The 12-point DFT is shown to decompose into a (4×3) two-dimensional DFT. The result is the same algorithm called the *nested* algorithm in Paper 1. Since there is some freedom in the ordering of the stages of the algorithm, the author shows that the number of additions can be minimized by the proper choice of ordering. Table 3 summarizes the operation count for the WFTA.

The last section of Paper 4 is devoted to topics that might be considered future theoretical directions for the development and application of Winograd's ideas. The application is for multidimensional DFT's. Notice that the construction of minimal cyclic convolution algorithms depends on the field of scalars over which one *chooses* to work (see Paper 13). Typically, the field is chosen to be Q, the set of rational numbers. Four-point cyclic convolution requires 5 multiplications over the rationals, but only 4 multiplications over the extension field $Q(i)$, the set of complex rationals. This fact can be exploited in the situation where the DFT is to be calculated for multiple data sets. The numbers of operations for a $120 \times 120 \times 120$ DFT are given. Other extension fields can be used to advantage, depending upon the length.

A more subtle application reduces the complexity of a 5×5 DFT from 36 multiplications to 33 by using a special construction from Paper 13. The savings in a $120 \times 120 \times 120$ DFT are dramatic. Although Winograd's two papers (4 and 13) are somewhat mathematical in nature, they contain the seeds of ideas that should spark interesting and profitable research in the digital signal processing community.

Paper 5—The Relationship Between Two Fast Fourier Transforms

Paper 5, by I. J. Good, deals explicitly with the difference between the mixed-radix FFT (developed by Cooley and Tukey) and the prime factor algorithm for decomposing the DFT (also known as *Good's algorithm*). Both techniques reduce the computation of a one-dimensional (1-*D*) DFT to the calculation of a number of very small DFT's. Good shows that both methods can be regarded as methods that reindex the 1-*D* data vector to form a multidimensional (*M-D*) data array.

For the radix-2 ($N = 2^L$ point) FFT algorithm the correspondence between the 1-*D* sequence $x[n]$ and the L-dimensional array $\tilde{x}[n_1, n_2, \ldots n_L]$ is given in terms of the binary expansion of the index n as

$$n = \sum_{i=1}^{L} n_i 2^{i-1} \tag{8.1}$$

The necessary correspondence in the transform domain is different. If $\tilde{X}[k_1, k_2, \ldots k_L]$ is the L-dimensional transform array and $X[k]$ is the 1-*D* DFT of $x[n]$ then,

$$k = \sum_{i=1}^{L} k_i 2^{L-i} \tag{8.2}$$

Equation (8.1) displays, of course, the bit-reversal associated with a radix-2 FFT.

Equations (8.1) and (8.2) can be generalized to the mixed-radix case where $N = t_1 t_2 \ldots t_L$—see equation (7) of Good's paper. Good points out that the relationship between $\tilde{x}[n_1, n_2, \ldots, n_L]$ and $\tilde{X}[k_1, k_2, \ldots k_L]$ is NOT that of the multidimensional DFT equation (8), but on the other hand, there are no restrictions on the factors of N in the FFT method.

In the prime factor algorithm the index mapping from one to L dimensions is based on the residue arithmetic defined by the Chinese Remainder Theorem. Good calls this correspondence *the Sino representation* and applies it to the index mapping in the transform between $\tilde{X}[k_1, k_2, \ldots k_L]$ and $X[k]$. Once again a different mapping is required in the other domain. Good calls this mapping the *Ruritanian correspondence* (equation (4)).

Using the Sino and Ruritanian correspondences, the relationship between $\hat{x}[n_1, n_2, \ldots n_L]$ and $\hat{X}[k_1, k_2, \ldots k_L]$ is precisely that of an L-dimensional DFT of size $t_1 \times t_2 \times \ldots t_L$ where N factors as $N = t_1 t_2 \ldots t_L$. From our knowledge of the Chinese Remainder Theorem, we know that the prime factor algorithm depends on the constraint that the factors t_i be mutually prime. Although not explictly stated in the article, the Sino correspondence may be used in the transform domain and the Ruritanian in the time domain, or vice versa.

Good proceeds to develop a representation of the FFT and the prime factor algorithm in terms of a *direct-product* of sparse matrices. Such a representation, although unfamiliar to many engineers, allows a simple derivation of the WFTA, for example. Good also points out that the Hadamard transform of length 2^n is, in fact, an n-dimensional $2 \times 2 \times 2 \times \ldots \times 2$ DFT. Finally, he suggests the existence of a number-theoretic Fourier transform and more generally a Fourier-Galois transform.

Paper 6—The Fast Fourier Transform in a Finite Field

Number-theoretic transforms (NNT's) were discovered independently by a number of researchers around 1970–71. One of the early papers on the subject was J. M. Pollard's "The Fast Fourier Transform in a Finite Field." Pollard showed that "transforms" analogous to the DFT could be defined in a Galois field (finite field) of p^n elements, where p is a prime number. The definition of an N-point transform requires the existence of a field element r of order N in the Galois field GF(p^n). This leads to the constraint that the transform length N divide $p^n - 1$. Many of the properties of the NTT are exactly the same as properties of the DFT, the most important being the "cyclic convolution property" (CCP). This property states that the multiplication of the NTT's of two N-point vectors $a[n]$ and $b[n]$ corresponds to the N-point cyclic (or circular) convolution $a[n] \circledN b[n]$. For the DFT this property is the basis of high-speed convolution algorithms using the FFT. Proof of the property rests on the orthogonality relation given in equation (8), Paper 6.

Pollard also indicated the appropriate extension necessary to define a transform in the ring of integers modulo M. The requirement on the transform length N is that N divide the greatest common divisor of $p_1 - 1, p_2 - 1, \ldots p_L - 1$, where M factors as $p_1^{r_1} p_2^{r_2} \ldots p_L^{r_L}$.

The rest of the paper is devoted to discussions of practical methods of performing the transform in GF(p^n), extensions to multidimensional transforms, and practical applications of the transforms. Two practical issues with the transforms are the determination of the element(s) r in GF(p^n) that have order N, and the implementation of addition and multiplication to conform with the algebraic laws of GF(p^n). Among the applications mentioned are multiplication of polynomials over GF(p^n), aperiodic convolution of integer sequences, multiplication of very large integers, and division of polynomials over GF(p). As a final item, a "chirp-z-transform" algorithm for NTT's is outlined, similar to Bluestein's work for the DFT.

The theoretical framework of the NTT permits a wide choice of possible transforms depending upon the choice of the finite field or ring, the transform length, and the root of unity.

Paper 7—Fast Convolution Using Fermat Number Transforms with Applications to Digital Filtering

The engineering aspects of the NTT require a transform whose parameters lead to efficient structures either in hardware or software. An in-depth study of this aspect of NTT's was undertaken by Agarwal and Burrus in Paper 7, "Fast Convolution Using Fermat Number Transforms with Applications to Digital Filtering." Drawing on the theoretical work of Pollard and others and some earlier work by Rader on Mersenne number transforms, Agarwal and Burrus showed that, among all NTT's, the Fermat number transform (FNT) was a particularly attractive transform from a hardware standpoint, involving on the order of $N \log_2 N$ additions, bit shifts, and subtractions but no multiplications.

Paper 7 examined the cyclic convolution property for invertible transforms with the "DFT structure" and gave the necessary and sufficient conditions that a transform of the form

$$\sum_{n=0}^{N-1} x[n]\alpha^{\langle nk \rangle}$$

have the cyclic convolution property when defined over the finite ring of integers modulo M. As with Paper 6, the conditions on N, M, and α are that N divide the GCD$\{p_1 - 1, p_2 - 1, \ldots p_L - 1\}$ where $M = \prod_{i=1}^{L} p_i^{r_i}$ and that α be an Nth root of unity modulo M.

Since hardware implementations of NTT's would have to perform addition and multiplication modulo M, the choice of M as 2^b, $2^b - 1$ or $2^b + 1$ would simplify matters. The choice $M = 2^b$ does not admit a useful transform (N must be one). The choice $M = 2^b - 1$ gives the Mersenne number transform, when b is prime. The root of unity is $\alpha \equiv -2 \bmod M$ which is of order $2b$. While α is simple, the transform length is not highly composite and FFT-type algorithms cannot be employed. The choice $M = 2^b + 1$ with $b = 2^t$ gives the Fermat number transforms (FNT's). A transform of length $N = 2^{t+1}$ with $\alpha \equiv 2 \bmod F_t$ (where F_t denotes the tth Fermat number $F_t = 2^{2^t} + 1$) can be defined. Furthermore, Paper 7 shows that the length can be extended to $N = 2^{2^{t+2}}$ by using $\hat{\alpha} \equiv 2^{2^{t-2}}(2^{2^{t-1}} - 1) \bmod F_t$. Since $\hat{\alpha}^2 \equiv 2 \bmod F_t$, the notation $\sqrt{2}$ is usually adopted for $\hat{\alpha}$. Since $\sqrt{2}$ has a two-bit binary representation in the ring of integers mod F_t, it is of interest for hardware implementation. Table I, Paper 7, summarizes the possible FNT's that can be defined over F_t for $t = 3, 4, 5$, and 6.

Implementation issues associated with the FNT are considered next. It is shown that arithmetic modulo $2^b + 1$ involves a simple correction stage to take care of overflows. This is based on the observation that $2^b \equiv -1 \bmod (2^b + 1)$. If the FNT is employed to do circular convolution of integer sequences, the result is only known modulo F_t. That is, a dynamic range constraint is imposed by the modular arithmetic. One must be able to bound a priori the result of the convolution in order to determine the true answer from the answer modulo F_t. The worst case bound is determined by the following simple procedure: If $y[n] = x[n] \circledN h[n]$ and $|x[n]| \leq 2^{b_x}$ and $|h[n]| \leq 2^{b_h}$, then $|y[n]| \leq N2^{b_x+b_h}$. For example, the 64-point circular convolution of two sequences requiring 8 bits plus sign would require at most 22 bits plus sign to represent the output sequence. The result is that F_4, F_5, and F_6 seem to offer the most promise for digital-filtering applications. The authors present some simulation results from a program written in assembly language for the IBM 370.

Last to be considered is the problem that the word length of the transform, dictated by the Fermat number, is tied to the transform length N. For example, with a 32-bit word using $F_5 = 2^{32} + 1$ it is possible to define at most a 128-point FNT using $\alpha \equiv \sqrt{2}$. One method for alleviating this constraint is to map the one-dimensional convolution problem to a two-dimensional convolution problem. This technique and some extensions are the subject of Agarwal and Burrus' companion paper, "Fast One-Dimensional Digital Convolution by Multidimensional Techniques."

Paper 8—Fast One-Dimensional Digital Convolution by Multidimensional Techniques

Paper 8 discusses two techniques for fast convolution by multidimensional techniques. In the first technique the two N-point sequences $x[n]$ and $h[n]$ are mapped to two $(2L - 1) \times M$ arrays (with $N = LM$) $\hat{X}[l, m]$ and $\hat{H}[l, m]$ such that the 2-D cyclic convolution of $\hat{X}$ and $\hat{H}$ contains, as part of its output, the correct result of the 1-D N-point cyclic convolution $x[n] \circledN h[n]$. When compared to the mappings discussed by Good (for example the Sino-Ruritanian correspondence) this procedure is inefficient, but it does have the advantage that L and M are unrestricted (that is, they need not be mutually prime). This fact is important for the definition of two-dimensional FNT's in which both dimensions will be powers of two. Table I, Paper 8, summarizes the word-length–transform-length tradeoff for 1-D and 2-D techniques. For example, the ring of integers modulo F_5 will allow the linear (noncyclic) convolution of two 64-point sequences when $\alpha \equiv \sqrt{2}$ is employed; this length is extended to 4096 with the 2-D mapping.

The technique is also shown to extend easily to higher-dimensional convolutions. The authors consider the case where the 1-D circular convolution is of length $N = 2^M$ and define M-dimensional arrays to be convolved. Combined with a special algorithm for two-point aperiodic convolution, this approach yields a convolution algorithm requiring 3^M multiplications rather than $N^2 = 4^M$ as in the direct approach. It is interesting to note that the special two-point and three-point convolution algorithms used in Paper 8 can be developed systematically using the methods introduced by Winograd (see Papers 4 and 13). Finally, a comparison of this technique with the FFT implementation of high-speed convolution is presented in Table II, Paper 8.

Paper 9—Hardware Realization of a Fermat Number Transform

The special attributes of the Fermat number transform led several researchers to consider seriously special-purpose hardware implementations of the transform. Two papers in this subject area are by McClellan and Leibowitz. In Paper 9, "Hardware Realization of a Fermat Number Transform," McClellan describes a hardware prototype built at M.I.T. Lincoln Laboratory in 1975. The machine was an implementation of a 64-point, 16-bit FNT (modulo F_4). An important element in the system design was a new coding scheme developed for the data representation to facilitate the implementation of arithmetic operations modulo the Fermat number. The use of this new coding scheme and the inherent simplicity of the FNT "butterfly" resulted in an FNT butterfly that could be clocked at 38 MHz.

One objective of this study was evaluating the total system cost of an FNT convolver versus a pipeline FFT convolver, primarily in the speed regime applicable to radar signal processing. Although the FNT has a significantly simpler arithmetic structure than the FFT, the implementation of long convolutions is penalized by the multidimensional mapping developed in Paper 8. Thus Paper 9 concludes that the FNT is a useful alternative to the FFT if the signal to be filtered is real and the computational elements are a major part of the overall system cost as in a pipeline architecture. Furthermore, for short length convolution (for example, length 64) and for two-dimensional convolution, the savings may be significant.

Paper 10—A Simplified Binary Arithmetic for the Fermat Number Transform

The coding scheme introduced in Paper 9 for FNT arithmetic was generalized by Leibowitz in Paper 10, "A Simplified Binary Arithmetic for the Fermat Number Transform." In place of the usual binary expansion, Leibowitz proposed that the integers in the range $0, 1, 2, 3, \ldots, 2^t - 1, 2^t$ be represented with the binary expansion of $(Ax - 1)$ mod $(2^t + 1)$, where A is a nonzero constant and x is the integer to be coded. The code of Paper 9 is the case $A = 2$; the simplest case is obviously $A = 1$.

Using this code, the rules for addition, negation, and shifting become the following:

1. Zero is a special case represented by a one in the tth bit and zeros in the lower t bits. If zero is an operand (for example, addition of $x + 0$) then it can be detected and the operation is trivial.
2. Addition is performed as on binary coded numbers with the added proviso that any overflow into the most significant bit (the tth place) is complemented and added to the least significant bit.
3. A nonzero number is negated by complementing its representation, bit by bit.
4. Multiplication by two (shifting) is accomplished by a circular shift of the bits one place to the left. In addition, the bit leaving the most significant place and entering the least significant position must be complemented.

The net result is an arithmetic system similar to ones-complement in its complexity and thus quite simple to implement in hardware.

The general problem of multiplying two numbers modulo F_t was also considered in Paper 10. This operation can be an important part of an FNT convolution system because the multiplications typically involve 16-, 32-, or 64-bit numbers.

Paper 11—The Use of Finite Fields to Compute Convolutions

The extensions of number-theoretic transforms to a more general setting that includes complex-valued transforms is the subject of the next two papers: Reed and Truong's "The Use of Finite Fields to Compute Convolutions" (Paper 11) considers the definition of NTT's over the Galois field of q^2 elements, GF(q^2), where q is a prime. This finite field is analogous to the field of complex numbers when (-1) is not a quadratic residue (that is, no element x satisfies $x^2 + 1 = 0 \bmod q$). The cyclic convolution of two complex-valued sequences can be performed naturally with such transforms. The special case where q is a Mersenne prime (for example, $2^p - 1$, $p = 13, 17, 19, 31, 61 \ldots$) is shown to yield a promising transform and is treated in some detail. The binary arithmetic needed is essentially ones-complement, hence easy to implement in hardware. Note that if q is a Fermat number, then (-1) is a quadratic residue and a similar derivation does not work.

The definition of the NTT over GF(q^2) follows the work of Pollard (Paper 6) in a straightforward manner. The length of the transform must divide $q^2 - 1 = (2^p - 1)^2 - 1 = 2^{p+1}(2^{p-1} - 1)$; hence transforms of length 2^k are possible, where $1 \leq k \leq p + 1$. In order to define the transform, a primitive element of order 2^{p+1} must be found. Theorems 1 and 2 (Paper 11) are helpful for finding the possible primitive elements. Since there are many primitive roots, a search can be made for those elements that have a simple binary expansion in order to simplify the ultimate hardware implementation. For example, in GF(31^2) there are four primitive elements of order 8 as listed in Table I.

Since the length of the transform is a power of two, an FFT-type algorithm can be developed to calculate the transform in $N \log_2 N$ operations. The authors also consider the dynamic range problem associated with the modulo-q arithmetic.

Paper 12—Digital Filtering Using Complex Mersenne Transforms

The twelfth paper, Nussbaumer's "Digital Filtering Using Complex Mersenne Transforms," also considers the definition of a complex Mersenne number transform. The definition is extended to include the ring of integers modulo a pseudo-Mersenne number. The pseudo-Mersenne number transform is attractive because it is more amenable to an FFT-type decomposition than the Mersenne number transform (MNT).

The definition of a complex MNT follows from the observation that $2j$ is an element of order $4q$ and $1 \pm j$ are elements of order $8q$ modulo $(2^q - 1)$. Although not explicitly stated in Paper 12, the author is essentially working in the Galois field GF(p^2) as was the case in Paper 11. The difference is that in Paper 12 the kernel is simple and the transform length is composite but contains a large prime factor q; whereas, in Paper 11 the kernel is complicated but the transform length is a power of two. Since $1 \pm j$ has order $8q$ the $8q$-point complex MNT can be decomposed, in a manner similar to the FFT, into eight q-point transforms in order to speed up the calculation.

The pseudo-Mersenne number transform is defined over the ring of integers modulo $(2^q - 1)/(p_i^{d_i})$ where $p_i^{d_i}$ is a divisor of $2^q - 1$. The transform length is governed by the greatest common divisor of $p_k - 1$ for $k \neq i$. Thus when $2^q - 1$ contains small prime factors, these can be divided out in the pseudo-Mersenne number to yield longer transform lengths. A factorization of $2^q - 1$ and of its factors $(p_i - 1)$ is presented in Table I (Paper 12). Table II gives the transform lengths and roots for some promising pseudo-MNT's. The arithmetic implementation of the pseudo-MNT can be performed by doing all calculations modulo $(2^q - 1)$, followed by a final residue reduction modulo $\dfrac{(2^q - 1)}{(p_i^{d_i})}$.

Paper 13—Some Bilinear Forms Whose Multiplicative Complexity Depends on the Field of Constants

The final paper in the collection is a mathematical paper on the general problem of the multiplicative complexity of bilinear forms and its dependence on the field of constants used in the derivation of the minimum complexity algorithms. In this paper,

Winograd presents the proof of the theorem that the minimum number of multiplications to compute $X(z)H(z) \bmod P(z)$ is $2N - k$, where N is the degree of $P(z)$ and k is the number of distinct irreducible factors of $P(z)$ over the field of interest. Paper 13 also contains a proof that the minimal algorithm must be based on the Chinese Remainder Theorem for polynomials. This paper is more mathematical than the others in the collection, but it should offer a stimulus to those interested in exploring the theoretical side of the complexity of computation as it has been studied for bilinear forms.

A Prime Factor FFT Algorithm Using High-Speed Convolution

DEAN P. KOLBA, STUDENT MEMBER, IEEE, AND THOMAS W. PARKS, MEMBER, IEEE

Abstract—Two recently developed ideas, the conversion of a discrete Fourier transform (DFT) to convolution and the implementation of short convolutions with a minimum of multiplications, are combined to give efficient algorithms for long transforms. Three transform algorithms are compared in terms of the number of multiplications and additions. Timing for a prime factor fast Fourier transform (FFT) algorithm using high-speed convolution, which was programmed for an IBM 370 and an 8080 microprocessor, is presented.

I. Introduction

THE calculation of the discrete Fourier transform (DFT)

$$X(k) = \sum_{n=0}^{N-1} x(n)\, e^{-j(2\pi/N)nk}$$

is one of the central operations in digital signal processing. The development and wide spread use of the fast Fourier transform (FFT), stimulated by the paper of Cooley and Tukey [1], has had a major impact on signal processing.

Recently, several new ideas have emerged which lead to new algorithms for the DFT. One key idea described by Rader [2] in 1968 was the observation that computation of the DFT can be changed into a circular convolution by rearranging the data when N is prime. Thus, if one has a fast way to do convolution, he now has a fast way to do the DFT.

Winograd [3] has shown the minimum number of multiples required for circular convolution. New convolution algorithms which often achieve this minimum are being developed by Agarwal and Cooley [4].

In a concise paper Winograd [5] combines the conversion of a DFT to convolution, for prime and prime power lengths, with these new convolution algorithms for short transforms. He proposes that long transforms be computed by nesting these short, high-speed transforms and presents a table comparing the number of operations required with the conventional FFT.

This paper first reviews the two central ideas; conversion of a DFT to circular convolution and convolution with the minimum number of multiples, then presents a study of various implementations of long transforms.

An alternative to the nested algorithm proposed by Winograd, a prime factor FFT algorithm using high-speed convolution for individual factors, is singled out as a promising approach and programmed for two machines; an IBM 370 and an 8080 microprocessor. Tables compare this approach with Winograd's nesting and with the conventional power of 2 FFT. While the idea of breaking up a one-dimensional transform into a multidimensional transform with prime factors is not new (see Good [6] and Thomas [7]), the combination of short, high-speed convolution algorithms with this multidimensional expansion appears to be a promising new way to implement the DFT.

Manuscript received November 1, 1976; revised March 1, 1977. This work was supported by the National Science Foundation under Grant ENG 70-01349 A03.

The authors are with the Department of Electrical Engineering, Rice University, Houston, TX 77001.

Printed in U.S.A. Annals No. 708AS002

II. Doing DFT's with Convolution

The discrete Fourier transform (DFT)

$$X(k) = \sum_{n=0}^{N-1} x(n) W^{nk}, \quad k = 0, \cdots, N-1$$
$$W = e^{-j(2\pi/N)} \tag{1}$$

is a linear transformation of the N-dimensional data vector

$$x = \begin{bmatrix} x(0) \\ x(1) \\ \vdots \\ x(N-1) \end{bmatrix}$$

into the vector

$$X = \begin{bmatrix} X(0) \\ X(1) \\ \vdots \\ X(N-1) \end{bmatrix}$$

of frequency samples.

The matrix representation

$$\begin{bmatrix} X(0) \\ X(1) \\ \vdots \\ X(N-1) \end{bmatrix} = \begin{bmatrix} 1 & 1 & \cdots & 1 \\ 1 & W^1 & W^2 \cdots & W^{N-1} \\ \vdots & W^2 & W^4 \cdots & W^{2(N-1)} \\ \vdots & \vdots & \vdots & \vdots \\ 1 & W^{N-1} & W^{2(N-1)} \cdots & W^{(N-1)(N-1)} \end{bmatrix} \begin{bmatrix} x(0) \\ x(1) \\ \vdots \\ x(N-1) \end{bmatrix} \tag{2}$$

shows that the complexity of the algorithm arises from the $(N-1)$ by $(N-1)$ lower right portion of the matrix. If

$$\bar{X}(k) = \sum_{n=1}^{N-1} x(n) W^{kn}, \quad k = 1, \cdots N-1 \tag{3}$$

can be computed efficiently, then we will have a fast algorithm for computing (1), the DFT, since from (3), (1) can easily be calculated:

$$X(0) = \sum_{n=0}^{N-1} x(n)$$
$$X(k) = x(0) + \bar{X}(k), \quad k = 1, 2, \cdots, N-1. \tag{4}$$

To see how (3) may be converted to circular convolution, consider the matrix representation of (3), for $N = 5$ with exponents taken modulo 5:

$$\begin{bmatrix} \bar{X}(1) \\ \bar{X}(2) \\ \bar{X}(3) \\ \bar{X}(4) \end{bmatrix} = \begin{bmatrix} W^1 & W^2 & W^3 & W^4 \\ W^2 & W^4 & W^1 & W^3 \\ W^3 & W^1 & W^4 & W^2 \\ W^4 & W^3 & W^2 & W^1 \end{bmatrix} \begin{bmatrix} x(1) \\ x(2) \\ x(3) \\ x(4) \end{bmatrix}. \tag{5}$$

If we interchange the last two columns of (5), and then interchange the last two rows we get

$$\begin{bmatrix} \bar{X}(1) \\ \bar{X}(2) \\ \bar{X}(4) \\ \bar{X}(3) \end{bmatrix} = \begin{bmatrix} W^1 & W^2 & W^4 & W^3 \\ W^2 & W^4 & W^3 & W^1 \\ W^4 & W^3 & W^1 & W^2 \\ W^3 & W^1 & W^2 & W^4 \end{bmatrix} \begin{bmatrix} x(1) \\ x(2) \\ x(4) \\ x(3) \end{bmatrix}. \tag{6}$$

We now have something like backwards circular convolution or circular correlation. By fixing $x(1)$ and reversing the remaining input vector, we obtain the conventional circular convolution:

$$\begin{bmatrix} \bar{X}(1) \\ \bar{X}(2) \\ \bar{X}(4) \\ \bar{X}(3) \end{bmatrix} = \begin{bmatrix} W^1 & W^3 & W^4 & W^2 \\ W^2 & W^1 & W^3 & W^4 \\ W^4 & W^2 & W^1 & W^3 \\ W^3 & W^4 & W^2 & W^1 \end{bmatrix} \begin{bmatrix} x(1) \\ x(3) \\ x(4) \\ x(2) \end{bmatrix}. \tag{7}$$

The computation of (3) (and thus, the DFT) has been changed to a circular convolution (7) by permuting the input and the output indices. If we have a fast way of doing convolutions, we now have a fast way of doing a DFT.

This idea was first presented by Rader [2]. He showed how the permutation can always be done when N is a prime number. Since $W^N = 1$, we are dealing with the product of integers modulo N in the exponent of W in (3). To change the DFT into a circular convolution a mapping of the indices is used to change multiplication of indices modulo N to addition of indices modulo $N-1$. The set of integers $(1, 2, \cdots, N-1)$ forms a cyclic group with the operation of multiplication modulo N [8]. We can always find at least one integer α in the group with the property that any integer in the group may be expressed as some power of α. By ordering the data according to the exponent of α we can always change (3) to circular convolution for prime N. The relation between the new index m and the original n is

$$n = \alpha^m \text{ modulo } N, \quad \begin{matrix} n = 1, 2, \cdots, N-1 \\ m = 0, 1, \cdots, N-2 \end{matrix} \tag{8}$$

where

$$\alpha^k \neq 1, \quad \text{for } 0 < k < N-1,$$
$$\alpha^{N-1} = 1. \tag{9}$$

The permutation map (8) is an isomorphism between the multiplicative group of $N-1$ integers $\{1, 2, \cdots, N-1\}$ and the additive group of $N-1$ integers $\{0, 1, \cdots, N-2\}$. An integer α with property (9) is called a primitive $(N-1)$st root of unity and is said to generate the group since any element can be written as some power of α. Just as logarithms change multiplication to addition, (8) changes multiplication into addition of indices. When (8) is used on both input and output indices, (3) becomes

$$\bar{X}(\alpha^l) = \sum_{m=0}^{N-2} x(\alpha^m)\, W^{\alpha^{(l+m)}}, \qquad l = 0, \cdots, N-2. \tag{10}$$

In (10) the exponents of α are taken modulo $N-1$. This gives for $N = 5$ the backward circular convolution of (6) when $\alpha = 2$. To obtain conventional circular convolution we change the sign of m in (10) which corresponds to fixing $x(\alpha^\circ)$ and reversing the remaining input sequence.

$$\bar{X}(\alpha^l) = \sum_{m=0}^{N-2} x(\alpha^{-m})\, W^{\alpha^{(l-m)}}, \qquad l = 0, \cdots, N-2. \tag{11}$$

Again, the indices in (11) (the exponents of α) are taken modulo $N-1$. By combining (11) with (4), we can always convert the computation of a DFT of prime length into a circular convolution.

Winograd [5] and McClellan and Rader [9] have shown that the DFT may also be converted to convolution when the transform length N is a prime power, i.e., $N = p^r$ for a prime $p \neq 2$. The conversion is a bit more complicated since we must first remove all integers which contain a factor p from the set $\{1, 2, \cdots, N-1\}$ to get a cyclic group with $p^{r-1}(p-1)$ elements. This cyclic group leads to a circular convolution of length $p^{r-1}(p-1)$ as before. The remaining computation consists of two DFT's of length p^{r-1}. For example with $N = 9 = 3^2$, we delete the integers 3 and 6 to obtain the set $\{1, 2, 4, 5, 7, 8\}$ which forms a cyclic group under multiplication modulo 9 and is isomorphic to the additive group of integers $\{0, 1, 2, 3, 4, 5\}$ under addition modulo 6. The integer 2 will generate the multiplicative group since the powers of 2 modulo 9, $2^m \bmod 9$, $m = 0, \cdots, 5$, are 1, 2, 4, 8, 7, 5.

In terms of the matrix representation

$$\begin{bmatrix} X(0)\\ X(1)\\ X(2)\\ X(3)\\ X(4)\\ X(5)\\ X(6)\\ X(7)\\ X(8) \end{bmatrix} = \begin{bmatrix} 1 & 1 & 1 & 1 & 1 & 1 & 1 & 1 & 1\\ 1 & W^1 & W^2 & W^3 & W^4 & W^5 & W^6 & W^7 & W^8\\ 1 & W^2 & W^4 & W^6 & W^8 & W^1 & W^3 & W^5 & W^7\\ 1 & W^3 & W^6 & 1 & W^3 & W^6 & 1 & W^3 & W^6\\ 1 & W^4 & W^8 & W^3 & W^7 & W^2 & W^6 & W^1 & W^5\\ 1 & W^5 & W^1 & W^6 & W^2 & W^7 & W^3 & W^8 & W^4\\ 1 & W^6 & W^3 & 1 & W^6 & W^3 & 1 & W^6 & W^3\\ 1 & W^7 & W^5 & W^3 & W^1 & W^8 & W^6 & W^4 & W^2\\ 1 & W^8 & W^7 & W^6 & W^5 & W^4 & W^3 & W^2 & W^1 \end{bmatrix} \begin{bmatrix} x(0)\\ x(1)\\ x(2)\\ x(3)\\ x(4)\\ x(5)\\ x(6)\\ x(7)\\ x(8) \end{bmatrix} \tag{12}$$

we remove rows and columns corresponding to indices 0, 3, and 6 and compute the remaining length 6 transformation

$$\begin{bmatrix} \bar{X}(1)\\ \bar{X}(2)\\ \bar{X}(4)\\ \bar{X}(5)\\ \bar{X}(7)\\ \bar{X}(8) \end{bmatrix} = \begin{bmatrix} W^1 & W^2 & W^4 & W^5 & W^7 & W^8\\ W^2 & W^4 & W^8 & W^1 & W^5 & W^7\\ W^4 & W^8 & W^7 & W^2 & W^1 & W^5\\ W^5 & W^1 & W^2 & W^7 & W^8 & W^4\\ W^7 & W^5 & W^1 & W^8 & W^4 & W^2\\ W^8 & W^7 & W^5 & W^4 & W^2 & W^1 \end{bmatrix} \begin{bmatrix} x(1)\\ x(2)\\ x(4)\\ x(5)\\ x(7)\\ x(8) \end{bmatrix} \tag{13}$$

using the permutation

$$n = 2^m \bmod 9, \qquad \begin{matrix} n = 1, 2, 4, 8, 7, 5\\ m = 0, 1, 2, 3, 4, 5 \end{matrix}$$

to obtain the circular convolution (with input reversed as before)

$$\begin{bmatrix} \bar{X}(1)\\ \bar{X}(2)\\ \bar{X}(4)\\ \bar{X}(8)\\ \bar{X}(7)\\ \bar{X}(5) \end{bmatrix} = \begin{bmatrix} W^1 & W^5 & W^7 & W^8 & W^4 & W^2\\ W^2 & W^1 & W^5 & W^7 & W^8 & W^4\\ W^4 & W^2 & W^1 & W^5 & W^7 & W^8\\ W^8 & W^4 & W^2 & W^1 & W^5 & W^7\\ W^7 & W^8 & W^4 & W^2 & W^1 & W^5\\ W^5 & W^7 & W^8 & W^4 & W^2 & W^1 \end{bmatrix} \begin{bmatrix} x(1)\\ x(5)\\ x(7)\\ x(8)\\ x(4)\\ x(2) \end{bmatrix}. \tag{14}$$

In addition to (14) we must complete the computation for the rows and columns removed from (12). For the deleted rows we have

$$\begin{bmatrix} X(0)\\ X(3)\\ X(6) \end{bmatrix} = \begin{bmatrix} 1 & 1 & 1 & 1 & 1 & 1 & 1 & 1 & 1\\ 1 & W^3 & W^6 & 1 & W^3 & W^6 & 1 & W^3 & W^6\\ 1 & W^6 & W^3 & 1 & W^6 & W^3 & 1 & W^6 & W^3 \end{bmatrix} \begin{bmatrix} x(0)\\ x(1)\\ x(2)\\ x(3)\\ x(4)\\ x(5)\\ x(6)\\ x(7)\\ x(8) \end{bmatrix}. \tag{15}$$

Since

$$W^3 = e^{-j(2\pi/9)\cdot 3} = e^{-j(2\pi/3)} = W_3,$$

(15) is simply a length 3 DFT of added data

$$\begin{bmatrix} X(0)\\ X(3)\\ X(6) \end{bmatrix} = \begin{bmatrix} 1 & 1 & 1\\ 1 & W_3 & W_3^2\\ 1 & W_3^2 & W_3 \end{bmatrix} \begin{bmatrix} x(0)+x(3)+x(6)\\ x(1)+x(4)+x(7)\\ x(2)+x(5)+x(8) \end{bmatrix}.$$

For the deleted columns we have

$$\begin{bmatrix} Y(0) \\ Y(1) \\ Y(2) \\ Y(3) \\ Y(4) \\ Y(5) \\ Y(6) \\ Y(7) \\ Y(8) \end{bmatrix} = \begin{bmatrix} 1 & 1 & 1 \\ 1 & W^3 & W^6 \\ 1 & W^6 & W^3 \\ 1 & 1 & 1 \\ 1 & W^3 & W^6 \\ 1 & W^6 & W^3 \\ 1 & 1 & 1 \\ 1 & W^3 & W^6 \\ 1 & W^6 & W^3 \end{bmatrix} \begin{bmatrix} x(0) \\ x(3) \\ x(6) \end{bmatrix} \tag{16}$$

For (16) a second length 3 DFT can be computed.

$$\begin{bmatrix} Y(0) \\ Y(1) \\ Y(2) \end{bmatrix} = \begin{bmatrix} Y(3) \\ Y(4) \\ Y(5) \end{bmatrix} = \begin{bmatrix} Y(6) \\ Y(7) \\ Y(8) \end{bmatrix} = \begin{bmatrix} 1 & 1 & 1 \\ 1 & W_3 & W_3^2 \\ 1 & W_3^2 & W_3 \end{bmatrix} \begin{bmatrix} x(0) \\ x(3) \\ x(6) \end{bmatrix}. \tag{17}$$

Only the last two entries $Y(1)$ and $Y(2)$ are needed from (17) to complete (12) using (14)

$$\begin{bmatrix} X(1) \\ X(2) \\ X(4) \\ X(5) \\ X(7) \\ X(8) \end{bmatrix} = \begin{bmatrix} \bar{X}(1) \\ \bar{X}(2) \\ \bar{X}(4) \\ \bar{X}(5) \\ \bar{X}(7) \\ \bar{X}(8) \end{bmatrix} + \begin{bmatrix} Y(1) \\ Y(2) \\ Y(1) \\ Y(2) \\ Y(1) \\ Y(2) \end{bmatrix}.$$

This length 9 example thus requires a length 6 circular convolution and 2 length 3 DFT's, which can be computed using length 2 circular convolutions. If $N = 3^3$, we have a length $3^2 \cdot 2 = 18$ circular convolution and two length $3^2 = 9$ DFT's. These can be reduced to 2 length 6 convolutions and 4 length 3 DFT's, which are calculated with length 2 circular convolutions. If $N = p^r$ the length N transform is computed with 1 length $p^{r-1}(p-1)$ convolution, 2 length $p^{r-2}(p-1)$ convolutions, 4 length $p^{r-3}(p-1)$ convolutions, 8 length $p^{r-4}(p-1)$ convolutions $\cdots$, terminating with 2^{r-1} length $p-1$ convolutions.

III. Convolution with the Minimum Number of Multiplications

The prime and prime power DFT algorithms are means of converting the calculations required in the DFT into circular convolution. Then, special fast cyclic convolution techniques may be used to perform the calculations. An algorithm for computing a short length cyclic convolution in the minimum number of multiples for small values of N is based on recent work by Winograd [3].

Winograd's theorem on the minimum number of multiplications is explained in terms of polynomial multiplication. To cyclicly convolve the sequences $h_0, h_1, \cdots h_{N-1}$ and $x_0, x_1, \cdots x_{N-1}$ we need only find the N coefficients of the polynomial

$$Y(z) = H(z) \cdot X(z) \bmod (z^N - 1) \tag{18}$$

where

$$H(z) = \sum_{k=0}^{N-1} h_k z^k \quad X(z) = \sum_{k=0}^{N-1} x_k z^k. \tag{19}$$

If $z^N - 1$ is written in terms of the K factors which are irreducible over the rationals

$$z^N - 1 = \prod_{i=1}^{K} Q_i(z) \tag{20}$$

with no common factors in the $Q_i(z)$ polynomials, Winograd's theorem states that the minimum number of multiplications required to compute the circular convolution of two length N sequences is $2N - K$. This theorem does not count multiplication by rational numbers.

In order to reduce the computation required for (18), $Y(z)$ is decomposed into K simpler parts using the polynomial version of the Chinese remainder theorem [4].

$$Y(z) = \left[\sum_{i=1}^{K} Y_i(z) S_i(z)\right] \bmod (z^N - 1) \tag{21}$$

$$Y_i(z) = H_i(z) X_i(z) \bmod (Q_i(z)), \quad i = 1, \cdots K \tag{22}$$

$$X_i(z) = X(z) \bmod (Q_i(z)), \quad i = 1, \cdots K$$

$$H_i(z) = H(z) \bmod (Q_i(z)), \quad i = 1, \cdots K. \tag{23}$$

The polynomials $S_i(z)\ i = 1, \cdots, K$ play the role of a Kronecker delta

$$S_i(z) \equiv 1 \bmod Q_i(z), \quad i = 1, \cdots K$$

$$S_i(z) \equiv 0 \bmod Q_j(z)\ \forall j \neq i. \tag{24}$$

The $S_i(z)$ may be found by applying Euclid's algorithm to polynomials [4]. For example, to do cyclic convolution of the length 6 sequences $h_0, h_1, \cdots h_5$ and $x_0, x_1, \cdots x_5$, form

$$H(z) = h_0 + h_1 z + \cdots h_5 z^5$$

$$X(z) = x_0 + x_1 z + \cdots + x_5 z^5.$$

First, the irreducible factors $Q_i(z)$ are found:

$$z^6 - 1 = (z+1)(z-1)(z^2+z+1)(z^2-z+1)$$

$$Q_1(z) = z + 1, Q_2(z) = z - 1, Q_3(z) = z^2 + z + 1,$$

$$Q_4(z) = z^2 - z + 1.$$

Next, the intermediary polynomials X_i, H_i, and Y_i are formed:[1]

$$X_1(z) = X(z) \bmod (z+1) = x_0^1$$

$$= x_0 - x_1 + x_2 - x_3 + x_4 - x_5$$

$$X_2(z) = X(z) \bmod (z-1) = x_0^2$$

$$= x_0 + x_1 + x_2 + x_3 + x_4 + x_5$$

$$X_3(z) = X(z) \bmod (z^2+z+1) = x_0^3 + x_1^3 z$$

[1] Superscripts are used to identify the ith polynomial.

$$= (x_0 - x_2 + x_3 - x_5) + (x_1 - x_2 + x_4 - x_5)z$$

$$X_4(z) = X(z) \bmod (z^2 - z + 1) = x_0^4 + x_1^4 z$$

$$= (x_0 - x_2 - x_3 + x_5) + (x_1 + x_2 - x_4 - x_5)z. \quad (25)$$

(The $H_i(z)$ polynomials are the same in form.)

$$Y_1(z) = H_1(z)X_1(z) \bmod (z + 1) = y_0^1 = h_0^1 x_0^1$$

$$Y_2(z) = H_2(z)X_2(z) \bmod (z - 1) = y_0^2 = h_0^2 x_0^2$$

$$Y_3(z) = H_3(z)X_3(z) \bmod (z^2 + z + 1) = y_0^3 + y_1^3 z$$

$$= (h_0^3 x_0^3 - h_1^3 x_1^3) + (h_1^3 x_0^3 + h_0^3 x_1^3 - h_1^3 x_1^3)z$$

$$Y_4(z) = H_4(z)X_4(z) \bmod (z^2 - z + 1) = y_0^4 + y_1^4 z$$

$$= (h_0^4 x_0^4 - h_1^4 x_1^4) + (h_1^4 x_0^4 + h_0^4 x_1^4 + h_1^4 x_1^4)z. \quad (26)$$

Now, the $Y_1(z)$ and $Y_2(z)$ polynomials are formed directly with one multiplication each. So, we have the intermediate variables

$$m_1 = h_0^1 x_0^1 \quad \text{and} \quad m_2 = h_0^2 x_0^2$$

giving

$$y_0^1 = m_1 \quad \text{and} \quad y_0^2 = m_2 \quad . \quad (27)$$

The polynomials $Y_3(z)$ and $Y_4(z)$ require only three multiplications each—similar to complex multiplication done in three real multiples. Thus, for $Y_3(z)$ we need

$$m_3 = (h_0^3 - h_1^3)(x_1^3 - x_0^3), \quad (28)$$

$$m_4 = h_0^3 x_0^3, \quad \text{and} \quad m_5 = h_1^3 x_1^3 \quad (29)$$

from which we obtain

$$y_1^3 = m_3 + m_4 \quad \text{and} \quad y_0^3 = m_4 - m_5.$$

For $Y_4(z)$ we need

$$m_6 = (h_0^4 + h_1^4)(x_0^4 + x_1^4), m_7 = h_0^4 x_0^4, \quad \text{and} \quad m_8 = h_1^4 x_1^4 \quad (30)$$

which we use to form $y_1^4 = m_6 - m_7$ and $y_0^4 = m_7 - m_8$. At this point, we have the 4 component polynomials: $Y_1(z) = y_0^1$, $Y_2(z) = y_0^2$, $Y_3(z) = y_0^3 + y_1^3 z$, and $Y_4(z) = y_0^4 + y_1^4 z$.

The final step is to express the polynomial $Y(z)$ in terms of the components $Y_i(z)$. We have

$$Y(z) = [Y_1(z)S_1(z) + Y_2(z)S_2(z) + Y_3(z)S_3(z) + Y_4(z)S_4(z)] \bmod (z^6 - 1). \quad (31)$$

The $S_i(z)$ polynomials satisfying (24) are

$$S_1(z) = -\tfrac{1}{6}(z^5 - z^4 + z^3 - z^2 + z - 1)$$

$$S_2(z) = \tfrac{1}{6}(z^5 + z^4 + z^3 + z^2 + z + 1)$$

$$S_3(z) = \tfrac{1}{6}(z^6 - z^5 - z^4 + 2z^3 - z^2 - z + 1)$$

$$S_4(z) = \tfrac{1}{6}(z^6 + z^5 - z^4 - 2z^3 - z^2 + z + 1). \quad (32)$$

In order to show the exact operations on the original $\{x_i\}$ and $\{h_i\}$, the vector of coefficients of $Y(z)$,

$$y = \begin{bmatrix} y_0 \\ y_1 \\ \vdots \\ y_{N-1} \end{bmatrix}$$

may be expressed in terms of

$$x = \begin{bmatrix} x_0 \\ x_1 \\ \vdots \\ x_{N-1} \end{bmatrix} \quad \text{and} \quad \boldsymbol{h} = \begin{bmatrix} h_0 \\ h_1 \\ \vdots \\ h_{N-1} \end{bmatrix}$$

with two equations using $\otimes$ to indicate point by point multiplication of column vectors:

$$\boldsymbol{m} = [B]\,\boldsymbol{h} \otimes [A]\,\boldsymbol{x} \quad (33)$$

$$\boldsymbol{y} = C\boldsymbol{m}. \quad (34)$$

A length N circular convolution requiring M multiplies can always be expressed this way with A and B $M \times N$ matrices and C an $N \times M$ matrix [4].

In order to put our example for $N = 6$ in this matrix form, we first identify the intermediate m parameters from (26), as shown in (27), (28), (29), and (30). The terms in (26) involving h and x are expanded according to (25) to obtain the matrices A and B.

$$m = \begin{bmatrix} 1 & -1 & 1 & -1 & 1 & -1 \\ 1 & 1 & 1 & 1 & 1 & 1 \\ 1 & -1 & 0 & 1 & -1 & 0 \\ 1 & 0 & -1 & 1 & 0 & -1 \\ 0 & 1 & -1 & 0 & 1 & -1 \\ 1 & 1 & 0 & -1 & -1 & 0 \\ 1 & 0 & -1 & -1 & 0 & 1 \\ 0 & 1 & 1 & 0 & -1 & -1 \end{bmatrix} \begin{bmatrix} h_0 \\ h_1 \\ h_2 \\ h_3 \\ h_4 \\ h_5 \end{bmatrix} \otimes \begin{bmatrix} 1 & -1 & 1 & -1 & 1 & -1 \\ 1 & 1 & 1 & 1 & 1 & 1 \\ -1 & 1 & 0 & -1 & 1 & 0 \\ 1 & 0 & -1 & 1 & 0 & -1 \\ 0 & 1 & -1 & 0 & 1 & -1 \\ 1 & 1 & 0 & -1 & -1 & 0 \\ 1 & 0 & -1 & -1 & 0 & 1 \\ 0 & 1 & 1 & 0 & -1 & -1 \end{bmatrix} \begin{bmatrix} x_0 \\ x_1 \\ x_2 \\ x_3 \\ x_4 \\ x_5 \end{bmatrix}. \quad (35)$$

To obtain the C matrix, rewrite (26) in terms of the m's given by (27)-(30).

$$Y_1(z) = m_1$$

$$Y_2(z) = m_2$$

$$Y_3(z) = (m_4 - m_5) + (m_3 + m_4)z$$

$$Y_4(z) = (m_7 - m_8 + (m_6 - m_7)z. \tag{36}$$

Substitute (36) and (32) into (31), collect powers of z to obtain

$$y = C\,m$$

$$\begin{bmatrix} y_0 \\ y_1 \\ y_2 \\ y_3 \\ y_4 \\ y_5 \end{bmatrix} = \frac{1}{6}\begin{bmatrix} 1 & 1 & -1 & 1 & -2 & 1 & 1 & -2 \\ -1 & 1 & 2 & 1 & 1 & 2 & -1 & -1 \\ 1 & 1 & -1 & -2 & 1 & 1 & -2 & 1 \\ -1 & 1 & -1 & 1 & -2 & -1 & -1 & 2 \\ 1 & 1 & 2 & 1 & 1 & -2 & 1 & 1 \\ -1 & 1 & -1 & -2 & 1 & -1 & 2 & 1 \end{bmatrix}\begin{bmatrix} m_1 \\ m_2 \\ m_3 \\ m_4 \\ m_5 \\ m_6 \\ m_7 \\ m_8 \end{bmatrix}. \tag{37}$$

We now have a length 6 cyclic convolution computed with 8 multiplications, which is the minimum number as given by Winograd's theorem:

$$2N - K = 2(6) - 4 = 8.$$

The multiplications by the rational numbers in A, B, and C are not counted.

The calculation of $Y_3(z)$ and $Y_4(z)$ may be done with a number of different algorithms. These will give values for m_3 through m_8 different from (28)-(30), and different evaluations of $Y_3(z)$ and $Y_4(z)$ in terms of the m's. This freedom in the choice of calculating $Y_3(z)$ and $Y_4(z)$ may be used to try to minimize the number of adds for the convolution algorithm. No attempt has been made here to minimize the number of adds.

Convolution algorithms which achieve $2N - K$ multiplies and have simple A, B, and C matrices are only known for short lengths. For longer convolutions it is difficult to see how to keep the number of additions under control.

IV. Application to the DFT

The polynomial method of generating a circular convolution algorithm which requires the minimum number of multiplications has been used effectively for short lengths. Winograd [5] states that all known algorithms for computing circular convolution in the minimum number of multiplications require a large number of additions when $z^N - 1$ has large irreducible factors. Therefore, to be of practical interest, the DFT length will be kept small to take advantage of the change to a circular convolution. DFT algorithms for small values of N have been written with these methods for use in a prime factor FFT, to be described in Section V. These algorithms are different from those used for the nested DFT proposed by Winograd and described by Silverman [10] which is also described in Section V. Table I shows the number of operations required for short transforms intended for use in the prime factor FFT and for short transforms intended for use in the nested algorithm. Explicit formulas for short transforms to be used with a prime factor FFT are given for lengths 3, 5, 7, and 9 in the Appendix. These formulas come from combining the correction terms with special convolution algorithms described in Section III. Short transforms for use with the nesting algorithm are derived by modifying the transforms in the Appendix to reduce the number of W^0 multiplies as in (45)-(47).

TABLE I
Short Length DFT Operations Count

	Prime Factor FFT			Nested Algorithm		
N	Multiplies	Shifts	Adds	Multiplies	W^0 multiplies	Adds
2	0	0	2	0	2	2
3	1	1	6	2	1	6
4	0	0	8	0	4	8
5	4	2	17	5	1	17
7	8	0	36	8	1	36
8	2	0	26	2	6	26
9	8	2	49	10	2	49

To see how a DFT is implemented with a convolution algorithm, consider the following length 3 example:

$$\begin{bmatrix} X(0) \\ X(1) \\ X(2) \end{bmatrix} = \begin{bmatrix} W^0 & W^0 & W^0 \\ W^0 & W^1 & W^2 \\ W^0 & W^2 & W^1 \end{bmatrix}\begin{bmatrix} x(0) \\ x(1) \\ x(2) \end{bmatrix}.$$

The convolution

$$\begin{bmatrix} \bar{X}(1) \\ \bar{X}(2) \end{bmatrix} = \begin{bmatrix} W^1 & W^2 \\ W^2 & W^1 \end{bmatrix}\begin{bmatrix} x(1) \\ x(2) \end{bmatrix} \tag{38}$$

provides $\bar{X}(1), \bar{X}(2)$, and

$$X(0) = W^0\,[x(0) + x(1) + x(2)]$$

$$X(1) = W^0\,x(0) + \bar{X}(1)$$

$$X(2) = W^0\,x(0) + \bar{X}(2). \tag{39}$$

To get an explicit formula for the DFT, the convolution (38) is written in matrix form as in (33) and (34).

$$\begin{bmatrix} m_1 \\ m_2 \end{bmatrix} = \begin{bmatrix} 1 & 1 \\ 1 & -1 \end{bmatrix}\begin{bmatrix} W^1 \\ W^2 \end{bmatrix} \otimes \begin{bmatrix} 1 & 1 \\ 1 & -1 \end{bmatrix}\begin{bmatrix} x(1) \\ x(2) \end{bmatrix} \tag{40}$$

$$\begin{bmatrix} \bar{X}(1) \\ \bar{X}(2) \end{bmatrix} = \frac{1}{2}\begin{bmatrix} 1 & 1 \\ 1 & -1 \end{bmatrix}\begin{bmatrix} m_1 \\ m_2 \end{bmatrix}. \tag{41}$$

In applying (40) and (41) to the length 3 DFT we absorb the factor of $\frac{1}{2}$ into the B matrix. Then, combining this convolution with (39) gives the following length 3 DFT algorithm for use in the prime factor FFT:

$$a_1 = x(1) + x(2),$$

$$a_2 = x(1) - x(2),$$

$$a_3 = x(0) + a_1; \tag{42}$$

$$m_1 = -\frac{1}{2}a_1,$$

$$m_2 = -j\frac{\sqrt{3}}{2}a_2; \tag{43}$$

$$c_1 = x(0) + m_1,$$
$$X(0) = a_3,$$
$$X(1) = c_1 + m_2,$$
$$X(2) = c_1 - m_2. \tag{44}$$

Now, for the algorithm to be used in the nested DFT method, the multiplies by W^0 must be accounted for as explained in Section V. Therefore, we wish to minimize the multiplies by W^0 as well. This may be done by modifying the above length 3 DFT as shown below:

$$a_1 = x(1) + x(2),$$
$$a_2 = x(1) - x(2),$$
$$a_3 = x(0) + a_1; \tag{45}$$
$$m_1 = (-\tfrac{1}{2} - 1)\, a_1 = -\tfrac{3}{2} a_1,$$
$$m_2 = -j\frac{\sqrt{3}}{2} a_2,$$
$$m_3 = W^0 a_3 = 1 \cdot a_3; \tag{46}$$
$$c_1 = m_3 + m_1,$$
$$X(0) = m_3,$$
$$X(1) = c_1 + m_2,$$
$$X(2) = c_1 - m_2. \tag{47}$$

The algorithm used for the prime factor FFT has 1 multiply, 1 shift (multiplication by $\frac{1}{2}$), and 6 adds, as shown in Table I. The algorithm used for the nested transform has 2 multiplies, 1 multiply by W^0, and 6 additions. For complex data, the values in Table I are only doubled because the coefficients formed from the BW portion of the convolution are pure real or pure imaginary numbers. This occurs since the B matrix is such that the W's occur as sums or differences of conjugate pairs.

These short length DFT algorithms may be written in a matrix form as

$$X = O\,D\,I\,x \tag{48}$$

where I is the $\mu \times N$ matrix representation of the input adds in (42) or (45), D is a $\mu \times \mu$ diagonal matrix of the multiply coefficients in (43) or (46), and O is the $N \times \mu$ matrix representation of the output adds in (44) or (47). For our length 3 example in (45)-(47),

$$\begin{bmatrix} X(0) \\ X(1) \\ X(2) \end{bmatrix} = \begin{bmatrix} 1 & 0 & 0 \\ 1 & 1 & 1 \\ 1 & 1 & -1 \end{bmatrix} \begin{bmatrix} 1 & 0 & 0 \\ 0 & -\frac{3}{2} & 0 \\ 0 & 0 & -j\frac{\sqrt{3}}{2} \end{bmatrix} \begin{bmatrix} 1 & 1 & 1 \\ 0 & 1 & 1 \\ 0 & 1 & -1 \end{bmatrix} \begin{bmatrix} x(0) \\ x(1) \\ x(2) \end{bmatrix}.$$

The summation form of these matrices gives another way to represent these algorithms:

$$X(k) = \sum_{l=0}^{\mu-1} o_{kl}\, d_l \sum_{n=0}^{N-1} i_{ln}\, x(n). \tag{49}$$

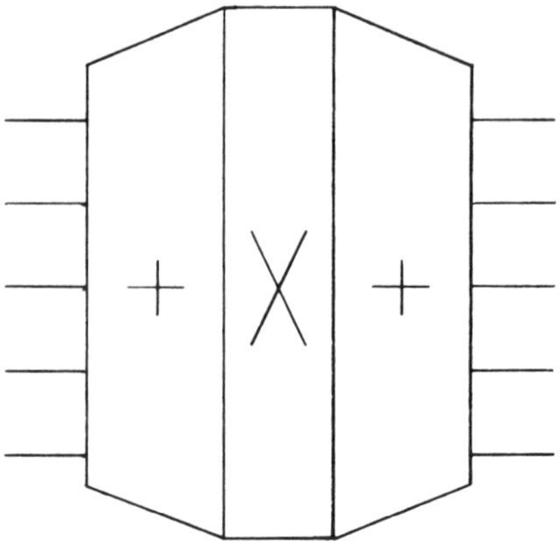

Fig. 1. DFT implemented with convolution showing expansion caused by more than 1 multiply per point.

In general, $\mu > N$ as shown by the expansion of the block labeled "X" in Fig. 1. The input and output additions are indicated by blocks labeled with a "+".

V. Long Transforms from Short

The short transforms described above can be combined in several ways to provide a long transform of length N. The idea is to convert a one-dimensional length $N = M_1 M_2 \cdots M_l$ transform into an l-dimensional transform requiring computation of l shorter, length M_k transforms for $k = 1, \cdots, l$. In this paper, we use a mapping from one to l dimensions which requires that the M_k factors be relatively prime [11]. Conventional FFT algorithms map one dimension to many dimensions. The Cooley-Tukey algorithm [1] allows common factors in N, while algorithms proposed by Good [6] and Thomas [7] use a mapping based on the Chinese remainder theorem [12] which requires relatively prime factors. We will use the Chinese remainder mapping which will be described for two factors M_1 and M_2 which are relatively prime.

In the DFT

$$X(k) = \sum_{n=0}^{N-1} x(n)\, W^{nk},$$

the index n of the input sequence $x(n)$ will be called the "input index," and the index k of the output sequence $X(k)$ will be called the "output index." The mapping from one-to-two dimensions maps the input index n into a pair of indices (n_1, n_2).

$$n_1 = r_1 n \bmod M_1 \quad n_1 = 0, 1, \cdots, M_1 - 1 \quad r_1 = M_2 \bmod M_1$$
$$n_2 = r_2 n \bmod M_2 \quad n_2 = 0, 1, \cdots, M_2 - 1 \quad r_2 = M_1 \bmod M_2.$$

The output index is similarly mapped

$$k_1 = k \bmod M_1 \quad k_1 = 0, 1, \cdots, M_1 - 1$$
$$k_2 = k \bmod M_2 \quad k_2 = 0, 1, \cdots, M_2 - 1.$$

The inverse mapping from two dimensions to one dimension for the output index is

$$k = (s_1 k_1 + s_2 k_2) \bmod N \tag{50}$$

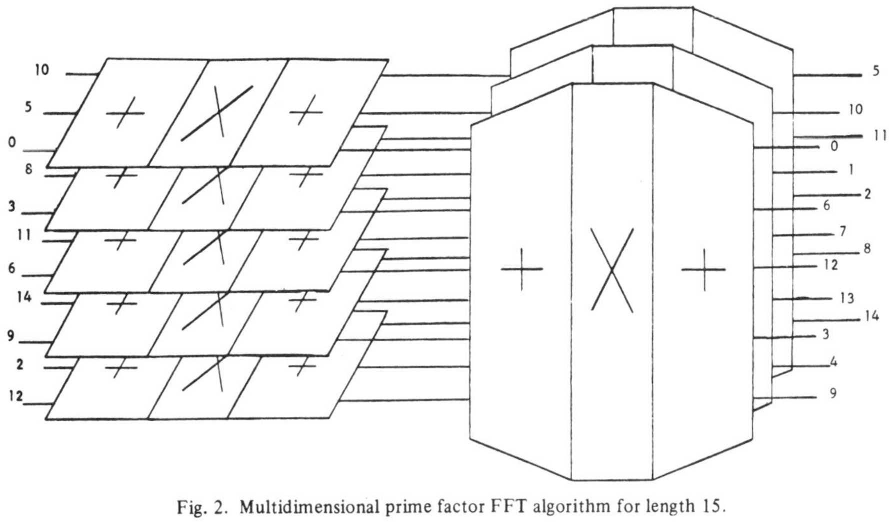

Fig. 2. Multidimensional prime factor FFT algorithm for length 15.

where

$$s_1 \equiv 1 \bmod M_1 \quad \text{and} \quad s_2 \equiv 0 \bmod M_1$$

$$s_1 \equiv 0 \bmod M_2 \qquad s_2 \equiv 1 \bmod M_2.$$

While the same inverse mapping could be used for the input index, it is more convenient to use

$$n = (M_2 n_1 + M_1 n_2) \bmod N. \tag{51}$$

When these mappings are used, the DFT becomes

$$X(k_1, k_2) = \sum_{n_1=0}^{M_1-1} \sum_{n_2=0}^{M_2-1} x(n_1, n_2)\, W_{M_2}^{n_2 k_2}\, W_{M_1}^{n_1 k_1} \tag{52}$$

where

$$W_{M_1} = e^{-j(2\pi/M_1)}$$

and

$$W_{M_2} = e^{-j(2\pi/M_2)}.$$

The two-dimensional transform in (52) may be implemented by first calculating M_1 length M_2 DFT's,

$$y(n_1, k_2) = \sum_{n_2=0}^{M_2-1} x(n_1, n_2) W^{n_2 k_2}, \tag{53}$$

then calculating M_2 length M_1 DFT's

$$X(k_1, k_2) = \sum_{n_1=0}^{M_1-1} y(n_1, k_2) W^{n_1 k_1} \tag{54}$$

The short transforms in (53) and (54) can be implemented using convolution methods as in (49). This procedure will be called the prime factor FFT algorithm and is illustrated in Fig. 2.

Fig. 2 shows a length 15 transform implemented by first calculating 5 length 3 transforms as in (50), then calculating 3 length 5 transforms as in (50). As in Fig. 1, blocks labeled "+" indicate addition, blocks labeled "X" indicate multiplication in the convolution DFT of (48). Fig. 2 is based on similar diagrams in Gold and Rader [13].

Winograd [5] has proposed another implementation of (52), which uses the special structure (49) of the short transforms to nest all multiplications inside of input and output additions. When the length M_1 short transform is written in terms of input additions $i^{(1)}$, output additions $o^{(1)}$, and multiplications $d^{(1)}$ and the length M_2 transform uses $i^{(2)}$, $o^{(2)}$, and $d^{(2)}$ as in (49), (53) becomes

$$y(n_1, k_2) = \sum_{l=0}^{\mu_2-1} o_{k_2 l}^{(2)} d_l^{(2)} \sum_{n_2=0}^{M_2-1} i_{l n_2}^{(2)}\, x(n_1, n_2). \tag{55}$$

Since $X(k_1, k_2)$ in (52) is a length M_1 transform of $y(n_1, k_2)$ which can also be implemented as in (49), (54) becomes

$$X(k_1, k_2) = \sum_{m=0}^{\mu_1-1} o_{k_1 m}^{(1)} d_m^{(1)} \sum_{n_1=0}^{M_1-1} i_{m n_1}^{(1)}\, Y(n_1, k_2). \tag{56}$$

Substituting (55) into (56), we get

$$X(k_1, k_2) = \sum_{m=0}^{\mu_1-1} o_{k_1 m}^{(1)} d_m^{(1)} \sum_{n_1=0}^{M_1-1} i_{m n_1}^{(1)} \sum_{l=0}^{\mu_2-1} o_{k_2 l}^{(2)} d_l^{(2)} \sum_{n_2=0}^{M_2-1} i_{l n_2}^{(2)}\, x(n_1, n_2). \tag{57}$$

The summations in (57) are an explicit representation of the operations indicated in Fig. 2. The order of summation may be changed as shown by McClellan and Rader [9] to nest all multiplies in the center giving

$$X(k_1, k_2) = \sum_{l=0}^{\mu_2-1} o_{k_2 l}^{(2)} \sum_{m=0}^{\mu_1-1} o_{k_1 m}^{(1)} d_m^{(1)} d_l^{(2)} \sum_{n_1=0}^{M_1-1} i_{m n_1}^{(1)} \sum_{n_2=0}^{M_2-1} i_{l n_2}^{(2)}\, x(n_1, n_2). \tag{58}$$

As shown in Fig. 3, (58) corresponds to first doing input adds on the rows of $x(n_1, n_2)$, then input adds on columns. Rows and columns are then multiplied as indicated by $d_m^{(1)}$

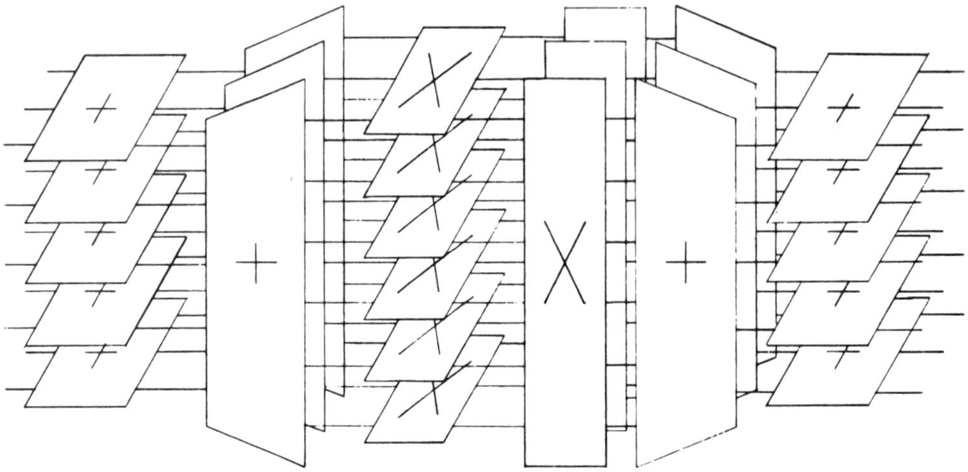

Fig. 3. Rearrangement of operations to nest multiplies inside of adds.

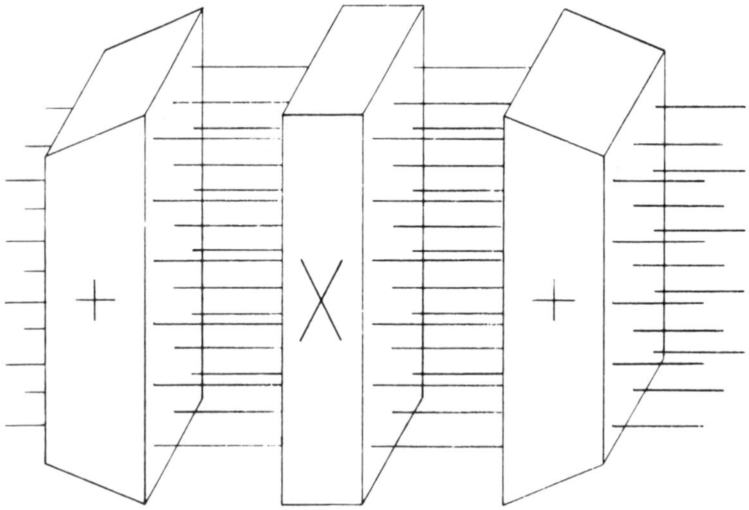

Fig. 4. Length 15 DFT implemented by the nested algorithm.

and $d_l^{(2)}$. Finally, output additions are performed on columns and rows. This algorithm, proposed by Winograd [5], will be called the nested algorithm in this paper because all multiplies are nested inside of additions as shown in Fig. 4.

The number of multiplications required by the nested algorithm is essentially the product of the total number of multiplications for each factor in the DFT, when implemented as in (58). However, some savings may be made for the case where the DFT being computed is at a point by point multiplication involving a product of W^0 coefficients. This corresponds to both d's in (58) being unity. The number of multiplications saved is the product of the number of W^0 multiplies in each short transform. Thus, the equation for the number of multiplications for 2 factors is # multiplies $= \mu_1\mu_2 - \kappa_1\kappa_2$, where

κ_i # of multiplies by W^0 for length M_i DFT,
μ_i total # of multiplies for length M_i DFT,
α_i # of additions.

For three factors we have # multiplies $= \mu_1\mu_2\mu_3 - \kappa_1\kappa_2\kappa_3$, and the pattern continues for more factors.

An equation for the number of additions is based on structures of the nested algorithms in (58) and on Fig. 3. The horizontal addition planes in Fig. 3 correspond to length M_2 transforms and there are M_1 of them (the indices n_1 and k_1 take on M_1 values). This contributes $M_1\alpha_2$ additions. The vertical addition planes in Fig. 3 correspond to length M_1 transforms and there are μ_2 of them (the index l takes on μ_2 values). This contributes $\mu_2\alpha_1$ additions. Thus, for two factors we have

$$\# \text{ additions} = M_1\alpha_2 + \mu_2\alpha_1.$$

For three factors we have

$$\# \text{ additions} = M_1M_2\alpha_3 + \mu_3[M_1\alpha_2 + \mu_2\alpha_1].$$

In the same manner, we have with four factors

TABLE II
COMPARISON OF DFT ALGORITHMS

N	Nested Algorithm Factors	Multiplies	Adds	Prime Factor FFT Multiplies	Adds	Radix 2 FFT Multiplies	Adds
30	$5 \cdot 3 \cdot 2$	68	384	68	384		
32	2^5					102	490
126	$9 \cdot 7 \cdot 2$	424	3312	512	2920		
128	2^7					774	3082
252	$9 \cdot 7 \cdot 4$	848	7128	1024	6344		
256	2^8					1926	7306
315	$9 \cdot 5 \cdot 7$	1292	11 286	1784	8812		
360	$9 \cdot 5 \cdot 8$	1152	9492	1396	8708		
504	$9 \cdot 7 \cdot 8$	1704	15 516	2300	13 948		
512	2^9					4614	16 906
840	$5 \cdot 7 \cdot 8 \cdot 3$	2592	24 804	4244	23 172		
1024	2^{10}					10 758	38 410
1260	$9 \cdot 5 \cdot 7 \cdot 4$	5168	50 184	7136	40 288		
2048	2^{11}					24 582	86 026
2520	$9 \cdot 5 \cdot 7 \cdot 8$	10 344	106 667	15 532	86 876		

$$\#\text{ additions} = M_1M_2M_3\alpha_4 + \mu_4 [M_1M_2\alpha_3 + \mu_3 [M_1\alpha_2 + \mu_2\alpha_1]]. \tag{59}$$

For the number of multiplications, the ordering of the factors is unimportant. However, the number of additions required depends on the ordering of the factors. For the nested transform additions given in Table II, the best ordering was used and is indicated by the order of factors. For complex data, the number of real multiplications and real additions is twice the numbers given by the above equations (multiplications of the complex data are with pure real or pure imaginary coefficients).

To compare the operations count of the nested algorithm with the operations required by the Cooley-Tukey and the prime factor FFT algorithm, we need to look at the latter two algorithms and determine the number of operations they require. In order to make the comparison more realistic, the radix 2 Cooley-Tukey algorithm will perform complex multiplies in three real multiplications and will not count multiplies by W^0. Multiplications by $\pm j$ will not be counted. We have for $N = 2^M$ ($\log N = M$) [14]:

$$\#\text{ multiplies} = 3\left[\frac{N}{2}\log N - \frac{3N}{2} + 2\right]$$

$$\#\text{ adds} = 2N\log N + \frac{5}{3}[\#\text{ multiplies}].$$

For the prime factor FFT algorithm we use the special short length transforms intended for this algorithm. See the first section of Table I. With $N = M_1M_2M_3$, M_2M_3 length M_1 transforms are computed, M_1M_3 length M_2 DFT's and M_1M_2 length M_3 DFT's. We have

$$\#\text{ multiplies} = 2[M_2M_3(\mu_1) + M_1M_3(\mu_2) + M_1M_2(\mu_3)] \tag{60}$$

$$\#\text{ adds} = 2[M_2M_3(\alpha_1) + M_1M_3(\alpha_2) + M_1M_2(\alpha_3)] \tag{61}$$

for complex data. Using these equations, the three algorithms are compared in Table II for several values of N.

VI. COMPARISON OF DFT ALGORITHMS—DISCUSSION OF TABLE II

The comparison in Table II shows that the prime factor FFT algorithm requires from 0–64 percent more multiplies than the nested transform. However the nested transform requires from 0–28 percent more additions than the prime factor FFT algorithm. In fact, if additions "cost" at least one-half as much as multiplications, then the multiply-add cost for the prime factor FFT algorithm is smaller for all lengths shown in Table II, except for lengths 30 and 840.

To develop an understanding for how these two transforms are related for various choices of factors, we will derive expressions for the number of operations required per output point.

From (60), the number of multiplies per output point for the prime factor FFT algorithm is simply the sum of the number of multiplies per point for each factor.

$$\#\text{ multiplies per point} = \sum\left(\frac{\mu_i}{M_i}\right). \tag{62}$$

Similarly, for the prime factor FFT algorithm,

$$\#\text{ adds per point} = \sum\left(\frac{\alpha_i}{M_i}\right). \tag{63}$$

For the nested algorithm, the number of multiplies per point is approximately the product of the number of multiplies per point for each factor.

$$\#\text{ multiplies per point} = \prod_i \left(\frac{\mu_i}{M_i}\right). \tag{64}$$

From (59), the number of additions per point for the nested transform is

$$\#\text{ adds per point} = \frac{\alpha_1}{M_1} + \left(\frac{\mu_1}{M_1}\right)\frac{\alpha_2}{M_2} + \left(\frac{\mu_1}{M_1}\right)\left(\frac{\mu_2}{M_2}\right)\frac{\alpha_3}{M_3} + \cdots. \tag{65}$$

For the factors used in this study, the number of operations per point is shown in Table III.

With k factors and an average number of multiplies per point, μ, the nested algorithm requires μ^k multiplies per point. The prime factor FFT algorithm requires $k\mu$ multiplies per point. When $\mu^k > k\mu$ or when $\mu > \sqrt[k-1]{k}$, the prime factor FFT algorithm requires fewer multiplies than the nested algorithm.

Since $\sqrt[k-1]{k}$ becomes smaller with increasing k (more factors) and μ_i increases for the extra factors which must be large, the prime factor FFT algorithm will have fewer multiplies per point than the nested algorithm when more factors are used.

With k factors and average number of additions per point, α, the prime factor FFT algorithm requires $k\alpha$ additions per point. The nested algorithm requires $\alpha + \mu\alpha + \cdots \mu^{k-1}\alpha = (\mu^k - 1)/(\mu - 1) \cdot \alpha$ additions per point for $\mu > 1$. In the special case $N = 2^r \cdot 3 \cdot p$ where p is a prime other than 2 or 3 and $r = 1, 2,$ or 3, both algorithms require the same number additions since $\mu = 1$ for the factors 2^r and 3. With other factors the prime factor FFT algorithm will have fewer additions. As shown above the difference in the number of additions will also increase rapidly when more factors are used.

A good strategy would be to use nesting for a few factors until μ began to grow, then combine, using the prime factor FFT algorithm, with another composite length short transform which was done with nesting.

VII. Programs for Comparing Transform Methods

The prime factor FFT algorithm was used to program a mixed radix DFT in which the short length DFT's are calculated using the fast convolution method previously described. A flowchart of the program is shown below.

Prime Factor FFT Flowchart

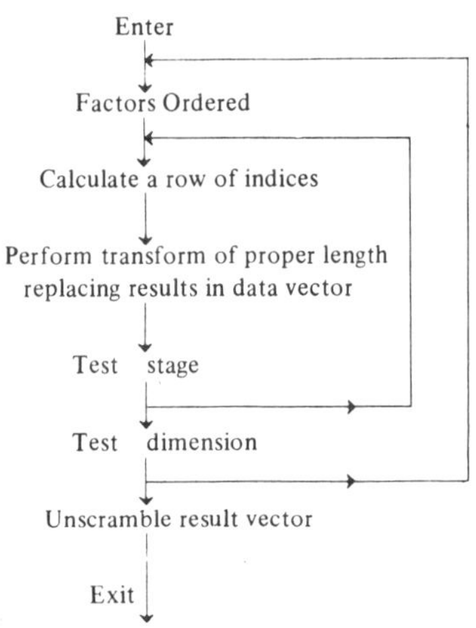

For $N = M_1 \cdot M_2 \cdot M_3$, there are three dimensions in the transform. First, $M_2 \cdot M_3$ length M_1 transforms are calculated. The indices of the data required for each length M_1 transform are calculated using the input mapping equation for the prime factor FFT algorithm (51). Then, the data corresponding to these indices is transformed by a length M_1 DFT using the fast convolution technique. The results of this transform are replaced in the data vector. Then, if more length M_1 transforms are required, the new indices are calculated. If all the length M_1 DFT's are finished, the program moves on to calculating the $M_1 \cdot M_3$ length M_2 DFT's. Finally, when the DFT's along all the dimensions have been calculated, the vector of the results must be unscrambled. This is done using the output mapping for the Good algorithms given by (50). The length N transform is then complete.

TABLE III

	Prime Factor FFT		Nested Algorithm	
N	μ/N	α/N	μ/N	α/N
2	0	1.0	1.0	1.0
3	0.33	2.0	1.0	2.0
4	0	2.0	1.0	2.0
5	0.8	3.4	1.2	3.4
7	1.14	5.14	1.28	5.14
8	0.25	3.25	1.0	3.25
9	0.89	5.44	1.33	5.44

TABLE IV
Time Test on IBM 370 155

	Times in Seconds			
N	Prime Factor FFT	Singleton's FFT	Radix 2 FFT	Percentage of Time Savings
32			0.013	
60	0.017	0.025		47 percent
64			0.027	
128			0.059	
210	0.080	0.119		49 percent
256			0.129	
315	0.111	0.179		61 percent
504	0.168	0.288		71 percent
512			0.280	
840	0.344	0.509		48 percent
1024			0.609	
1260	0.540	0.809		50 percent
2048			1.323	
2520	1.115	1.714		54 percent

This prime factor FFT algorithm for calculating the DFT was programmed in Fortran and in 8080 microprocessor assembly language. The Fortran prime factor FFT was compared in speed to a mixed radix FFT program written by Singleton [15], which uses a Cooley-Tukey mapping. The FFT subroutine of Singleton is 50 percent longer than the prime factor FFT subroutine. However, the prime factor FFT uses storage of two vectors of length N, while the Singleton FFT subroutine requires one length N vector. The results of the time tests for several transform lengths are given in Table IV. These tests were run on an IBM 370 computer for which the ratio of multiply to add time was 3. The power of 2 algorithm was taken from Rabiner and Gold [14]. It may be 15 percent slower than an algorithm which stores all powers of W.

The timing for the subroutines was accomplished using an interval timer on the IBM 370. The percent saving in time given is the percent by which the Singleton FFT subroutine is slower than the prime factor FFT subroutine.

The timing results of Table IV are for calculating the frequency response of a length 32 finite impulse response digital bandpass filter and are taken from a single run.

The times for the prime factor FFT may be calculated from the Table II values for the total number of operations by the following formula:

$$\text{time (in ms)} = (NF(0.052) + 0.028)N + 0.0096\,(\#\,\text{multiplies}) + 0.0045\,(\#\,\text{adds}).$$

The input indexing for the program took 52 μs per point for each dimension of the transform (NF equals the number of factors). The output indexing took 28 μs per point. The code generated by the Fortran add and multiply statements took 4.5 and 9.6 μs, respectively, to run. In the program, the shifts in the short DFT algorithms for the prime factor FFT were done as multiplies.

Next, the 8080 microprocessor assembly language version of the prime factor FFT subroutine was compared in speed with a radix 2 FFT subroutine. The radix 2 FFT was also written in assembly language, used three real multiplies for each complex multiply, and did not multiply by W^0. In addition, the FFT used precalculated values of W_N^i which were stored in a table. The FFT program was much shorter than the prime factor FFT program. The ratio of multiply to add times on the 8080 was approximately 30. A length 252 Good DFT requires 3.20 seconds. A length 256 radix 2 FFT requires 5.42 seconds. So, the radix 2 FFT subroutine is 70 percent slower than the prime factor FFT subroutine. The savings occur in both the multiplications and additions. The fewer multiplications required make up approximately 80 percent of the savings, the remainder of which is addition savings.

VIII. Conclusions

The conversion of a DFT into a circular convolution leads to new methods for computation of the DFT.

For short transforms these algorithms require few multiplications and additions as shown in Table I and as shown in the explicit formulas given in the Appendix.

Long transforms are built up from these short transforms in several ways which are compared in Table II. The prime factor FFT algorithm was chosen as the most attractive approach for several reasons. The prime factor FFT algorithm has about the same combined total of multiply-adds as the nested algorithm. However, it is easier to write a general prime factor FFT program. The prime factor FFT can be calculated using less memory than is required for the nested algorithm. It requires less data storage and probably less program memory. Since the prime factor FFT algorithm is done in small pieces it might run faster on machines with small high speed memory blocks. Special hardware for parallel computation will probably be simpler for the prime factor FFT algorithm.

A general prime factor FFT program was written for an IBM 370 in Fortran and for an 8080 microprocessor in assembly language, and the running time for this new algorithm was compared with a conventional FFT. In the 370 comparison the new algorithm was compared with the mixed radix algorithm of Singleton [15], since the prime factor FFT algorithm is a mixed radix algorithm. A reduction of approximately 50 percent was observed (see Table IV). Much larger savings may be expected if special hardware is constructed for the short convolution-based algorithms.

Many open questions remain. How should one combine nesting and prime factor FFT techniques to obtain long transforms from short ones? Should the multidimensional expansion always be done at the transform level, or should the convolutions contained within transforms also be implemented in multidimensional expansions? How can one improve on the indexing schemes required for these new transforms? It is likely that continuing development of longer and more efficient convolution algorithms will make implementations of the DFT using convolution even more attractive.

Appendix

3 Point DFT Algorithm

$a_1 = x(1) + x(2)$ $\quad m_1 = \frac{1}{2} a_1$ $\quad c_1 = x(0) - m_1$

$a_2 = x(1) - x(2)$ $\quad m_2 = 0.86603\, a_2$

$a_3 = x(0) + a_1$ $\quad X(0) = a_3$

$X(1) = c_1 - jm_2$

1 Multiply, 6 Adds, 1 Shift. $\quad X(2) = c_1 + jm_2$

5 Point DFT Algorithm

$a_1 = x(1) + x(4)$ $\quad m_1 = 0.95106\, a_5$ $\quad c_1 = x(0) - m_5$

$a_2 = x(1) - x(4)$ $\quad m_2 = 1.53884\, a_2$ $\quad c_2 = c_1 + m_4$

$a_3 = x(2) + x(3)$ $\quad m_3 = 0.36327\, a_4$ $\quad c_3 = c_1 - m_4$

$a_4 = x(2) - x(3)$ $\quad m_4 = 0.55902\, a_6$ $\quad c_4 = m_1 - m_3$

$a_5 = a_2 + a_4$ $\quad m_5 = \frac{1}{4} a_7$ $\quad c_5 = m_2 - m_1$

$a_6 = a_1 - a_3$

$a_7 = a_1 + a_3$

$a_8 = x(0) + a_7$

4 Multiplies, 17 Adds, 2 Shifts.

$X(0) = a_8$

$X(1) = c_2 - jc_4$

$X(2) = c_3 - jc_5$

$X(3) = c_3 + jc_5$

$X(4) = c_2 + jc_4$

7 Point DFT Algorithm

$a_1 = x(1) + x(6)$	$m_1 = 0.16667\, a_7$	$c_1 = x(0) - m_1$
$a_2 = x(1) - x(6)$	$m_2 = 0.79016\, a_8$	$c_2 = c_1 + m_2 + m_3$
$a_3 = x(2) + x(5)$	$m_3 = 0.05585\, a_9$	$c_3 = c_1 - m_2 - m_4$
$a_4 = x(2) - x(5)$	$m_4 = 0.73430\, a_{10}$	$c_4 = c_1 - m_3 + m_4$
$a_5 = x(3) + x(4)$	$m_5 = 0.44096\, a_{11}$	$c_5 = m_5 + m_6 - m_7$
$a_6 = x(3) - x(4)$	$m_6 = 0.34087\, a_{12}$	$c_6 = m_5 - m_6 - m_8$
$a_7 = a_1 + a_3 + a_5$	$m_7 = 0.53397\, a_{13}$	$c_7 = -m_5 - m_7 - m_8$
$a_8 = a_1 - a_5$	$m_8 = 0.87484\, a_{14}$	
$a_9 = -a_3 + a_5$		$X(0) = a_{15}$
$a_{10} = -a_1 + a_3$		$X(1) = c_2 - jc_5$
$a_{11} = a_2 + a_4 - a_6$		$X(2) = c_3 - jc_6$
$a_{12} = a_2 + a_6$		$X(3) = c_4 - jc_7$
$a_{13} = -a_4 - a_6$		$X(4) = c_4 + jc_7$
$a_{14} = -a_2 + a_4$		$X(5) = c_3 + jc_6$
$a_{15} = x(0) + a_7$		$X(6) = c_2 + jc_5$

8 Multiplies, 36 Adds.

9 Point DFT Algorithm

$a_1 = x(1) + x(8)$	$m_1 = 0.19740\, a_9$	$c_1 = x(0) - m_7$
$a_2 = x(1) - x(8)$	$m_2 = 0.56858\, a_{10}$	$c_2 = m_2 - m_3$
$a_3 = x(2) + x(7)$	$m_3 = 0.37111\, a_{11}$	$c_3 = m_1 + m_3$
$a_4 = x(2) - x(7)$	$m_4 = 0.54253\, a_{12}$	$c_4 = m_1 + m_2$
$a_5 = x(4) + x(5)$	$m_5 = 0.10026\, a_{13}$	$c_5 = c_1 + c_2 - c_3$
$a_6 = x(4) - x(5)$	$m_6 = 0.44228\, a_{14}$	$c_6 = c_1 + c_3 + c_4$
$a_7 = x(3) + x(6)$	$m_7 = \frac{1}{2}\, a_7$	$c_7 = c_1 - c_2 - c_4$
$a_8 = x(3) - x(6)$	$m_8 = 0.86603\, a_8$	$c_8 = m_4 - m_6$
$a_9 = -a_1 + a_5$	$m_9 = \frac{1}{2}\, a_{15}$	$c_9 = m_5 - m_6$
$a_{10} = a_1 - a_3$	$m_{10} = 0.86603\, a_{16}$	$c_{10} = m_4 - m_5$
$a_{11} = -a_3 + a_5$		$c_{11} = c_8 + c_9 + m_8$
$a_{12} = a_2 - a_6$		$c_{12} = c_8 + c_{10} - m_8$
$a_{13} = a_2 + a_4$		$c_{13} = -c_9 + c_{10} + m_8$
$a_{14} = -a_4 - a_6$		$c_{14} = x(0) + a_7 - m_9$
$a_{15} = a_1 + a_3 + a_5$		$X(0) = a_{17}$
$a_{16} = a_2 - a_4 + a_6$		$X(1) = c_5 - jc_{11}$

$$a_{17} = x(0) + a_{15} + a_7$$

8 Multiplies, 49 Adds, 2 Shifts.

$$X(2) = c_6 - jc_{12}$$
$$X(3) = c_{14} - jm_{10}$$
$$X(4) = c_7 - jc_{13}$$
$$X(5) = c_7 + jc_{13}$$
$$X(6) = c_{14} + jm_{10}$$
$$X(7) = c_6 + jc_{12}$$
$$X(8) = c_5 + jc_{11}$$

Acknowledgment

The authors would like to acknowledge the suggestions and helpful comments of C. M. Rader and J. H. McClellan.

References

[1] J. W. Cooley and J. W. Tukey, "An algorithm for the machine calculation of complex Fourier series," *Math. Comput.*, vol. 19, pp. 297-301, Apr. 1965.

[2] C. M. Rader, "Discrete Fourier transforms when the number of data samples is prime," *Proc. IEEE*, vol. 56, pp. 1107-1108, June 1968.

[3] S. Winograd, "Some bilinear forms whose multiplicative complexity depends on the field of constants," IBM T. J. Watson Res. Ctr., Yorktown Heights, NY, IBM Res. Rep., RC 5669, Oct. 10, 1975.

[4] R. Agarwal and J. W. Cooley, "New algorithms for digital convolution," presented at the IEEE Arden House Workshop, Feb. 1976; also to appear in this Transactions.

[5] S. Winograd, "On computing the discrete Fourier transform," *Proc. Nat. Acad. Sci. USA*, vol. 73, no. 4, pp. 1005-1006, Apr. 1976.

[6] I. J. Good, "The interaction algorithm and practical Fourier analysis," *J. Royal Statist. Soc., ser. B*, vol. 20, pp. 361-372, 1958; "Addendum," vol. 22 (MR 21 1674; MR 23 A4231), pp. 372-375, 1960.

[7] L. H. Thomas, "Using a computer to solve problems in physics," in *Applications of Digital Computers*. Boston, MA: Ginn and Co., 1963.

[8] C. D. Mostow, J. H. Sampson, and J. Meyer, *Fundamental Structures of Algebra*. New York: McGraw-Hill, 1963.

[9] J. McClellan and C. M. Rader, "There is something much faster than the fast Fourier transform," Seminar Notes, Oct. 21, 1976.

[10] H. F. Silverman, "An introduction to programming the Winograd Fourier transform algorithm (WFTA)," *IEEE Trans. Acoust., Speech, Signal Processing*, vol. ASSP-25, pp. 152-165, Apr. 1977.

[11] C. S. Burrus, "Index mappings for multidimensional formulation of the DFT and convolution," *IEEE Trans. Acoust., Speech, Signal Processing*, vol. ASSP-25, pp. 239-242, June 1977.

[12] D. E. Knuth, "Seminumerical algorithms," in *The Art of Computer Programming*, vol. 2. Reading, MA: Addison-Wesley, 1971.

[13] B. Gold and C. M. Rader, *Digital Processing of Signals*. New York: McGraw-Hill, 1969, p. 194.

[14] L. R. Rabiner and B. Gold, *Theory and Application of Digital Signal Processing*. Englewood Cliffs, NJ: Prentice-Hall, 1975.

[15] R. C. Singleton, "An algorithm for computing the mixed radix fast Fourier transform," *IEEE Trans. Audio Electroacoust.*, vol. AU-17, pp. 93-103, June 1969.

Reprinted from the PROCEEDINGS OF THE IEEE
VOL. 56, NO. 6, JUNE, 1968
pp. 1107-1108

Discrete Fourier Transforms When the Number of Data Samples Is Prime

***Abstract*—The discrete Fourier transform of a sequence of N points, where N is a prime number, is shown to be essentially a circular correlation. This can be recognized by rearranging the members of the sequence and the transform according to a rule involving a primitive root of N. This observation permits the discrete Fourier transform to be computed by means of a fast Fourier transform algorithm, with the associated increase in speed, even though N is prime.**

To compute the discrete Fourier transform of a long sequence of data samples, one of the fast Fourier transform (FFT) algorithms should be used if possible.[1],[2] The limitation common to all of these algorithms is that the number of data samples, N, must be highly composite. In this letter we show how FFT techniques can be applied to the computation of a discrete Fourier transform when N is prime.

I. Notation and Definitions

We basically restrict our attention to the discrete Fourier transform. Let the data to be transformed be a sequence of N numbers $\{a_i\}$, $i=0, 1, \cdots, N-1$, where the braces indicate the entire sequence, while the symbol a_i represents only the ith member of the sequence. The discrete Fourier transform (DFT) is the sequence $\{A_k\}$, $k=0, 1, \cdots, N-1$, whose members are given by

$$A_k = \sum_{i=0}^{N-1} a_i \exp(-j(2\pi/N)ik). \tag{1}$$

The values of A_k are samples of the z-transform of the finite length sequence $\{a_i\}$ at N points equally spaced on the unit circle.

We need a notation for an integer modulo N. For brevity we will indicate this by superfluous parentheses:

$$((X)) = X \text{ modulo } N. \tag{2}$$

If N is prime there is some number g, not necessarily unique, such that there is a one-to-one mapping of the integers $i=1, \cdots, N-1$ to the integers $j=1, 2, \cdots, N-1$, given by

$$j = ((g^i)). \tag{3}$$

For example, let $N=7$ and $g=3$. The table below gives the mapping of i onto j.

i	1	2	3	4	5	6
j	3	2	6	4	5	1

Manuscript received March 4, 1968.

In number theory, g is called a primitive root of N. A table of primitive roots of primes less than 10^4 is given by Abramowitz and Stegun.[3]

II. Computation of $\{A_k\}$ when N Is Prime

In (1) we have an expression for A_k for all k. The expression for A_0 is particularly simple,

$$A_0 = \sum_{i=0}^{N-1} a_i, \tag{4}$$

and is to be computed directly. For the other A_k we observe that a_0 is not to be multiplied, and we choose to add it last into the summation. We are left with the sequence $\{A_k - a_0\}$, $k=1, 2, \cdots, N-1$ to compute, given by

$$A_k - a_0 = \sum_{i=1}^{N-1} a_i \exp\left(-j\frac{2\pi}{N} ik\right). \tag{5}$$

We permute the terms in the summation, and change the order of the equations via transformations

$$\begin{aligned} i &\to ((g^i)) \\ k &\to ((g^k)) \end{aligned} \tag{6}$$

and, noting that $((g^{N-1}))=((g^0))$, we can see that

$$(A_{((g^k))} - a_0) = \sum_{i=0}^{N-2} a_{((g^i))} \exp\left(-j\frac{2\pi}{N} g^{(i+k)}\right). \tag{7}$$

We are now able to recognize that the sequence $\{A_{((g^k))} - a_0\}$ is the circular correlation of the sequence $\{a_{((g^i))}\}$ and the sequence $\{\exp(-j(2\pi/N)g^i)\}$. But Stockham has shown how circular (or ordinary) correlation functions can be computed with a greatly reduced number of operations by making use of FFT algorithms.[4] There are two ways we can proceed. Since N is prime, $N-1$ must be composite. Suppose that it is highly composite. Then the $N-1$ point circular correlation (7) may be recognized as the inverse DFT of the product of the DFT of $\{a_{((g^{-i}))}\}$ and the DFT of $\{\exp(-j(2\pi/N)g^i\}$.[1] All the DFT operations called for are performed by an FFT algorithm.

$$\{A_{((g^k))} - a_0\} = \mathrm{DFT}^{-1}\left\{(\mathrm{DFT}\{a_{((g^{-i}))}\})\left(\mathrm{DFT}\left\{\exp\left(-j\frac{2\pi}{N} g^i\right)\right\}\right)\right\}. \tag{8}$$

[1] It may be shown that this DFT has magnitude $\sqrt{N}$ at all frequencies but the zeroth, where it is -1.

Proceeding in this first way we will be successful only if $N-1$ is highly composite. If $N-1$ is only modestly composite, as with $N=563$, the savings of the FFT algorithm will be overcome by the fact that more than one DFT must be computed. However, there is another way we can proceed which is not subject to these limitations. The second method is based on the observation that a circular correlation or convolution where the number of points is not highly composite can be computed as a part of a circular convolution with a larger number of points. Letting N' be any highly composite integer greater than $2N-4$, we create an N' point sequence $\{b_i\}$ by inserting $(N'-N+1)$ zeros between the zeroth and first points of $\{a_{((g^{-i}))}\}$ and we create a second N' point sequence $\{c_i\}$ by periodically repeating the $N-1$ point sequence $\{\exp(-j(2\pi/N)g^i)\}$ until N' points are present. Then the inverse DFT of the product of the DFTs of $\{b_i\}$ and $\{c_i\}$ contains $\{A_{((g^k))}-a_0\}$ as a subsequence—the first $N-1$ points. Since N' can be chosen to be highly composite, even a power of two, an FFT algorithm can be used to compute the DFTs.

Using either technique, about one-third of the computation can be saved if the transform of $\{\exp(-j(2\pi/N)g^i)\}$ is precomputed. One method requires a computation proportional to $(N-1)$ times the sum of the factors of $(N-1)$ whereas the second method requires a computation proportional to $N'\log N'$. Furthermore, the summation called for in (4) and the addition of a_0 to each other A_k can each be performed with negligible additional computation by operating on intermediate quantities available when the correlation is done by FFT techniques.

III. Conclusions

While the restriction that N be a highly composite number for FFT techniques to be useful has not proved severe, it is interesting to know that it can be removed. On the other hand, the recognition that a DFT can be expressed as a convolution may be useful in itself, as this implies that a single network with fixed parameters can compute all the points of a DFT. It is expected that such diverse applications as radar beam forming and modem design may profitably use this result..

Charles M. Rader
M.I.T. Lincoln Lab.[2]
Lexington, Mass. 02173

References

[1] J. W. Cooley and J. W. Tukey, "An algorithm for the machine computation of complex Fourier series," *Math. Comput.*, vol. 19, pp. 297–301, April 1965.

[2] "What is the fast Fourier transform?" *Proc. IEEE*, vol. 55, pp. 1664–1674, October 1967.

[3] M. Abramowitz and I. Stegun, *Handbook of Mathematical Functions*. New York: Dover, 1965, p. 827 and pp. 864–869.

[4] T. G. Stockham, "High speed convolution and correlation," *1966 Spring Joint Computer Conf., AFIPS Proc.*, vol. 28. Washington, D.C.: Spartan, 1966, pp. 229–233.

[2] Operated with support from the U. S. Air Force.

New Algorithms for Digital Convolution

RAMESH C. AGARWAL, MEMBER, IEEE, AND JAMES W. COOLEY, MEMBER, IEEE

Abstract—It is shown how the Chinese Remainder Theorem (CRT) can be used to convert a one-dimensional cyclic convolution to a multidimensional convolution which is cyclic in all dimensions. Then, special algorithms are developed which compute the relatively short convolutions in each of the dimensions. The original suggestion for this procedure was made in order to extend the lengths of the convolutions which one can compute with number-theoretic transforms. However, it is shown that the method can be more efficient, for some data sequence lengths, than the fast Fourier transform (FFT) algorithm. Some of the short convolutions are computed by methods in an earlier paper by Agarwal and Burrus. Recent work of Winograd, consisting of theorems giving the minimum possible numbers of multiplications and methods for achieving them, are applied to these short convolutions.

I. Introduction and Background

THE calculation of the finite digital convolution

$$y_i = \sum_{k=0}^{N-1} h_{i-k} x_k \tag{1.1}$$

has extensive applications in both general-purpose computers and specially constructed digital processing devices. It is used to compute auto and cross correlation functions, to design and implement finite impulse response (FIR) and infinite impulse response digital filters, to solve difference equations, and to compute power spectra.

While the direct calculation of the convolution according to the defining formula (1.1) would require a number of multiplications and additions proportional to N^2 for large N [which we denote by $0(N^2)$], use of the fast Fourier transform algorithm (FFT) (see [5]) has been able to reduce this to $0(N \log N)$ operations when N is a power of 2. To be more specific, we consider the problem where $h_i, i = \cdots -1, 0, 1, \cdots$ is a periodic sequence of period N so that $h_i = h_{N+i}$. Then the discrete Fourier transform (DFT)

$$X_n = \sum_{k=0}^{N-1} x_k e^{-2\pi i n k/N} \tag{1.2}$$

has the property that the DFT's H_n, X_n, and Y_n, $n = 0, 1, 2, \cdots, N-1$, of the three sequences h_k, x_k, and y_k, $k = 0, 1, \cdots, N-1$, respectively, are related by

$$Y_n = H_n X_n, \qquad n = 0, 1, \cdots, N-1. \tag{1.3}$$

If (1.1) is regarded as a multiplication of a vector x by a matrix H whose i, k element is h_{i-k}, then the DFT (1.2) is seen to be a transformation which diagonalizes H. This is a transformation to the frequency domain where the computationally expensive convolution operation in (1.1) corresponds to the N complex multiplications in (1.3). The DFT is, therefore, said to have the cyclic convolution property (CCP). Since the FFT algorithm enables one to calculate the DFT in $0(N \log N)$ operations, the entire convolution requires $0(N \log N)$ operations.

A seemingly paradoxical situation arises here when one considers that all numbers in (1.1) may be integers making exact calculation of the convolution possible. However, the computationally efficient DFT method involves intermediate quantities, i.e., sines and cosines, which are irrational numbers, thereby making exact results impossible on a digital machine. This, as shown by Agarwal and Burrus [2], is a consequence of the fact that, in order to have the CCP, a transformation must have the form

$$X_n = \sum_{k=0}^{N-1} x_k \alpha^{nk}, \qquad n = 0, 1, \cdots, N-1 \tag{1.4}$$

where, in the ring in which the calculation takes place, α must be a primitive Nth root of unity. There is no primitive Nth root of unity in the ring of integers where the calculation may be considered to be defined, or even in the field of rational numbers. However, $e^{-2\pi i/N}$ is a primitive Nth root of unity in the complex number field, so the whole calculation is, therefore, carried out in the complex number field with $\alpha = e^{-2\pi i/N}$ when applying the DFT method.

The theories of DFT's and the FFT algorithm were investigated in finite fields and rings by Nicholson [11] and Pollard [12]. The FFT algorithm applications to Fourier, Walsh, and Hadamard transforms were shown to be special cases of Fourier transforms in algebras over fields or rings. Pollard described applications where the DFT is defined in finite (Galois) fields. This led Rader [13] to suggest performing the calculations in the ring of integers modulo a Mersenne number, $M_p = 2^p - 1$, i.e., in remainder arithmetic modulo M_p. In this ring, $2^p \equiv 1$ so that 2 is a pth primitive root of unity and -2 is a $2p$th primitive root of unity. Thus, a Mersenne transform is defined which has the CCP for sequences of length $N = 2p$, with -2 replacing $e^{-2\pi i/N}$ as the Nth primitive root of unity, and with all calculations done in remainder arithmetic modulo M_p. Rader advocated such a transform since using 2 or -2 as a root of unity would necessitate only shift and add operations in computing the transforms. The only multiplications required would be the N multiplications of the values of the transforms. If one takes $N = p$, a prime, the FFT algorithm cannot be used and the number of shift and add operations would be $0(N^2)$. Rader also mentioned the possibility of using Fermat numbers as moduli so that N would be a power of 2, permitting the use of the FFT algorithm.

Agarwal and Burrus [2] made a thorough investigation of the necessary and sufficient conditions on the modulus, word

Manuscript received December 2, 1976; revised March 31, 1977.

The authors are with IBM Thomas J. Watson Research Center, Yorktown Heights, NY 10598.

length, and sequence lengths for number-theoretic transforms (NTT's) to have the CCP and to permit use of the FFT algorithm. Their results show the rather stringent limitation on the sequence lengths which can be used. They show that the use of the Fermat numbers $F_b = 2^t + 1$ where $t = 2^b$ and particularly F_5, offer some of the best choices as moduli for the NTT. In this case too, however, the sequence length is severely limited. It is proportional to the number of bits in the modulus.

A number of suggestions have arisen for lengthening the sequences which can be handled by the NTT. One suggestion is to perform the calculation modulo several mutually prime moduli and then obtain the desired result by using the CRT. Reed and Truong [15] have also shown how one can extend the method to Galois fields over complex integers modulo Mersenne primes to enable one to use the FFT algorithm to compute convolutions of complex sequences, and to lengthen the sequences which the method can handle. But, in that case, the resulting primitive Nth root of unity is not simple and, therefore, the computation of the complex Mersenne transform would require general multiplications.

One of the most promising methods for lengthening the sequences one can handle was suggested by Rader [13], and then developed by Agarwal and Burrus [1]. This consisted of mapping the one-dimensional sequences into multidimensional sequences and expressing the convolution as a multidimensional convolution. Then, the Fermat Number Transform (FNT) is suggested for the computation of the convolution in the longest dimension. For the convolutions in the other dimensions, Agarwal and Burrus devised special algorithms which reduced the number of multiplications considerably. The number of additions usually increased slightly, but when considering the NTT, one is already considering either a special-purpose machine or a computer which favors integer arithmetic, in which case multiplication is considerably more expensive than addition.

The mapping of the one-dimensional array considered by Agarwal and Burrus was to simply assign the elements lexicographically to the multidimensional array. This meant that the multidimensional array was cyclic in only one dimension, and, to employ cyclic convolution algorithms in the other dimensions, one would have to double the number of points in all dimensions except one. The result of this effect was to show that a variety of short convolutions, combined with FNT's, could reduce the amount of computation considerably. It was also shown how, even without NTT's, multidimensional techniques can compute convolutions faster for N less than or equal to 128 as compared with the use of the FFT algorithm.

One innovation of the present paper consists of an extension and improvement of the general idea of the Agarwal and Burrus [1] paper, i.e., to compute a convolution in terms of a multidimensional convolution in which the short convolutions in some of the dimensions are done by special efficient algorithms. The second innovation is to let the dimensions of the multidimensional arrays be mutually prime numbers, and then use the CRT to map the sequences into multidimensional arrays. This makes the data cyclic in all dimensions and avoids the necessity of appending zeros in order to use cyclic convolution algorithms. Although this method was also originally conceived with the idea that the convolution in the longest dimension would be done by the NTT, it is shown that it is efficient even when the NTT is not used. In fact, the crossover N-value, below which the present method is more efficient than FFT methods, is much higher and, in some cases, is around 400.

The algorithms developed by Agarwal and Burrus [1] were generally developed by skillful, but tedious manipulations which, however, lacked systematic methods for doing longer convolutions or for examining the many possible such algorithms for an optimal choice. Since then, Winograd [18] has applied computational complexity theory to the problem of computing convolutions. He has developed one theorem which gives the minimum number of multiplications required for computing a convolution and another theorem, which describes the general form of any algorithm which computes the convolution in the minimum number of multiplications. He has also developed a theoretical framework which can be used to find the best algorithms in terms of both numbers of multiply/adds and complexity. For the present purposes, his important theorems will be cited and algorithms resulting from them will be compared with the algorithms used here. Actually, it is not necessary, in the multidimensional technique, to have optimal algorithms for more than a few powers of small primes. Some of these have already appeared in the Agarwal and Burrus paper [1], and some of the additional ones given here were worked out by the same methods. After work on the present paper was well under way, the authors became acquainted with Winograd's methods and used them for simplifying the derivation of the longer convolutions and for developing several algorithms from which to choose. It was also found that Winograd had worked out many of the algorithms for the same convolutions.

In what follows, we will show how some of the long tedious parts of the derivations of the algorithms by Winograd's methods were done with SCRATCHPAD [8], a computer system at the IBM Watson Research Center for doing formula manipulation. This not only permitted the derivation of algorithms for longer convolutions, but simplified the choice of the best from a number of algorithms.

A rather simple matrix formulation is shown to be satisfied by the convolution algorithms developed here. The algorithm is then made to resemble, in a loose sense, other transform techniques having the CCP, i.e., the ability to replace the convolution operation by element by element multiplication in the transform domain. This is called the "rectangular transform," since the matrices defining it are rectangular instead of square.

II. Algorithms for Short Convolutions

A. The Cook-Toom Algorithm

In order to show the general idea of how complexity theory is applied and what type of algorithms are being developed, the Cook-Toom algorithm (see [9]) for noncyclic convolution will be explained in detail. This yields algorithms with the minimum number of multiplications, but with greater complexity than the ones developed in the following subsection.

In any case, it yields algorithms having the general form of those we are treating.

The noncyclic convolution being considered here is of the form

$$w_i = \sum_{k=\max(0,i-N+1)}^{\min(N-1,i)} h_{i-k}x_k, \quad i = 0, 1, \cdots, 2N-2. \tag{2.1}$$

The sequence length N, in this and the following sections, is the number of points in one dimension in the multidimensional arrays mentioned above. We will consider algorithms for both the cyclic and noncyclic cases. The first theorem we consider is described by Knuth [9]. We give it in a slightly different form to make it resemble the formulas in the next section.

Theorem 1: (The Cook-Toom Algorithm)

The noncyclic convolution (2.1) can be computed in $2N-1$ multiplications.

The proof is given by constructing the algorithm. Let us define the generating polynomial[1] of a sequence x_i, $i = 0, 1, \cdots, N-1$ by

$$X(z) = \sum_{i=0}^{N-1} x_i z^i. \tag{2.2}$$

We will assume similar definitions for $H(z)$, $W(z)$, and $Y(z)$ as generating polynomials of the h_i, w_i, and y_i sequences, respectively. It is easily seen that

$$W(z) = H(z)\, X(z) \tag{2.3}$$

where $W(z)$ is a $2N-2$ degree polynomial. Let the x_i's and h_i's be treated as indeterminates in terms of which we will obtain formulas for the w_i's.

To determine the $2N-1$ w_i's, one selects $2N-1$ distinct numbers α_j, $j = 0, 1, \cdots, 2N-2$, and substitutes them for z in (2.3) to obtain the $2N-1$ products

$$m_j = W(\alpha_j) = H(\alpha_j)\, X(\alpha_j), \quad j = 0, 1, \cdots, 2N-2 \tag{2.4}$$

of linear combinations of the h_i's and x_i's. The Lagrange interpolation formula may be used to uniquely determine the $2N-2$ degree polynomial

$$W(z) = \sum_{j=0}^{2N-2} m_j \prod_{k \neq j} \frac{(z-\alpha_k)}{(\alpha_j - \alpha_k)}. \tag{2.5}$$

Thus, the convolution (2.1) is obtained at the cost of the $2N-1$ multiplications in (2.4). This completes the proof of Theorem 1.

The Cook-Toom algorithm is then formulated as follows: since the $H(\alpha_j)$'s and $X(\alpha_j)$'s are linear combinations of the h_i's and x_i's, respectively, we can, therefore, write (2.4) in the matrix-vector form

$$\boldsymbol{m} = (A\boldsymbol{h}) \times (A\boldsymbol{x}) \tag{2.6}$$

where $\boldsymbol{h}$ and $\boldsymbol{x}$ are N-element column vectors with elements h_i and x_i, respectively, and where $\times$ denotes element by element multiplication. The elements of $\boldsymbol{m}$ are the $W(\alpha_j)$'s and

$$A = \begin{bmatrix} 1 & \alpha_0 & \alpha_0^2 & \cdots & \alpha_0^{N-1} \\ 1 & \alpha_1 & \alpha_1^2 & \cdots & \alpha_1^{N-1} \\ \vdots & \vdots & \vdots & \cdots & \vdots \\ 1 & \alpha_{2N-2} & \cdot & \cdots & \alpha_{2N-2}^{N-1} \end{bmatrix}. \tag{2.7}$$

Therefore, from (2.5) we see that the coefficients of $W(z)$ will be linear combinations of the m_j's and may be written as

$$\boldsymbol{w} = C^*\boldsymbol{m} \tag{2.8}$$

where C^* is a $2N-1$ by $2N-1$ matrix. If the α_j's are rational numbers, the elements of C^* will be rational numbers. To apply the above to the calculation of cyclic convolutions, it remains only to compute

$$Y(z) = W(z) \bmod (z^N - 1). \tag{2.9}$$

Since $z^N \equiv 1 \bmod (z^N - 1)$, this means simply that

$$\begin{aligned} y_0 &= w_0 + w_N \\ y_1 &= w_1 + w_{N+1} \\ &\vdots \\ y_{N-2} &= w_{N-2} + w_{2N-2} \\ y_{N-1} &= w_{N-1} \end{aligned} \tag{2.10}$$

which leads to

$$\boldsymbol{y} = C\boldsymbol{m} \tag{2.11}$$

where C is an N by $2N-1$ matrix obtained from C^* by performing the row operations on C^* corresponding to (2.10). Here, and in what follows, we seek algorithms of the general form (2.6) and (2.8) or (2.11), except that we will not require that $\boldsymbol{x}$ be multiplied by the same matrix as $\boldsymbol{h}$ and consider, instead, algorithms of a more general form,

$$\boldsymbol{m} = (A\boldsymbol{h}) \times (B\boldsymbol{x}). \tag{2.12}$$

We will usually consider applications where a fixed impulse response sequence $\boldsymbol{h}$ is convolved with many $\boldsymbol{x}$ sequences so that $A\boldsymbol{h}$ will be precomputed and the operations required for computing $A\boldsymbol{h}$ will not be counted.

Although we write the algorithms in terms of matrices, it will be shown that, for efficiency, one does not store the matrices as such and does not perform full matrix-vector multiplications. In what follows, however, we will refer to A, B, and C as either matrices or as operators, interchangeably.

If derived as described above, with integers for the α_j's, A and B will have integer coefficients and C will have rational coefficients. Since $A\boldsymbol{h}$ is precomputed, we usually redefine A and C so that the denominators in C appear in a redefined A and the redefined C has integer elements. Therefore, in the methods and theorems which are given below, the operators B and C are considered to involve no multiplications. The only multiplications counted are in the element by element multiplication of $A\boldsymbol{h}$ by $B\boldsymbol{x}$. However, the Cook-Toom algorithm yields rather large integer coefficients in the A, B, and C matrices which can be as costly as multiplication. The ob-

[1] This is the familiar z transform, except for the fact that we have chosen to use positive instead of negative powers of z.

jective in the following section will be to obtain algorithms with as few multiplications as possible while still keeping B and C simple.

To give an example, suppose we wish to calculate the noncyclic 2-point convolution

$$\begin{aligned} w_0 &= h_0 x_0 \\ w_1 &= h_0 x_1 + h_1 x_0 \\ w_2 &= h_1 x_1. \end{aligned} \tag{2.13}$$

In terms of z transforms, this is equivalent to

$$w_0 + w_1 z + w_2 z^2 = (h_0 + h_1 z)(x_0 + x_1 z). \tag{2.14}$$

Letting $\alpha_j = -1, 0, 1$ for $j = 0, 1, 2$ in (2.4),

$$\begin{aligned} m_0 &= (h_0 - h_1)(x_0 - x_1) \\ m_1 &= h_0 x_0 \\ m_2 &= (h_0 + h_1)(x_0 + x_1) \end{aligned} \tag{2.15}$$

and, for (2.5) we obtain

$$W(z) = m_2 \frac{z(z+1)}{1 \cdot 2} + m_1 \frac{(z+1)(z-1)}{1 \cdot (-1)} + m_0 \frac{(z-1)z}{(-2) \cdot (-1)} \tag{2.16}$$

so that

$$\begin{aligned} w_0 &= m_1 \\ w_1 &= (m_2 - m_0)/2 \\ w_2 &= (m_0 + m_2)/2 - m_1. \end{aligned} \tag{2.17}$$

To illustrate what was said above about transferring denominators from the C to the A matrix, we combine the factor $\frac{1}{2}$ with the h_j's and store the precomputed constants

$$\begin{aligned} a_0 &= (h_0 - h_1)/2 \\ a_1 &= h_0 \\ a_2 &= (h_0 + h_1)/2 \end{aligned} \tag{2.18}$$

so that the algorithm becomes, in terms of the a_j's and redefined m_j's,

$$\begin{aligned} m_0 &= a_0(x_0 - x_1) \\ m_1 &= a_1 x_0 \\ m_2 &= a_2(x_0 + x_1) \end{aligned} \tag{2.19}$$

$$\begin{aligned} w_0 &= m_1 \\ w_1 &= m_2 - m_0 \\ w_2 &= m_0 + m_2 - m_1. \end{aligned} \tag{2.20}$$

Thus, only 3 multiplications and 5 additions are required instead of the 4 multiplications and 1 addition appearing in the defining formula.

Finally, if one were multiplying two complex numbers $x = x_0 + ix_1$ and $h = h_0 + ih_1$, the result would be $w_0 - w_2 + iw_1$. The above derivation, therefore, gives one of several ways of multiplying complex numbers in 3 instead of 4 real multiplications.

It is seen here that one can generate as many algorithms as one wishes by using different choices of α_j-values in (2.4). For example, if one uses $\alpha_j = 0, 1, 2$, one obtains

$$\begin{aligned} m_0 &= h_0 x_0 \\ m_1 &= (h_0 + h_1)(x_0 + x_1) \\ m_2 &= (h_0 + 2h_1)(x_0 + 2x_1) \end{aligned} \tag{2.21}$$

and

$$\begin{aligned} w_0 &= m_0 \\ w_1 &= (-3m_0 - m_2)/2 + 2m_1 \\ w_2 &= (m_0 + m_2)/2 - m_1. \end{aligned} \tag{2.22}$$

The first algorithm, (2.18)-(2.20), may be preferable due to its simpler coefficients.

B. Optimal Short Convolution Algorithms

The general form of the algorithm (2.11) and (2.12) is

$$y = C(Ah) \times (Bx). \tag{2.23}$$

This suggests a similarity with the general class of algorithms having the CCP. The rectangular matrices A and B transform h and x, respectively, to a higher dimensional manifold in which the transforms are multiplied. Then, the rectangular matrix C transforms the products back to the data space. Agarwal and Burrus [2] showed that if the transformation is into a manifold of the same dimension as the data and $A = B = C^{-1}$, the elements of the transform would have to be powers of the roots of unity. By allowing the transform space to be of a higher dimension and permitting $A \neq B \neq C^{-1}$, the consequent increase in the number of degrees of freedom permits a great simplification in the transform.

In this section, two theorems of Winograd [18] will be stated in a form relevant to the present context. Then, a procedure using the CRT, which was also suggested by Winograd for helping to derive optimal and near-optimal algorithms, will be described.

Theorem 2:

Let

$$Y(z) = H(z)\, X(z) \bmod P_n(z) \tag{2.24}$$

where $P_n(z)$ is an irreducible polynomial of degree n, and $H(z)$ and $X(z)$ are any polynomials of degree $n - 1$ or greater. Then the minimum number of multiplications required to compute $Y(z)$ is $2n - 1$.

We refer the reader to Winograd [18] for the proof of this theorem and only point out that the Cook-Toom algorithm gives a method for achieving this minimum number of multiplications.

Theorem 3:

The minimum number of multiplications required for computing the convolution (2.26) is $2N - K$ where K is the number of divisors of N, including 1 and N.

The following method for finding optimal algorithms will prove Theorem 3 and prove that the minimum $2N - K$ can be achieved.

Let

$$W(z) = H(z)\,X(z) \tag{2.25}$$

and

$$Y(z) = W(z) \bmod (z^N - 1). \tag{2.26}$$

The polynomial $z^N - 1$ is factored into a product of irreducible polynomials with integer coefficients

$$z^N - 1 = P_{d_1}(z)\,P_{d_2}(z) \cdots P_{d_K}(z). \tag{2.27}$$

These factors are well known in the literature on number theory (see Nagell [10]) as *cyclotomic* polynomials. There is one $P_{d_j}(z)$ for each divisor d_j of N, including $d_1 = 1$ and $d_K = N$. The roots of the polynomial $P_{d_j}(z)$ are the primitive d_jth roots of unity. The number of such roots is $n_j = \varphi(d_j)$ where $\varphi(d_j)$ is Euler's φ function and is equal to the number of positive integers smaller than d_j which are prime to d_j. Therefore, the degree of $P_{d_j}(z)$ is $n_j = \varphi(d_j)$. The degree of the product is the sum of the degrees of the $P_{d_j}(z)$'s, so one obtains the relation familiar to number theorists,

$$\sum_{j=1}^{K} n_j = \sum_{j=1}^{K} \varphi(d_j) = N \tag{2.28}$$

where the sum is over all divisors d_j of N. The properties of the $P_{d_j}(z)$'s which are important here are that they are irreducible and have simple coefficients. In fact (see [10], prob. 116, p. 185) if d_j has no more than two distinct odd prime factors, the coefficients will be ± 1 or 0. The smallest integer d with three prime factors is $d = 105 = 3 \cdot 5 \cdot 7$. Using SCRATCHPAD, we have found that of the nonzero coefficients of $P_{105}(z)$, 31 are ± 1 and two are equal to -2. Therefore, we say that reduction mod $P_{d_j}(z)$ generally involves only simple additions.

A reduction of the calculation of a convolution to a set of smaller convolutions is accomplished by the use of the CRT applied to the ring of polynomials with rational coefficients. The statement of the theorem in this context is that the set of congruences

$$Y_j(z) = Y(z) \bmod P_{d_j}(z), \qquad j = 1, 2, \cdots, K \tag{2.29}$$

has the unique solution

$$Y(z) = \sum_{j=0}^{K} Y_j(z)\,S_j(z) \bmod (z^N - 1) \tag{2.30}$$

where

$$S_j(z) \equiv 1 \bmod P_{d_j}(z) \tag{2.31}$$

and

$$S_j(z) \equiv 0 \bmod P_{d_k}(z), \qquad k \neq j. \tag{2.32}$$

The reader may be more familiar with the CRT as applied to rings of integers in residue class arithmetic as described in Section III, below.

The calculation of the convolution algorithm is easily carried out by using SCRATCHPAD [8], the computer-based formula manipulation system at the IBM Watson Research Center. To compute the polynomials $S_j(z)$, all one has to do is give a command to factor $z^N - 1$ and then, in three more lines of SCRATCHPAD commands, compute

$$T_j(z) = (z^N - 1)/P_{d_j}(z) \tag{2.33}$$

$$Q_j(z) = [T_j(z)]^{-1} \bmod P_{d_j}(z) \tag{2.34}$$

$$S_j(z) = T_j(z)\,Q_j(z). \tag{2.35}$$

The inverse in (2.34) is, by definition, the solution $Q_j(z)$ of the congruence relation

$$S_j(z) = T_j(z)\,Q_j(z) \equiv 1 \bmod P_{d_j}(z). \tag{2.36}$$

The reduction in calculation should now be apparent since the $Y_j(z)$'s in (2.30) can be obtained from

$$Y_j(z) = H_j(z)\,X_j(z) \bmod P_{d_j}(z) \tag{2.37}$$

where

$$H_j(z) = H(z) \bmod P_{d_j}(z) \tag{2.38}$$

$$X_j(z) = X(z) \bmod P_{d_j}(z). \tag{2.39}$$

The coefficients of the product polynomial $H_j(z)\,X_j(z)$ give the values of the noncyclic n_j-point convolution of the coefficients of $H_j(z)$ and $X_j(z)$. Then, according to (2.37), $Y_j(z)$ is the result of reducing this polynomial mod $P_{d_j}(z)$. The Cook-Toom algorithm shows that $H_j(z)\,X_j(z)$ can be computed by multiplying linear combinations of the coefficients h_i^j of the $H_j(z)$'s by linear combinations of the coefficients x_i^j of the $X_j(z)$'s. These coefficients are, in turn, linear combinations of the h_i's and x_i's, respectively. The set of products so formed is, therefore, of the form (2.6)

$$m = (Ah) \times (Bx).$$

Substituting the $Y_j(z)$'s in the CRT (2.30) results in formulas for the y_i's as linear combinations of the above-mentioned products. Thus, one obtains the form (2.11),

$$y = Cm.$$

The minimum number of multiplications required for computing $Y_j(z)$ is, according to Theorem 2, equal to $2n_j - 1$, so, summing over j and using (2.28), we have

$$\sum_{j=1}^{K} (2n_j - 1) = 2N - K. \tag{2.40}$$

This concludes the proof of Theorem 3.

It is seen from the above how convolution calculations can be described in terms of operations with polynomials. In so doing, the CRT for polynomials is used to reduce the problem of computing the N-point cyclic convolution, which, in terms of polynomials is

$$Y(z) = H(z)\,X(z) \bmod (z^N - 1) \tag{2.41}$$

to the problem of computing the set of K smaller convolutions

$$Y_j(z) = H_j(z)\,X_j(z) \bmod P_{d_j}(z). \tag{2.42}$$

The Cook-Toom algorithm, other systematic procedures, or even manual manipulation can then be used to obtain an algorithm for computing $H_j(z)\,X_j(z)$. While it is important to know the minimum number of multiplications and how to obtain them from the above theory, it is, due to the complexity of the A, B, and C matrices, well worth developing slightly less than optimal algorithms for the small convolutions (2.42). In many cases, the algorithms developed by Agarwal and Burrus [1] did this but it was not known, when they were written, how close they were to being optimal.

Evidently, the manipulations to be carried out in deriving the A, B, and C operators are quite tedious and fraught with opportunities for errors. Therefore, SCRATCHPAD [8] was of enormous help in deriving and checking error-free expressions for a sequence of calculations of intermediate quantities leading to expressions for the final results. The authors of SCRATCHPAD added a few commands to the language which made the entire procedure quite simple. At first, SCRATCHPAD was used interactively to develop concepts and expressions which helped to minimize the number of additions and to yield formulas convenient for programming. Then, the resulting set of commands was run in a batch mode to develop alternate formulas for each N and to go up to higher N. In using SCRATCHPAD for the above calculations, all one had to do was to define the various polynomials recursively and request the printing of various formulas at appropriate points.

The program then printed out expressions for

1) the x_i^j's in terms of the x_i's (formulas for the h_i^j's are the same),
2) the y_i^j's in terms of the products of the h_i^j's and the x_i^j's, and
3) the y_i's in terms of the y_i^j's.

Other quantities such as the factors of $z^N - 1$ were also given, but not really needed to describe the final algorithms. .

The numbers of operations for some of the convolution formulas derived by the above methods are given in Table I where K is the number of divisors of N, $2N - K$ is the minimum number of multiplications required for an N-point convolution, and M and A are the number of multiplications and additions, respectively, required for the algorithms given in Appendix A.

C. An Example with $N = 4$

The derivation of an optimal algorithm for a cyclic $N = 4$ convolution will be given here in detail, according to the methods in Section II-B. The convolution is defined by

$$y_i = \sum_{k=0}^{3} h_{i-k} x_k, \quad i = 0, 1, 2, 3. \tag{2.43}$$

In terms of polynomials whose coefficients are the sequences involved, this corresponds to

$$Y(z) = H(z)\, X(z) \bmod (z^4 - 1). \tag{2.44}$$

The factors of 4 are $d_j = 1, 2$, and 4, so the irreducible factors of $z^4 - 1$ are the cyclotomic polynomials

$$\begin{aligned} P_1(z) &= z - 1 \\ P_2(z) &= z + 1 \\ P_4(z) &= z^2 + 1. \end{aligned} \tag{2.45}$$

From these we compute

$$\begin{aligned} T_1(z) &= (z + 1)(z^2 + 1) \\ T_2(z) &= (z - 1)(z^2 + 1) \\ T_3(z) &= z^2 - 1 \end{aligned} \tag{2.46}$$

TABLE I
THEORETICAL MINIMUM NUMBER OF MULTIPLICATIONS FOR CONVOLUTION AND NUMBER OF MULTIPLICATIONS AND ADDITIONS FOR ALGORITHMS OF APPENDIX A

N	K	$2N - K$	M	A
2	2	2	2	4
3	2	4	4	11
4	3	5	5	15
5	2	8	10	35
6	4	8	8	44
7	2	12	19	72
8	4	12	14	46
9	3	15	22	98
10	4	16		
11	2	20		
12	6	18		

and

$$\begin{aligned} Q_1(z) &= [T_1(z)]^{-1} \bmod (z - 1) = \tfrac{1}{4} \\ Q_2(z) &= [T_2(z)]^{-1} \bmod (z + 1) = -\tfrac{1}{4} \\ Q_3(z) &= [T_3(z)]^{-1} \bmod (z^2 + 1) = -\tfrac{1}{2} \end{aligned} \tag{2.47}$$

giving

$$\begin{aligned} S_1(z) &= (z^3 + z^2 + z + 1)/4 \\ S_2(z) &= -(z^3 - z^2 + z - 1)/4 \\ S_3(z) &= -(z^2 - 1)/2. \end{aligned} \tag{2.48}$$

The reduced polynomials

$$H_j(z) = H(z) \bmod P_{d_j}(z) \tag{2.49}$$

are

$$\begin{aligned} H_1(z) &= h_0^1 = h_0 + h_1 + h_2 + h_3 \\ H_2(z) &= h_0^2 = h_0 - h_1 + h_2 - h_3 \\ H_3(z) &= h_0^3 + h_1^3 z = (h_0 - h_2) + (h_1 - h_3)z. \end{aligned} \tag{2.50}$$

As stated previously, the superscript j is put on the coefficients of the polynomials reduced mod $P_{d_j}(z)$. The equations for

$$X_j(z) = X(z) \bmod P_{d_j}(z) \tag{2.51}$$

are exactly the same form as those for $H_j(z)$. The relation

$$Y_j(z) = H_j(z)\, X_j(z) \bmod P_{d_j}(z) \tag{2.52}$$

is, in terms of the coefficients of $H_j(z)$ and $X_j(z)$,

$$\begin{aligned} y_0^1 &= h_0^1 x_0^1 \\ y_0^2 &= h_0^2 x_0^2 \\ y_0^3 &= h_0^3 x_0^3 - h_1^3 x_1^3 \\ y_1^3 &= h_0^3 x_1^3 + h_1^3 x_0^3. \end{aligned} \tag{2.53}$$

The calculation of $Y_3(z)$ is exactly like complex multiplication and is carried out as though $z = \sqrt{-1}$. Therefore, as shown in Section II-A, the Cook-Toom algorithm can be used to compute y_0^3 and y_1^3, in 3 instead of 4 multiplications. For the present purpose, however, we will use a slightly different complex number multiplication algorithm also requiring 3 multiplications, but requiring fewer additions involving the variable data x_i and y_i. The result is that we have to compute the five products

$$\begin{aligned} m_0 &= h_0^1 x_0^1 \\ m_1 &= h_0^2 x_0^2 \\ m_2 &= h_0^3 (x_0^3 + x_1^3) \\ m_3 &= (h_0^3 - h_1^3) x_0^3 \\ m_4 &= (h_0^3 + h_1^3) x_1^3. \end{aligned} \tag{2.54}$$

In terms of these, the y_i^j's in (2.53) are

$$\begin{aligned} y_0^1 &= m_0 \\ y_0^2 &= m_1 \\ y_0^3 &= m_2 - m_4 \\ y_1^3 &= m_2 - m_3. \end{aligned} \tag{2.55}$$

The polynomials $Y_j(z)$ whose coefficients are given by (2.55) are then substituted in the CRT

$$Y(z) = \sum_{j=1}^{3} Y_j(z)\, S_j(z) \tag{2.56}$$

to give the final result,

$$\begin{aligned} y_0 &= (m_0 + m_1)/4 + (m_2 - m_4)/2 \\ y_1 &= (m_0 - m_1)/4 + (m_2 - m_3)/2 \\ y_2 &= (m_0 + m_1)/4 - (m_2 - m_4)/2 \\ y_3 &= (m_0 - m_1)/4 - (m_2 - m_3)/2. \end{aligned} \tag{2.57}$$

As mentioned above, we assume that h_i is fixed and used repeatedly for many x_i sequences. Accordingly, we simplify the computation by redefining the m_k's and combining the $\frac{1}{2}$ and $\frac{1}{4}$ factors with the h_i's. The resulting algorithm, as described in Appendix A, is of the general form of (2.11) and (2.12).

The algorithms for N = 2, 3, 4, 5, 6, 7, 8, and 9 are given in Appendix A so as to show the grouping of terms, by means of parentheses, which hopefully minimizes the number of additions. With the above arrangement it is seen that for $N = 4$, not counting the calculation of Ah, there are 5 multiplications and 15 additions compared with the 16 multiplications and 12 additions required by direct use of the defining formula (2.43). It is interesting to note that, if the parentheses are grouped around intermediate quantities occurring as the coefficients of reduced polynomials, a grouping of additions is obtained which we have, in every case, been unable to improve upon in terms of the number of additions required. However, we know of no theorems about the minimum number of additions, or of systematic procedures for reducing the number of additions.

III. Composite Algorithms

A. The Two-Factor Algorithm

For large values of N, the optimal algorithms, i.e., those requiring the minimum number of multiplications, can become rather complicated. Some of the elements of the C matrix in (2.23) become too large to make it practical to multiply them by using successive additions and, in general, the number of additions becomes large. Furthermore, if one wishes to write a general computer program which can be used for a number of different N-values, it is more practical to write the convolution as a multidimensional convolution where the product of the dimensions is the given N.

Here, it will be shown that, instead of using the one-to-many-dimensional mapping suggested by Agarwal and Burrus [1], one can, by requiring that the chosen factors of N be mutually prime, use the mapping given by the CRT for integers mod N. This will yield a multidimensional convolution which is periodic in all dimensions without the necessity for appending zeros.[2]

In the following, a description of the CRT mapping and the general form of the resulting algorithm for composite N will be given. The formulation is designed so as to lead to effective ways of organizing computer programs for computing cyclic convolutions for all N, which can be formed from products of a fixed set of mutually prime factors. These factors will be the sequence lengths for which optimal algorithms are available.

Consider again, the problem of computing the cyclic convolution

$$y_i = \sum_{k=0}^{N-1} h_{i-k} x_k \tag{3.1}$$

where N is a composite number

$$N = r_1 r_2 \tag{3.2}$$

with mutually prime factors r_1 and r_2. This permits us to define the one-to-one mapping

$$i \longleftrightarrow (i_1, i_2) \tag{3.3}$$

where i_1 and i_2 are defined by the congruence relations

$$\begin{aligned} i_1 &\equiv i \bmod r_1, \quad 0 \leq i_1 < r_1 \\ i_2 &\equiv i \bmod r_2, \quad 0 \leq i_2 < r_2. \end{aligned} \tag{3.4}$$

The CRT says that there is a unique solution i to the congruences (3.4) which is given by

$$\begin{aligned} i &\equiv i_1 s_1 + i_2 s_2 \bmod N \\ &0 \leq i < N \end{aligned} \tag{3.5}$$

where

$$\begin{aligned} s_1 &\equiv 1 \bmod r_1 \\ s_2 &\equiv 1 \bmod r_2 \end{aligned} \tag{3.6}$$

$$\begin{aligned} s_1 &\equiv 0 \bmod r_2 \\ s_2 &\equiv 0 \bmod r_1. \end{aligned} \tag{3.7}$$

[2] This mapping was used by Good [7] and Thomas [17] for expressing the DFT as a multidimensional DFT, thereby reducing the amount of computation required. This procedure is described by Cooley, Lewis, and Welch [5].

Equation (3.7) implies that for some q_1 and q_2,

$$s_1 = q_1 r_2$$
$$s_2 = q_2 r_1 \tag{3.8}$$

which, with (3.6), requires that

$$q_1 = (r_2)_{r_1}^{-1}$$
$$q_2 = (r_1)_{r_2}^{-1}, \tag{3.9}$$

the notation denoting that q_1 is the inverse mod r_1 of r_2, and that q_2 is the inverse mod r_2 of r_1.

Let each of the vectors $\boldsymbol{y}$, $\boldsymbol{h}$, and $\boldsymbol{x}$, containing the elements y_i, h_i, and x_i, respectively, be indexed by the index pairs (i_1, i_2). Conceptually, one may think of this as a mapping of the one-dimensional arrays y_i, h_i, and x_i, $i = 0, 1, \cdots, N-1$, onto the respective two-dimensional arrays according to (3.4) and (3.5). Next, let us consider the elements of the vectors $\boldsymbol{y}$, $\boldsymbol{h}$, and $\boldsymbol{x}$ to be indexed lexicographically in i_1, i_2. Substituting (3.5) for i and a similar expression for k in terms of (k_1, k_2), the convolution (3.1) can be written

$$y_{i_1,i_2} = \sum_{k_2=0}^{r_2-1} \sum_{k_1=0}^{r_1-1} h_{i_1-k_1,i_2-k_2} x_{k_1,k_2} \tag{3.10}$$

where the indices of h_{i_1,i_2} are understood to be taken mod r_1 and r_2, respectively. In vector-matrix notation, this may be written

$$\boldsymbol{y} = \boldsymbol{H}\boldsymbol{x} \tag{3.11}$$

where the index of $\boldsymbol{y}$, which is also the row index of $\boldsymbol{H}$, is the sequence of pairs (k_1, k_2) in lexicographical order. Although $\boldsymbol{y}$, $\boldsymbol{h}$, and $\boldsymbol{x}$ are vectors, it will sometimes help to explain certain operations by thinking of them as two-dimensional arrays with row and column indices i_1 and i_2, respectively, or k_1 and k_2, respectively, whichever the case may be. Equation (3.10) represents a two-dimensional cyclic convolution where the first dimension is of length r_1 and the second dimension is of length r_2. It will be shown below that this two-dimensional cyclic convolution can be computed using a two-dimensional transformation having the CCP. Being a two-dimensional transformation, it can be expressed as a direct product of two one-dimensional transformations having the CCP for lengths r_1 and r_2. Let us assume that both these transformations are rectangular transforms of the type represented by (2.23).

With subscripts to denote which of the factors r_1 or r_2 the matrices refer to, we let $\boldsymbol{A}_1$, $\boldsymbol{B}_1$, and $\boldsymbol{C}_1$ represent a set of rectangular matrices of dimensions $M_1 \times r_1$, $M_1 \times r_1$, and $r_1 \times M_1$, respectively, having the CCP for length r_1 and requiring M_1 multiplications. Similarly, $\boldsymbol{A}_2$, $\boldsymbol{B}_2$, and $\boldsymbol{C}_2$ represent a set of rectangular matrices of dimensions $M_2 \times r_2$, $M_2 \times r_2$, and $r_2 \times M_2$, respectively, having the CCP for length r_2 and requiring M_2 multiplications. Then, the two-dimensional rectangular transformation having the CCP can be derived as follows.

For the moment, let $\boldsymbol{h}$ and $\boldsymbol{x}$ be regarded as two-dimensional arrays. The sum over k_1 in (3.10) is, for each fixed i_2 and k_2 a convolution of column $i_2 - k_2$ of the array $\boldsymbol{h}$ with column k_2 of the array $\boldsymbol{x}$. Each of these convolutions may be computed by the above transform methods, giving

$$y_{i_1,i_2} = \sum_{k_2=0}^{r_2-1} \sum_{n_1=0}^{M_1-1} C^1_{i_1,n_1} H'_{n_1,i_2-k_2} X'_{n_1,k_2} \tag{3.12}$$

where

$$H'_{n_1,i_2} = \sum_{k_1=0}^{r_1-1} A^1_{n_1,k_1} h_{k_1,i_2} \tag{3.13}$$

and

$$X'_{n_1,k_2} = \sum_{k_1=0}^{r_1-1} B^1_{n_1,k_1} x_{k_1,k_2}. \tag{3.14}$$

The superscript "1" is put on the elements of $\boldsymbol{A}_1$, $\boldsymbol{B}_1$, and $\boldsymbol{C}_1$. By changing the order of summation in (3.12), we obtain a sum over n_1, of convolutions with respect to k_2, of the sequences H'_{n_1,k_2} with X'_{n_1,k_2}, for $k_2 = 0, 1, \cdots, r_2 - 1$. These may be computed by the r_2-point rectangular transform algorithm yielding

$$y_{i_1,i_2} = \sum_{n_1=0}^{M_1-1} C^1_{i_1,n_1} \sum_{n_2=0}^{M_2-1} C^2_{i_2,n_2} H_{n_1,n_2} X_{n_1,n_2} \tag{3.15}$$

where

$$H_{n_1,n_2} = \sum_{i_2=0}^{r_2-1} A^2_{n_2,i_2} H'_{n_1,i_2}$$
$$= \sum_{i_2=0}^{r_2-1} \sum_{i_1=0}^{r_1-1} A^2_{n_2,i_2} A^1_{n_1,i_1} h_{i_1,i_2} \tag{3.16}$$

$$X_{n_1,n_2} = \sum_{k_2=0}^{r_2-1} B^2_{n_2,k_2} X'_{n_1,k_2}$$
$$= \sum_{k_2=0}^{r_2-1} \sum_{k_1=0}^{r_1-1} B^2_{n_2,k_2} B^1_{n_1,k_1} x_{k_1,k_2}. \tag{3.17}$$

In operator notation, the calculation can be described[3] by

$$\boldsymbol{y} = \boldsymbol{C}_1 \boldsymbol{C}_2 [(\boldsymbol{A}_2 \boldsymbol{A}_1 \boldsymbol{h}) \times (\boldsymbol{B}_2 \boldsymbol{B}_1 \boldsymbol{x})]. \tag{3.18}$$

The notation $\boldsymbol{B}_2\boldsymbol{B}_1\boldsymbol{x}$ means that one computes the transform $\boldsymbol{B}_1$ of the columns of $\boldsymbol{x}$ and then the transform $\boldsymbol{B}_2$ of the rows of the result. Since the ordering of the operators corresponds to the ordering of the summations, they commute. However, the ordering of the operators affects the sizes of intermediate arrays, the number of additions, and program organization. These will be discussed in Section V-A.

We have thus shown that the composite two-dimensional transform algorithm as described by (3.18) has the CCP.

Mapping the result y_{i_1,i_2} into the one-dimensional array y_i via the CRT (3.5) yields the one-dimensional convolution (3.1). Hence, the total transformation (3.18) has the one-

[3] Equation (3.18) can be written in Kronecker product notation as

$$\boldsymbol{y} = (\boldsymbol{C}_1 \times \boldsymbol{C}_2)[(\boldsymbol{A}_2 \times \boldsymbol{A}_1 \boldsymbol{h}) \times (\boldsymbol{B}_2 \times \boldsymbol{B}_1 \boldsymbol{x})],$$

where $\times$ denotes the Kronecker product and $*$ denotes element by element multiplication. However, this notation serves no useful purpose and can cause some confusion. Therefore, it will not be used here.

dimensional CCP with respect to the one-dimensional sequences y_i, h_i, and $x_i, i = 0, 1, \cdots, N-1$.

B. Number of Operations for Two-Factor Algorithms

As mentioned in Section II-A, the matrices are not stored and multiplied as matrices. Instead, to save storage and operations, the calculation is performed by explicit formulas which are arranged so that intermediate quantities are saved and reused. Some of the algorithms are written in Appendix A in this manner. We also mention again that it is assumed that $\boldsymbol{h}$ is to be used for many different $\boldsymbol{x}$ vectors and, therefore, operations involving $\boldsymbol{h}$ are not counted.

Let us consider the sizes of the arrays involved. Since B_1 is $M_1 \times r_1$ and x is $r_1 \times r_2$, B_1x is $M_1 \times r_2$, meaning that its columns are of length M_1 and are, in general, longer than those of x. Similarly, the effect of B_2, which is $M_2 \times r_2$, is to lengthen rows when it operates, producing the $M_1 \times M_2$ array $X = B_2B_1x$. In the same way, C_1C_2 is an operator which reduces the dimensionality, in reverse order, of the array on which it operates.

The number of multiplications involved is, therefore, the number of elements in X,

$$M(r_1, r_2) = M_1M_2 \tag{3.19}$$

and is seen to be independent of the ordering. On the other hand, the number of additions depends on the ordering. Let A_{B_j} and A_{C_j} be the number of additions required to apply the B_j and C_j operators, respectively, in a one-dimensional convolution. Let

$$A_1 = A_{B_1} + A_{C_1}$$

$$A_2 = A_{B_2} + A_{C_2}. \tag{3.20}$$

Then, since B_1x takes A_{B_1} additions when B operates on each of the r_2 columns of x, it takes $A_{B_1}r_2$ additions in all. But, B_2 operates on the M_1 rows of the $M_1 \times r_2$ array B_1x taking $A_{B_2}M_1$ additions. Next, C_2 operates on the M_1 rows of the array $Y = HX$, taking $A_{C_2}M_1$ additions. Then C_1 operates on the r_2 columns of C_2Y taking $A_{C_1}r_2$ additions. In all, we get

$$A(r_1, r_2) = A_{B_1}r_2 + A_{B_2}M_1 + A_{C_2}M_1 + A_{C_1}r_2$$

$$= A_1r_2 + A_2M_1 \tag{3.21}$$

operations. The reader may verify that if the C_j's were applied in the order C_2C_1, one would obtain

$$A^*(r_1, r_2) = A_{B_1}r_2 + A_{B_2}M_1 + A_{C_1}M_2 + A_{C_2}r_1. \tag{3.22}$$

This is more complicated than (3.21) and makes it more difficult to minimize the number of additions. Both of these formulas were tested with actual operation counts and, in only one case, was it found that (3.22) gave fewer additions. Therefore, we have adopted the convention of placing the C_j operators in the reverse order of that used for the B_j's in order to be able to use (3.21). As mentioned earlier, this ordering also simplifies programming.

Now let us consider reversing the order of the factors. If the transforms are computed first along index 2 and then along index 1, the total number of additions required will be

$$A(r_2, r_1) = A_2r_1 + A_1M_2. \tag{3.23}$$

TABLE II
TABLE OF VALUES OF $T(r_j) = M_j - r_j/A_j$

r_j	$T(r_j)$
2	0.000
3	0.091
4	0.066
5	0.142
6	0.045
7	0.166
8	0.130
9	0.131

For the ordering r_1, r_2 to take fewer operations, we must have

$$A(r_1, r_2) < A(r_2, r_1)$$

or

$$A_1r_2 + A_2M_1 < A_2r_1 + A_1M_2$$

from which it follows that

$$\frac{M_1 - r_1}{A_1} < \frac{M_2 - r_2}{A_2}. \tag{3.24}$$

Therefore, the transformation to perform first is the one for which the quantity

$$T(r_1) = \frac{M_1 - r_1}{A_1} \tag{3.25}$$

is smaller. Values of $T(r)$ for the algorithms of Appendix A are listed in Table II.

C. The General Multifactor Algorithm

A one-dimensional cyclic convolution can be implemented as a multidimensional cyclic convolution by factoring N into more than two mutually prime factors,

$$N = r_1r_2 \cdots r_t \tag{3.26}$$

where, as stated above, the r_j's are mutually prime. The multidimensional index mapping is defined by

$$i_j \equiv i \bmod r_j$$

$$0 \leqslant i_j < r_j, \quad j = 1, 2, \cdots, t \tag{3.27}$$

and the inverse mapping is

$$i \equiv i_1s_1 + i_2s_2 + \cdots + i_ts_t \bmod N \tag{3.28}$$

where

$$s_j = q_j(N/r_j),$$

in which q_j satisfies

$$q_j(N/r_j) \equiv 1 \bmod r_j. \tag{3.29}$$

With this t-dimensional mapping, the one-dimensional cyclic convolution (3.1) can be written as a t-dimensional cyclic convolution in a form which is a generalization of (3.10). The t-dimensional cyclic convolution can be computed using a t-dimensional transformation having the CCP. This t-dimensional transformation is a direct product of t one-dimensional transformations having the CCP for lengths $r_1, r_2, \cdots, r_t$, respectively. Computation for the t-dimensional transformation can be carried out by a simple generalization of the two-dimensional transformation (3.18) which can be written

$$y = C_1 C_2 \cdots C_t[(A_t \cdots A_2 A_1 h) \times (B_t \cdots B_2 B_1 x)]. \quad (3.30)$$

Letting x be regarded as a t-dimensional array with indices $k_1, k_2, \cdots, k_t$, $B_t \cdots B_2 B_1 x$ denotes a t-dimensional rectangular transform of x. This is obtained by first computing the r_1-point transform $B_1 x$, with respect to the first index k_1 of x for fixed values of all other indices. Note here that if the first transform is a Fourier transform or an NTT, $B_1 x$ will be of the same size as x. If B_1 is a rectangular transform, however, $B_1 x$ will be larger in the first dimension. Then, one computes the r_2-point transform with respect to k_2 for each fixed set of values of all other indices, increasing the length of the second dimension. The inverse operation with the C_j's is to be performed in a similar fashion where, as mentioned before, we apply the C_j's in reverse order as in the two-dimensional case above. Multiplication by each C_j is seen to reduce the length of the array in the k_jth dimension. Results on the computational requirements for a t-dimensional transformation can be easily generalized from the two-dimensional case.

D. Number of Operations for the General Multifactor Algorithm

Let A_j and M_j be the number of additions and multiplications, respectively, required for a length r_j one-dimensional convolution. Then, the number of multiplications required for the t-dimensional cyclic convolution is

$$M(r_1, r_2, \cdots, r_t) = M_1 M_2 \cdots M_t \quad (3.31)$$

and the number of additions required is

$$A(r_1, r_2, \cdots, r_t) = A_1 r_2 \cdots r_t + M_1 A_2 r_3 \cdots r_t + M_1 M_2 A_3 r_4 \cdots r_t + \cdots + M_1 \cdots M_{t-1} A_t. \quad (3.32)$$

As before, the ordering of the arguments of $A(\cdot \cdot \cdot)$ indicates the order in which the transforms are computed. Inverse transforms are computed in the reverse order. As in the two-dimensional case, the number of additions depends on the order in which the transforms are computed.

It is fairly simple to show that the ordering of the indices $r_1, r_2, \cdots, r_m$ which minimizes the number of additions is given by a generalization of the two-dimensional case treated above. Thus, the ordering should be according to the size of

$$T(r_j) = \frac{M_j - r_j}{A_j}, \quad (3.33)$$

i.e., such that

$$T(r_k) \leqslant T(r_j) \text{ when } k < j. \quad (3.34)$$

Appendix A lists explicitly or implicitly, the A, B, and C matrices for some basic short length cyclic convolution algorithms. These algorithms are the basic building blocks which may be used to obtain algorithms for computing convolutions of long sequences by multidimensional implementations. Table I lists the number of multiplications and additions required for these basic algorithms. Mutually prime factors from this list are selected to obtain algorithms for longer N. Table III lists the number of multiplications and additions required for some multidimensional implementations of one-dimensional convolutions with rectangular transforms. Both Tables I and III assume that the transform of h is precomputed and stored. The factors column lists factors of N in the order in which the transform of x is computed. The ordering listed gives the minimum number of additions. For comparison, Table IV lists the number of multiplications per point required for a length $N = 2^t$ cyclic convolution using the FFT algorithm. The FFT algorithm used is a very efficient radix 2, 4, 8 algorithm which also makes use of the fact that the data are real.

IV. Use with Fermat Number Transforms

The FNT provides an efficient and error-free means of computing cyclic convolutions. The computation of the FNT requires $0(N \log N)$ bit shifts and additions, but no multiplications. The only multiplications required for an FNT implementation of cyclic convolution are the N multiplications required to multiply the transforms. This is a very efficient technique for computing cyclic convolutions, but unfortunately, the maximum transform length for an FNT is proportional to the word length of the machine used. Agarwal and Burrus [2] showed that a very practical choice of a Fermat number for this application is $F_5 = 2^{32} + 1$, and that the FNT mod F_5 can be implemented on a 32-bit machine. For this choice of the Fermat number, the maximum transform length is 128. To compute the cyclic convolution of a one-dimensional sequence longer than 128, we write the one-dimensional sequence as a multidimensional sequence using the CRT mapping as in (3.4) and (3.5). The length of the first dimension is taken as 128, and the lengths of the other dimensions are taken as mutually prime odd numbers. Thus,

$$N = 128 r_2 r_3 \cdots r_t. \quad (4.1)$$

For the FNT, the matrices A, B, and C in (2.23) satisfy $A = B$ and $C = A^{-1}$ and they are 128 by 128 matrices. Since for FNT, $M = r$, (3.24) tells us that the first transform to compute is a length 128 FNT's. This is computed for each of the indices in the other dimensions and then followed by the computation of the rectangular transforms along all other dimensions. Finally, the transforms of h and x are multiplied and the inverse transforms, in all dimensions, are applied to the product in the reverse order, the last inverse transform being the FNT. All calculations, including those for the rectangular transforms must be done modulo F_5.

The total number of multiplications required is

$$M = 128 M_2 M_3 \cdots M_t \quad (4.2)$$

while the number of length 128 FNT's and inverse FNT's required is

$$F = 2 r_2 r_3 \cdots r_t. \quad (4.3)$$

The number of additions required in excess of those required for computing the FNT is

$$A(128, r_2, \cdots, r_t) = 128 A(r_2, \cdots, r_t) = 128(A_2 r_3 r_4 \cdots r_t + M_2 A_3 r_4 \cdots r_t + \cdots + M_2 M_3 \cdots M_{t-1} A_t). \quad (4.4)$$

TABLE III
NUMBER OF MULTIPLICATIONS AND ADDITIONS PER OUTPUT POINT FOR CONVOLUTION USING COMPOSITE ALGORITHMS FORMED FROM THE RECTANGULAR TRANSFORMS IN APPENDIX A

N	Factors	Total Number of Multiplications	Total Number of Additions	Multiplications per Point	Additions per Point
6	2, 3	8	34	1.33	5.67
12	4, 3	20	100	1.67	8.33
18	2, 9	44	232	2.44	12.89
20	4, 5	50	250	2.50	12.50
30	2, 3, 5	80	450	2.67	15.00
36	4, 9	110	625	3.06	17.36
60	4, 3, 5	200	1200	3.33	20.00
72	8, 9	308	1786	4.28	24.80
84	4, 3, 7	380	2140	4.52	25.48
120	3, 8, 5	560	3320	4.67	27.67
180	4, 9, 5	1100	6975	6.11	38.75
210	2, 3, 5, 7	1520	8910	7.24	42.42
360	8, 9, 5	3080	19 710	8.56	54.75
420	4, 3, 5, 7	3800	22 800	9.05	54.29
504	8, 9, 7	5852	34 678	11.61	68.81
840	3, 8, 5, 7	10 640	63 560	12.67	75.67
1260	4, 9, 5, 7	20 900	128 025	16.59	101.61
2520	8, 9, 5, 7	58 520	359 730	23.22	142.75

TABLE IV
NUMBER OF MULTIPLICATIONS AND ADDITIONS PER OUTPUT POINT FOR CONVOLUTION USING COMPOSITE FFT ALGORITHMS (RADICES 2, 4, 8)

N	Real Multiplications per Point	Real Additions per Point
4	2.00	7.00
8	2.50	9.50
16	4.25	12.37
32	5.12	14.81
64	6.06	17.53
128	8.03	20.51
256	9.01	23.00
512	10.00	25.75
1024	12.00	28.75
2048	13.00	31.25
4096	14.00	34.00

Note: (It is assumed that one will do two real transforms with each complex FFT.)

TABLE V
AMOUNT OF COMPUTATION FOR CONVOLUTION USING THE FNT IN MULTIDIMENSIONAL ALGORITHMS

N	Factors of N	Number of Multiplies per Point	Number of Extra Adds per Point
128	128 × 1	1.0	0.00
384	128 × 3	1.33	3.66
640	128 × 5	2.0	7.00
896	128 × 7	2.71	10.28
1152	128 × 9	2.44	10.88
1920	128 × 3 × 5	2.66	13.00

Table V lists the amount of computation required for multidimensional implementation of cyclic convolution using FNT's and rectangular transforms.

The data in Table V are to be compared with that in Tables III and IV, where comparable data for the computation of convolutions by rectangular transform and FFT methods are given. The comparison is difficult to make since the FNT does depend for its efficiency upon special machine hardware for the transformations. However, the data do show how much is to be gained if one has a machine with such hardware. The reduction in numbers of multiplications is quite impressive. For example, a mixed radix FFT algorithm (see [16]) for 1024 points takes 12 multiplications per output point to compute a cyclic convolution while the FNT, used with the present algorithms for a composite 896 point transform, takes only 2.71 multiplications per output point. The comparable figure for 840 points with the composite rectangular transform method is 12.67 multiplications per output point. For N = 1920, we have 2.66 multiplications per output point for the FNT method while for N = 2048, the FFT method takes 13 multiplications per output point.

V. MISCELLANEOUS CONSIDERATIONS

A. Programming of the Algorithm and Machine Organization

We first summarize the calculation in matrix operator notation. The two-dimensional convolution (3.10) may be written in the form

$$y = h^{**}x \tag{5.1}$$

where "**" denotes the fact that there are two convolutions of h with x, the first being a convolution of *columns*, the second a convolution of *rows*. Application of the rectangular transform algorithm to the r_1-point column convolutions gives (3.12)-(3.14) which we express in operator notation as

$$H' = A_1 h \tag{5.2}$$

$$X' = B_1 x \tag{5.3}$$

$$Y' = H' \times {}^*X' \tag{5.4}$$

$$y = C_1 Y'. \tag{5.5}$$

Equations (5.4) and (5.5) are defined by the result of changing the order of summation in (3.12). One may think of the "$\times$ *" in (5.4) as signifying an element by element multiplication with respect to the first index and a convolution with respect to the second index of the arrays H' and X', i.e., of the *rows* of H' with the respective *rows* of X'. These convolutions are calculated with the r_2-point convolution algorithm which can be written

$$H = A_2 H' \tag{5.6}$$

$$X = B_2 X' \tag{5.7}$$

$$Y = H \times\times X \tag{5.8}$$

$$Y' = C_2 Y \tag{5.9}$$

where the "$\times\times$" in (5.8) denotes element by element multiplication of all elements. The above formulation can be used to define the structure of a program for implementing the algorithm. Such a program would carry out the operations defined by (5.2)-(5.5) in that order. This would essentially be an r_1-point convolution program operating on vectors. In computing (5.4), however, the program would compute the convolutions by performing the operations defined by (5.6)-(5.9) in that order. The latter computation can be done by a subroutine having exactly the same structure as (5.2)-(5.5). This is essentially an r_2-point convolution subroutine also operating on vectors. On step (5.8), an element by element multiplication is performed. If there were a third factor, (5.8) would contain a convolution and would be computed by still another convolution subroutine operating on vectors. This could thus proceed for as many levels of subroutines as there are factors in N.

For convolutions of real sequences, the rectangular transform approach requires only real arithmetic as compared with complex arithmetic required by the FFT algorithm. This should reduce hardware complexity considerably.

It may appear that the CRT mapping of a one-dimensional sequence into a multidimensional array may require substantial computation. However, this is not so. To map a one-dimensional sequence of length N into a t-dimensional array of dimensions $r_1, r_2, \cdots, r_t$ [as given by (3.27)], we set up t address registers which give the t-dimensional array address for each data point. As the input data comes in sequentially, all address registers are updated by one. These address registers are so set up that when the contents of the jth register becomes r_j, it is automatically reset to zero. Using this scheme, no additional computation is required for the address mapping. After computing the convolution, removing the data from the machine using (3.28) will require a substantial amount of computation. We can get around this by removing the data sequentially in the form of a one-dimensional sequence $\boldsymbol{y}$. Again, we use the scheme as described above to give the t-dimensional array address where the output is residing. For both input and output we use the mapping (3.27) which is much simpler. If the $\boldsymbol{h}$ sequence is fixed, the rectangular transform of $\boldsymbol{h}$ can be precomputed and stored in a read-only memory (ROM).

For basic short length convolution algorithms, the $\boldsymbol{A}$, $\boldsymbol{B}$, and $\boldsymbol{C}$ matrices are very simple and require few additions. Furthermore, as mentioned above, a rectangular transformation with respect to one index is done for all values of the other indices and is, therefore, a vector operation which can be done simultaneously or in pipelined fashion for all vector elements. This can be done conveniently by an array processor where one may even consider hard-wiring the circuits which compute the rectangular transforms.

Also, since the computation involves multidimensional transforms, it can easily be adapted to a two-level memory hierarchy. A slow memory unit can be used to store all the data, and a fast memory unit can be used to compute on a part of the data at a time (usually on a row or a column).

B. Bounds on Intermediate Results

If a multidimensional convolution is implemented in modular arithmetic (for example when the FNT is used) then we do not have to worry about the intermediate values as long as the final output is correctly bounded. But if ordinary arithmetic is used, all the intermediate values should be correctly bounded so that no overflow of the intermediate values occurs. Below, we will give some simple bounds for the case where data are real and only rectangular transforms are used. It is assumed that the $\boldsymbol{h}$ sequence is predetermined and remains fixed. Results are given for the two-dimensional case, but they generalize easily to more than two dimensions.

Let

$$N = r_1 r_2 \tag{5.10}$$

and let

$$x_{\max} = \max_{k_1 k_2} |x_{k_1, k_2}|. \tag{5.11}$$

A bound $y_{\max}$ on the magnitudes of the elements of $\boldsymbol{y}$ in (5.1) satisfies

$$|y|_{\max} \leq x_{\max} \sum_{k_2=0}^{r_2-1} \sum_{k_1=0}^{r_1-1} |h_{k_1, k_2}|. \tag{5.12}$$

The above bound is also a least upper bound. For a particular $\boldsymbol{x}$ array it can be achieved. Equation (5.12) is a bound on the output, but we also need bounds on the intermediate results. Consider the X' array (5.3) obtained after computing the B_1 transform along the first dimension. A simple bound on the elements of X' satisfies

$$|X'_{n_1, j_2}| \leq x_{\max} B(r_1, n_1) \tag{5.13}$$

for all n_1, j_2 where here, and in what follows,

$$B(r_j, n_j) = \sum_{k_j=0}^{r_j-1} |B_{n_j, k_j}|, \quad j = 1, 2. \tag{5.14}$$

The absolute values of the elements of the X array, (5.7), are bounded by

$$|X_{n_1, n_2}| \leq |X'_{n_1, j_2}|_{\max} B(r_2, n_2) \tag{5.15}$$

where the "max" refers to the maximum with respect to j_2. This, with (5.13) gives

$$|X_{n_1, n_2}| \leq x_{\max} B(r_1, n_1) B(r_2, n_2)$$
$$n_1 = 0, 1, \cdots, M_1 - 1, \quad n_2 = 0, 1, \cdots, M_2 - 1. \tag{5.16}$$

Both bounds (5.13) and (5.15) are least upper bounds. We get a bound on the elements of the transform Y in (5.8) in terms of the known fixed H by substituting the bound (5.16) in

$$|Y_{n_1,n_2}| \leq |H_{n_1,n_2}|\,|X_{n_1,n_2}|_{\max} \tag{5.17}$$

to get

$$|Y_{n_1,n_2}| \leq x_{\max}\,|H_{n_1,n_2}| B(r_1, n_1)\, B(r_2, n_2). \tag{5.18}$$

Bounds on the elements of Y' are obtained directly from (5.4) giving

$$|Y'_{n_1,j_2}| \leq |X'_{n_1,j_2}|_{\max} \sum_{k_2=0}^{r_2-1} |H'_{n_1,k_2}| \tag{5.19}$$

where the "max" refers to the maximum over j_2. Substituting (5.13) we have

$$|Y'_{n_1,j_2}| \leq x_{\max}\, B(r_1, n_1) \sum_{k_2=0}^{r_2-1} |H'_{n_1,k_2}|. \tag{5.20}$$

To summarize, (5.12), (5.13), (5.16), (5.18), and (5.20) give least upper bounds on the elements of y, X', X, Y, and Y', respectively, in terms of $x_{\max}$ and known fixed values of h and its transforms H' and H. These bounds can easily be generalized to the multidimensional case.

C. The Effect of Roundoff Error

If the multidimensional convolution is implemented in modular arithmetic, there is no roundoff error introduced at any stage of the computation. Even if ordinary arithmetic is used, the rectangular transform implementation of cyclic convolution is likely to have less arithmetical roundoff noise (error) than an FFT implementation. There are several reasons for this. To compute convolutions of real sequences, the rectangular transform approach requires only real operations, but the FFT implementation requires complex operations. Complex arithmetic introduces more roundoff noise than real arithmetic. Moreover, for short length convolutions, the rectangular transform approach requires a smaller total number of arithmetical operations as compared to the FFT implementation. Fewer arithmetical operations generally result in smaller roundoff noise. Furthermore, if fixed point arithmetic is used, roundoff noise is introduced only during multiplications. Therefore, for a rectangular transform fixed point implementation, the only source of noise is in the multiplication of the transforms. All these factors should lead to substantially less roundoff noise for a rectangular transform than for an FFT.

D. Optimal Block Length for Noncyclic Convolution

In many digital signal processing applications, one of the sequences (the impulse response h of the filter) is fixed and of short length, say p, while the other sequence (the input sequence x) is much longer and can be considered to be infinitely long. The convolution of these sequences is obtained by blocking the input sequences in blocks of length L. Now, for each block, we have to convolve a sequence of length L with a sequence of length p. They can be convolved using a length N cyclic convolution if $L + p - 1 \leq N$. For each p there is an optimum N, depending on the cyclic convolution scheme used, which requires the minimum amount of computation per output point. Let $F_1(N)$ be the number of multiplications per point required for a length N cyclic convolution. Then $F_2(p, N)$, the number of multiplications per output point, is given by

$$F_2(p, N) = F_1(N)\, N/(N - p + 1) \tag{5.21}$$

for a fixed p, $N/(N - p + 1)$ is a decreasing function of N. For an FFT implementation, $F_1(N)$ is proportional to $\log N$, a slowly increasing function of N. Therefore, for the FFT, the optimum block length N for a given p is much larger than p. For a rectangular transform calculation of a cyclic convolution $F_1(N)$ is a rapidly increasing function of N. Thus, for this case, the optimum N is not much larger than p. Table VI lists optimum N and corresponding $F_1(N)$ and $F_2(p, N)$ for several values of p. The values of N selected are from Table III. For comparison, Table VII lists for the same p-values the corresponding data obtained by using the FFT algorithm with the multiplication count as given in Table IV.

VI. Conclusions

The multidimensional method for computing convolutions was investigated by Agarwal and Burrus [1] in order to permit the efficient use of FNT's. While this presented computational advantages for computers capable of the special arithmetic required for the FNT, it was also shown that even without the FNT, a general-purpose computer could compute convolutions by this method in fewer multiplications than others using the FFT for sequence lengths up to around 128.

The present paper suggests the use of the CRT for mapping into multidimensional sequences. This, with improved short convolution algorithms, makes the multidimensional method better than FFT methods for sequence lengths up to around 420. The present methods are also more attractive since they do not require complex arithmetic with sines and cosines. This means that the calculation can be carried in integer arithmetic without rounding errors.

Theoretical results from computational complexity theory showing how close the special algorithms are to optimal are cited. Some of this theory is used for developing systematic techniques for deriving optimal short convolution algorithms.

It is expected that these techniques, using computer-based formula manipulation systems, will be useful for developing tailor-made convolution algorithms which take advantage of the special properties of a given computer. For the same reasons, one may also expect such techniques to have an effect on the design of special-purpose digital processing systems.

Appendix A

Convolution Algorithms for $2 \leq N \leq 9$

Optimal and near-optimal algorithms for a number of short convolutions are given with the number of multiplications M and the number of additions A_B, A_C, and A. The operations involving h are not counted. The elements of Ah and Bx are denoted by a_k and b_k, $k = 0, \cdots, M - 1$, respectively.

The expressions for a_k and b_k are written with parentheses arranged so as to show the ordering of the operations, which

TABLE VI
OPTIMUM SIZE SEGMENTS OF LONG SEQUENCES WHEN CONVOLVING WITH A SHORT SEQUENCE BY RECTANGULAR TRANSFORM METHODS

Filter Tap Length		Number of Multiplications		Multiplications per Point
p	N	M	$F_1(N)$	$F_2(p,N)$
2	6	8	1.33	1.60
4	12	20	1.66	2.22
8	30	80	2.66	3.47
16	60	200	3.33	4.44
32	120	560	4.66	6.29
64	180	1100	6.11	9.40
128	420	3800	9.04	12.97
256	840	10 640	12.66	18.17

TABLE VII
OPTIMUM SIZE SEGMENTS OF LONG SEQUENCES WHEN CONVOLVING WITH A SHORT SEQUENCE BY FFT METHODS

Filter Tap Length		Number of Multiplications		Multiplications per Point
p	N	M	$F_1(N)$	$F_2(p,N)$
2	4	8	2.00	2.66
4	8	20	2.50	4.00
8	32	164	5.12	6.55
16	64	388	6.06	7.88
32	256	2308	9.01	10.27
64	512	5124	10.00	11.40
128	512	5124	10.00	13.30
256	2048	26 628	13.00	14.82

takes the number of additions given for each algorithm. We have done our best to minimize the number of additions, but have no proof that we have succeeded.

With the algorithms for $N = 6, 7$, and 8 we also give the A, B, and C matrices. Where possible, the A matrix is given in terms of B premultiplied by a diagonal matrix, written diag $(\cdots)$ with the diagonal elements within the parentheses.

$N = 2$ Algorithm–$M = 2, A_B = 2, A_C = 2, A = 4$:

$$
\begin{aligned}
a_0 &= (h_0 + h_1)/2\\
a_1 &= (h_0 - h_1)/2\\
b_0 &= x_0 + x_1\\
b_1 &= x_0 - x_1\\
m_k &= a_k b_k, \quad k = 0, 1\\
y_0 &= m_0 + m_1\\
y_1 &= m_0 - m_1.
\end{aligned}
$$

$N = 3$ Algorithm–$M = 4, A_B = 5, A_C = 6, A = 11$:

$$
\begin{aligned}
a_0 &= (h_0 + h_1 + h_2)/3\\
a_1 &= h_0 - h_2\\
a_2 &= h_1 - h_2\\
a_3 &= [(h_0 - h_2) + (h_1 - h_2)]/3\\
b_0 &= x_0 + x_1 + x_2\\
b_1 &= x_0 - x_2\\
b_2 &= x_1 - x_2\\
b_3 &= (x_0 - x_2) + (x_1 - x_2)\\
m_k &= a_k b_k, \quad k = 0, 1, 2, 3\\
y_0 &= m_0 + (m_1 - m_3)\\
y_1 &= m_0 - (m_1 - m_3) - (m_2 - m_3)\\
y_2 &= m_0 + (m_2 - m_3).
\end{aligned}
$$

$N = 4$ Algorithm–$M = 5, A_B = 7, A_C = 8, A = 15$:

$$
\begin{aligned}
a_0 &= [(h_0 + h_2) + (h_1 + h_3)]/4\\
a_1 &= [(h_0 + h_2) - (h_1 + h_3)]/4\\
a_2 &= (h_0 - h_2)/2\\
a_3 &= [(h_0 - h_2) - (h_1 - h_3)]/2\\
a_4 &= [(h_0 - h_2) + (h_1 - h_3)]/2\\
b_0 &= (x_0 + x_2) + (x_1 + x_3)\\
b_1 &= (x_0 + x_2) - (x_1 + x_3)\\
b_2 &= (x_0 - x_2) + (x_1 - x_3)\\
b_3 &= x_0 - x_2\\
b_4 &= x_1 - x_3
\end{aligned}
$$

$$
\begin{aligned}
m_k &= a_k b_k, \quad k = 0, 1, 2, 3, 4\\
y_0 &= (m_0 + m_1) + (m_2 - m_4)\\
y_1 &= (m_0 - m_1) + (m_2 - m_3)\\
y_2 &= (m_0 + m_1) - (m_2 - m_4)\\
y_3 &= (m_0 - m_1) - (m_2 - m_3).
\end{aligned}
$$

$N = 5$ Algorithm–$M = 10, A_B = 13, A_C = 22, A = 35$:

$$
\begin{aligned}
a_0 &= (h_0 + h_1 + h_2 + h_3 + h_4)/5\\
a_1 &= h_0 - h_4\\
a_2 &= h_1 - h_4\\
a_3 &= h_2 - h_4\\
a_4 &= h_3 - h_4\\
a_5 &= (h_0 - h_4) + (h_1 - h_4)\\
a_6 &= (h_2 - h_4) + (h_3 - h_4)\\
a_7 &= (h_0 - h_4) + (h_2 - h_4)\\
a_8 &= (h_1 - h_4) + (h_3 - h_4)\\
a_9 &= \{[(h_0 - h_4) + (h_1 - h_4)] + [(h_2 - h_4) + (h_3 - h_4)]\}/5\\
b_0 &= x_0 + x_1 + x_2 + x_3 + x_4\\
b_1 &= x_0 - x_4\\
b_2 &= x_1 - x_4\\
b_3 &= x_2 - x_4\\
b_4 &= x_3 - x_4\\
b_5 &= (x_0 - x_4) + (x_1 - x_4)\\
b_6 &= (x_2 - x_4) + (x_3 - x_4)\\
b_7 &= (x_0 - x_4) + (x_2 - x_4)\\
b_8 &= (x_1 - x_4) + (x_3 - x_4)\\
b_9 &= [(x_0 - x_4) + (x_1 - x_4)] + [(x_2 - x_4) + (x_3 - x_4)]\\
m_k &= a_k b_k, \quad k = 0, 1, \cdots, 9
\end{aligned}
$$

$$
C = \begin{bmatrix}
1 & 1 & 0 & -1 & -1 & 0 & 1 & 0 & 0 & -1\\
1 & -1 & -1 & 0 & 1 & 1 & 0 & 0 & 0 & -1\\
1 & -1 & 1 & -1 & 0 & 0 & 0 & 1 & 0 & -1\\
1 & 1 & 1 & 1 & 1 & -1 & -1 & -1 & -1 & 4\\
1 & 0 & -1 & 1 & -1 & 0 & 0 & 0 & 1 & -1
\end{bmatrix}
$$

$$
\begin{aligned}
y_0 &= (m_0 - m_9) + (m_1 - m_4) - m_3 + m_6\\
y_1 &= (m_0 - m_9) - (m_1 - m_4) - m_2 + m_5\\
y_2 &= (m_0 - m_9) + (m_2 - m_3) - m_1 + m_7\\
y_3 &= (m_0 + m_0) + 2m_0 + m_0 - y_0 - y_1 - y_2 - y_4\\
y_4 &= (m_0 - m_9) - (m_2 - m_3) - m_4 + m_8.
\end{aligned}
$$

$N = 6$ Algorithm–$M = 8, A_B = 18, A_C = 26, A = 44$:

Note that this is not as good as the composite algorithm for $N = 2 \times 3$ in Table III which also takes 8 multiplications, but takes only 34 additions.

$$A = \text{diag}\,(1\ \ 1\ \ -1\ \ 1\ \ 1\ \ 1\ \ 1\ \ 1) \cdot B/6$$

where

$$B = \begin{bmatrix} 1 & 0 & -1 & 1 & 0 & -1 \\ 0 & 1 & -1 & 0 & 1 & -1 \\ 1 & -1 & 0 & 1 & -1 & 0 \\ 1 & 0 & -1 & -1 & 0 & 1 \\ 0 & 1 & 1 & 0 & -1 & -1 \\ 1 & 1 & 0 & -1 & -1 & 0 \\ 1 & -1 & 1 & -1 & 1 & -1 \\ 1 & 1 & 1 & 1 & 1 & 1 \end{bmatrix}$$

$$\begin{aligned}
a_0 &= [(h_0 - h_2) + (h_3 - h_5)]/6 \\
a_1 &= [(h_1 - h_2) + (h_4 - h_5)]/6 \\
a_2 &= a_1 - a_0 \\
a_3 &= [(h_0 - h_2) - (h_3 - h_5)]/6 \\
a_4 &= [(h_1 + h_2) - (h_4 + h_5)]/6 \\
a_5 &= a_3 + a_4 \\
a_6 &= [(h_0 + h_2) - (h_1 + h_3) + (h_4 - h_5)]/6 \\
a_7 &= [(h_0 + h_2) + (h_1 + h_3) + (h_4 + h_5)]/6
\end{aligned}$$

$$\begin{aligned}
b_0 &= (x_0 - x_2) + (x_3 - x_5) \\
b_1 &= (x_1 - x_2) + (x_4 - x_5) \\
b_2 &= b_0 - b_1 \\
b_3 &= (x_0 - x_2) - (x_3 - x_5) \\
b_4 &= (x_1 + x_2) - (x_4 + x_5) \\
b_5 &= b_3 + b_4 \\
b_6 &= (x_0 + x_2) - (x_1 + x_3) + (x_4 - x_5) \\
b_7 &= (x_0 + x_2) + (x_1 + x_3) + (x_4 + x_5)
\end{aligned}$$

$$m_k = a_k b_k, \qquad k = 0, 1, \cdots, 7$$

$$C = \begin{bmatrix} 1 & -2 & -1 & 1 & -2 & 1 & 1 & 1 \\ 1 & 1 & 2 & -1 & -1 & 2 & -1 & 1 \\ -2 & 1 & -1 & -2 & 1 & 1 & 1 & 1 \\ 1 & -2 & -1 & -1 & 2 & -1 & -1 & 1 \\ 1 & 1 & 2 & 1 & 1 & -2 & 1 & 1 \\ -2 & 1 & -1 & 2 & -1 & -1 & -1 & 1 \end{bmatrix}$$

$$\begin{aligned}
y_0 &= \{(m_0 - m_1) - (m_1 + m_2)\} + \{(m_3 - m_4) - (m_4 - m_5)\} \\
&\quad + (m_6 + m_7) \\
y_1 &= \{(m_0 + m_2) + (m_1 + m_2)\} - \{(m_3 - m_5) + (m_4 - m_5)\} \\
&\quad - (m_6 - m_7) \\
y_2 &= -\{(m_0 - m_1) + (m_0 + m_2)\} - \{(m_3 - m_4) + (m_3 - m_5)\} \\
&\quad + (m_6 + m_7) \\
y_3 &= \{(m_0 - m_1) - (m_1 + m_2)\} - \{(m_3 - m_4) - (m_4 - m_5)\} \\
&\quad - (m_6 - m_7) \\
y_4 &= \{(m_0 + m_2) + (m_1 + m_2)\} + \{(m_3 - m_5) + (m_4 - m_5)\} \\
&\quad + (m_6 + m_7) \\
y_5 &= -\{(m_0 - m_1) + (m_0 + m_2)\} + \{(m_3 - m_4) + (m_3 - m_5)\} \\
&\quad - (m_6 - m_7).
\end{aligned}$$

$N = 7$ *Algorithm*–$M = 19, A_B = 23, A_C = 49, A = 72$:

$$A = \text{diag}\,(\tfrac{1}{7}\ 1\ 1\ 1\ 1\ 1\ 1\ 1\ 1\ 1\ 1\ 1\ 1\ 1\ 1\ 1\ 1\ 1\ \tfrac{1}{7})\, B$$

where

$$A = \begin{bmatrix} 1 & 1 & 1 & 1 & 1 & 1 & 1 \\ 1 & 0 & 0 & 0 & 0 & 0 & -1 \\ 0 & 1 & 0 & 0 & 0 & 0 & -1 \\ 0 & 0 & 1 & 0 & 0 & 0 & -1 \\ 0 & 0 & 0 & 1 & 0 & 0 & -1 \\ 0 & 0 & 0 & 0 & 1 & 0 & -1 \\ 0 & 0 & 0 & 0 & 0 & 1 & -1 \\ 1 & 0 & 0 & 1 & 0 & 0 & -2 \\ 0 & 1 & 0 & 0 & 1 & 0 & -2 \\ 0 & 0 & 1 & 0 & 0 & 1 & -2 \\ 1 & 1 & 0 & 0 & 0 & 0 & -2 \\ 0 & 1 & 1 & 0 & 0 & 0 & -2 \\ 1 & 0 & 1 & 0 & 0 & 0 & -2 \\ 0 & 0 & 0 & 1 & 1 & 0 & -2 \\ 0 & 0 & 0 & 0 & 1 & 1 & -2 \\ 0 & 0 & 0 & 1 & 0 & 1 & -2 \\ 1 & 1 & 0 & 1 & 1 & 0 & -4 \\ 0 & 1 & 1 & 0 & 1 & 1 & -4 \\ 1 & 1 & 1 & 1 & 1 & 1 & -6 \end{bmatrix}$$

$$\begin{aligned}
a_1 &= h_0 - h_6 \\
a_2 &= h_1 - h_6 \\
a_3 &= h_2 - h_6 \\
a_4 &= h_3 - h_6 \\
a_5 &= h_4 - h_6 \\
a_6 &= h_5 - h_6 \\
a_7 &= a_1 + a_4 \\
a_8 &= a_2 + a_5 \\
a_9 &= a_3 + a_6 \\
a_{10} &= a_1 + a_2 \\
a_{11} &= a_2 + a_3 \\
a_{12} &= a_1 + a_3 \\
a_{13} &= a_4 + a_5 \\
a_{14} &= a_5 + a_6 \\
a_{15} &= a_4 + a_6 \\
a_{16} &= a_{10} + a_{13} \\
a_{17} &= a_{11} + a_{14} \\
a_{18} &= (a_7 + a_{17})/7 \\
a_0 &= a_{18} + h_6
\end{aligned}$$

$$\begin{aligned}
b_1 &= x_0 - x_6 \\
b_2 &= x_1 - x_6 \\
b_3 &= x_2 - x_6 \\
b_4 &= x_3 - x_6 \\
b_5 &= x_4 - x_6 \\
b_6 &= x_5 - x_6 \\
b_7 &= b_1 + b_4 \\
b_8 &= b_2 + b_5 \\
b_9 &= b_3 + b_6 \\
b_{10} &= b_1 + b_2 \\
b_{11} &= b_2 + b_3 \\
b_{12} &= b_1 + b_3
\end{aligned}$$

$b_{13} = b_4 + b_5$
$b_{14} = b_5 + b_6$
$b_{15} = b_4 + b_6$
$b_{16} = b_{10} + b_{13}$
$b_{17} = b_{11} + b_{14}$
$b_{18} = b_7 + b_{17}$
$b_0 = b_{18} + x_6 + [(x_6 + x_6) + 2x_6] + 2x_6$

$m_k = a_k b_k, \quad k = 0, 1, \cdots, 18$

$$C = \begin{bmatrix} 1 & 1 & 0 & -1 & -1 & -1 & -1 & 0 & 0 & 1 & 0 & 0 & 0 & 1 & 0 & 0 & 0 & 0 & -1 \\ 1 & -1 & -1 & 0 & -1 & 1 & -1 & 0 & 0 & 0 & 1 & 0 & 0 & 0 & 0 & 1 & 0 & 0 & -1 \\ 1 & -1 & 1 & -1 & 0 & -1 & -1 & 0 & 0 & 0 & 0 & 0 & 1 & 0 & 1 & 0 & 0 & 0 & -1 \\ 1 & -1 & -1 & -1 & -1 & 0 & 1 & 1 & 0 & 0 & 0 & 1 & 0 & 0 & 0 & 0 & 0 & 0 & -1 \\ 1 & 1 & 1 & 1 & 1 & 1 & 0 & -1 & -1 & 0 & -1 & 0 & 0 & -1 & 0 & 0 & 1 & 0 & -1 \\ 1 & 1 & -1 & 1 & 1 & -1 & 1 & 0 & 2 & 0 & 0 & 0 & -1 & 0 & 0 & -1 & -1 & -1 & 6 \\ 1 & 0 & 1 & 1 & 1 & 1 & 1 & 0 & -1 & -1 & 0 & -1 & 0 & 0 & -1 & 0 & 0 & 1 & -1 \end{bmatrix}$$

$u_0 = m_0 - m_{18}$
$u_1 = m_1 - m_5$
$u_2 = m_4 + m_6$
$u_3 = m_1 + m_3$
$u_4 = m_2 - m_6$
$u_5 = m_2 + m_3 + m_4 + m_5 - m_8$
$u_6 = u_0 - u_3$
$u_7 = u_0 + u_5$
$y_0 = u_0 + u_1 - u_2 - m_3 + m_9 + m_{13}$
$y_1 = u_0 - u_1 - u_2 - m_2 + m_{10} + m_{15}$
$y_2 = u_6 + u_4 - m_5 + m_{12} + m_{14}$
$y_3 = u_6 - u_4 - m_4 + m_7 + m_{11}$
$y_4 = u_7 + m_1 - m_7 - m_{10} - m_{13} + m_{16}$
$y_5 = (m_0 + m_0) + (2m_0 + 2m_0) + m_0 - y_0 - y_1 - y_2 - y_3 - y_4 - y_6$
$y_6 = u_7 + m_6 - m_9 - m_{11} - m_{14} + m_{17}.$

$N = 8$ *Algorithm*–$M = 14, A_B = 20, A_C = 26, A = 46$:

$A = \mathrm{diag}\,(\tfrac{1}{2}\ \tfrac{1}{2}\ \tfrac{1}{2}\ \tfrac{1}{2}\ \tfrac{1}{2}\ \tfrac{1}{2}\ \tfrac{1}{2}\ \tfrac{1}{2}\ \tfrac{1}{2}\ \tfrac{1}{4}\ \tfrac{1}{4}\ \tfrac{1}{4}\ \tfrac{1}{8}\ \tfrac{1}{8})E$

where

$$E = \begin{bmatrix} 1 & 0 & 0 & -1 & -1 & 0 & 0 & 1 \\ 1 & 0 & 0 & 0 & -1 & 0 & 0 & 0 \\ 1 & 1 & 0 & 0 & -1 & -1 & 0 & 0 \\ 1 & 1 & 1 & -1 & -1 & -1 & -1 & 1 \\ 1 & 0 & 1 & 0 & -1 & 0 & -1 & 0 \\ 1 & 1 & 1 & 1 & -1 & -1 & -1 & -1 \\ -1 & 1 & 1 & 1 & 1 & -1 & -1 & -1 \\ -1 & 0 & 1 & 0 & 1 & 0 & -1 & 0 \\ -1 & -1 & 1 & 1 & 1 & 1 & -1 & -1 \\ 1 & 0 & -1 & 0 & 1 & 0 & -1 & 0 \\ 1 & 1 & -1 & -1 & 1 & 1 & -1 & -1 \\ -1 & 1 & 1 & -1 & -1 & 1 & 1 & -1 \\ 1 & -1 & 1 & -1 & 1 & -1 & 1 & -1 \\ 1 & 1 & 1 & 1 & 1 & 1 & 1 & 1 \end{bmatrix}.$$

Also,

$$B = \begin{bmatrix} 0 & 1 & 0 & 1 & 0 & -1 & 0 & -1 \\ 1 & -1 & 1 & -1 & -1 & 1 & -1 & 1 \\ 1 & 0 & 1 & 0 & -1 & 0 & -1 & 0 \\ 0 & 0 & 0 & 1 & 0 & 0 & 0 & -1 \\ 0 & 0 & 1 & -1 & 0 & 0 & -1 & 1 \\ 0 & 0 & 1 & 0 & 0 & 0 & -1 & 0 \\ 0 & 1 & 0 & 0 & 0 & -1 & 0 & 0 \\ 1 & -1 & 0 & 0 & -1 & 1 & 0 & 0 \\ 1 & 0 & 0 & 0 & -1 & 0 & 0 & 0 \\ 1 & 1 & -1 & -1 & 1 & 1 & -1 & -1 \\ 0 & 1 & 0 & -1 & 0 & 1 & 0 & -1 \\ 1 & 0 & -1 & 0 & 1 & 0 & -1 & 0 \\ 1 & -1 & 1 & -1 & 1 & -1 & 1 & -1 \\ 1 & 1 & 1 & 1 & 1 & 1 & 1 & 1 \end{bmatrix}.$$

In the following, the superscript j appears on the coefficients of polynomials mod $P_{d_j}(z)$.

$h_0^1 = (h_0 - h_4)/2$
$h_1^1 = (h_1 - h_5)/2$
$h_2^1 = (h_2 - h_6)/2$
$h_3^1 = (h_3 - h_7)/2$
$h_0^2 = [(h_0 + h_4) - (h_2 + h_6)]/4$
$h_1^2 = [(h_1 + h_5) - (h_3 + h_7)]/4$
$h_0^3 = \{[(h_0 + h_4) + (h_2 + h_6)] - [(h_1 + h_5) + (h_3 + h_7)]\}$
$h_0^4 = \{[(h_0 + h_4) + (h_2 + h_6)] + [(h_1 + h_5) + (h_3 + h_7)]\}$
$a_0 = h_0^1 - h_3^1$
$a_1 = h_0^1$
$a_2 = h_0^1 + h_1^1$
$a_3 = (h_0^1 + h_2^1) + (h_1^1 - h_3^1)$
$a_4 = h_0^1 + h_2^1$
$a_5 = (h_0^1 + h_2^1) + (h_1^1 + h_3^1)$
$a_6 = (-h_0^1 + h_2^1) + (h_1^1 + h_3^1)$
$a_7 = -h_0^1 + h_2^1$

$a_8 = (-h_0^1 + h_2^1) - (h_1^1 - h_3^1)$
$a_9 = h_0^2$
$a_{10} = h_0^2 + h_1^2$
$a_{11} = -h_0^2 + h_1^2$
$a_{12} = h_0^3$
$a_{13} = h_0^4$

$x_0^1 = x_0 - x_4$
$x_1^1 = x_1 - x_5$
$x_2^1 = x_2 - x_6$
$x_3^1 = x_3 - x_7$
$x_0^2 = (x_0 + x_4) - (x_2 + x_6)$
$x_1^2 = (x_1 + x_5) - (x_3 + x_7)$
$x_0^3 = [(x_0 + x_4) + (x_2 + x_6)] - [(x_1 + x_5) + (x_3 + x_7)]$
$x_0^4 = [(x_0 + x_4) + (x_2 + x_6)] + [(x_1 + x_5) + (x_3 + x_7)]$

$b_0 = x_1^1 + x_3^1$
$b_1 = (x_0^1 + x_2^1) - (x_1^1 + x_3^1)$
$b_2 = x_0^1 + x_2^1$
$b_3 = x_3^1$
$b_4 = x_2^1 - x_3^1$
$b_5 = x_2^1$
$b_6 = x_1^1$
$b_7 = x_0^1 - x_1^1$
$b_8 = x_0^1$
$b_9 = x_0^2 + x_1^2$
$b_{10} = x_1^2$
$b_{11} = x_0^2$
$b_{12} = x_0^3$
$b_{13} = x_0^4$

$m_k = a_k b_k, \quad k = 0, 1, \cdots, 13$

$$C = \begin{bmatrix} 1 & 1 & 0 & -1 & -1 & 0 & 0 & 0 & 0 & 1 & -1 & 0 & 1 & 1 \\ 0 & -1 & 1 & 0 & 1 & -1 & 0 & 0 & 0 & 1 & 0 & 1 & -1 & 1 \\ 1 & 1 & 0 & 0 & 0 & 0 & 1 & 1 & 0 & -1 & 1 & 0 & 1 & 1 \\ 0 & -1 & 1 & 0 & 0 & 0 & 0 & -1 & 1 & -1 & 0 & -1 & -1 & 1 \\ -1 & -1 & 0 & 1 & 1 & 0 & 0 & 0 & 0 & 1 & -1 & 0 & 1 & 1 \\ 0 & 1 & -1 & 0 & -1 & 1 & 0 & 0 & 0 & 1 & 0 & 1 & -1 & 1 \\ -1 & -1 & 0 & 0 & 0 & 0 & -1 & -1 & 0 & -1 & 1 & 0 & 1 & 1 \\ 0 & 1 & -1 & 0 & 0 & 0 & 0 & 1 & -1 & -1 & 0 & -1 & -1 & 1 \end{bmatrix}$$

$y_0 = [(m_0 + m_1) - (m_3 + m_4)] + [(m_9 - m_{10}) + (m_{12} + m_{13})]$
$y_1 = [(m_2 - m_1) + (m_4 - m_5)] + [(m_9 + m_{11}) + (m_{13} - m_{12})]$
$y_2 = [(m_0 + m_1) + (m_6 + m_7)] + [-(m_9 - m_{10}) + (m_{12} + m_{13})]$
$y_3 = [(m_2 - m_1) + (m_8 - m_7)] + [-(m_9 + m_{11}) + (m_{13} - m_{12})]$
$y_4 = -[(m_0 + m_1) - (m_3 + m_4)] + [(m_9 - m_{10}) + (m_{12} + m_{13})]$
$y_5 = -[(m_2 - m_1) + (m_4 - m_5)] + [(m_9 + m_{11}) + (m_{13} - m_{12})]$
$y_6 = -[(m_0 + m_1) + (m_6 + m_7)] + [-(m_9 - m_{10}) + (m_{12} + m_{13})]$
$y_7 = -[(m_2 - m_1) + (m_8 - m_7)] + [-(m_9 + m_{11}) + (m_{13} - m_{12})].$

$N = 9$ *Algorithm*–$M = 22, A_B = 29, A_C = 69, A = 98$:

$b_0 = x_0^3 = (x_0 + x_3 + x_6) + (x_2 + x_5 + x_8) + (x_1 + x_4 + x_7)$
$b_1 = x_0^2 = (x_0 + x_3 + x_6) - (x_2 + x_5 + x_8)$
$b_2 = x_1^2 = (x_1 + x_4 + x_7) - (x_2 + x_5 + x_8)$
$b_3 = x_0^2 + x_1^2$
$b_4 = x_0^1 = x_0 - x_6$
$b_5 = x_1^1 = x_1 - x_7$
$b_6 = x_2^1 = x_2 - x_8$
$b_7 = x_3^1 = x_3 - x_6$
$b_8 = x_4^1 = x_4 - x_7$
$b_9 = x_5^1 = x_5 - x_8$
$b_{10} = x_0^1 + x_3^1$
$b_{11} = x_1^1 + x_4^1$
$b_{12} = x_2^1 + x_5^1$
$b_{13} = x_0^1 + x_1^1$
$b_{14} = x_1^1 + x_2^1$
$b_{15} = x_2^1 + x_0^1$
$b_{16} = x_3^1 + x_4^1$
$b_{17} = x_4^1 + x_5^1$
$b_{18} = x_5^1 + x_3^1$
$b_{19} = b_{13} + b_{16}$
$b_{20} = b_{14} + b_{17}$
$b_{21} = b_{15} + b_{18}$

$$A = \text{diag}\left(\tfrac{1}{9}\ \tfrac{1}{3}\ \tfrac{1}{3}\ \tfrac{1}{9}\ 1\ 1\ 1\ 1\ 1\ 1\ \tfrac{1}{3}\ \tfrac{1}{3}\ \tfrac{1}{3}\ 1\ 1\ 1\ 1\ 1\ 1\ \tfrac{1}{3}\ \tfrac{1}{3}\ \tfrac{1}{3}\right) B$$

$m_k = a_k b_k, \quad k = 0, 1, \cdots, 21$

$u_0 = m_0 + (m_1 - m_3)$
$u_1 = m_0 - (m_1 - m_3) - (m_2 - m_3)$
$u_2 = m_0 + (m_2 - m_3)$
$u_3 = m_{20} - m_{11} - m_{12}$
$u_4 = m_{19} - m_{10} - m_{11}$
$u_5 = m_{21} - m_{10} + m_{11} - m_{12}$
$u_6 = u_3 + m_{10}$
$u_7 = u_4 + m_{12}$
$u_8 = u_0 - u_6$
$u_9 = u_1 - u_7$
$u_{10} = u_2 - u_5$
$u_{11} = m_4 - m_9$
$u_{12} = m_4 - m_5$
$u_{13} = m_8 - m_9$
$u_{14} = m_6 + m_7$
$u_{15} = m_4 + m_5 + u_{14}$
$u_{16} = m_8 + m_9 + u_{14}$
$y_0 = u_8 + u_{11} - m_8 + m_{17}$
$y_1 = u_9 - u_{11} - m_5 + m_{13}$
$y_2 = u_{10} - u_{12} - m_6 + m_{15}$
$y_3 = u_8 - u_{15} + 3m_{10} + m_{14}$

$$y_4 = u_9 + u_{15} + m_8 + 3u_4 - m_{13} - m_{16}$$
$$y_5 = u_2 + u_{12} - u_{13} + u_{14} + 2u_5 - m_{15} - m_{18}$$
$$y_6 = u_8 + u_{16} + m_5 + 3u_3 - m_{14} - m_{17}$$
$$y_7 = u_9 - u_{16} + 3m_{12} + m_{16}$$
$$y_8 = u_{10} + u_{13} - m_7 + m_{18}.$$

Appendix B
Rectangular Transforms Having the Cyclic Convolution Property

In this section, we will establish relationships between the A, B, and C matrices which are necessary and sufficient for y to be the cyclic convolution defined by (3.1). These relationships are very general and any square or rectangular transformation having the CCP must satisfy them.

The transforms of h and x are defined by

$$H = Ah \tag{B1}$$

$$X = Bx \tag{B2}$$

where A and B are rectangular matrices of dimensions $M \times N$ where N is the length of the cyclic convolution and M is the number of points in the transform domain. It is obvious that $M \geqslant N$.

The M multiplications required to multiply the transforms H and X arise in the calculation of

$$Y = H \times X \tag{B3}$$

where $\times$ denotes the element by element product.

The output vector y which is the cyclic convolution of x and h is obtained by another rectangular transformation

$$y = CY \tag{B4}$$

where C is an $N \times M$ matrix.

We would like to establish conditions on the A, B, and C matrices so that y is the cyclic convolution of x and h. Equations (B1) and (B2) can be written in terms of their elements,

$$H_k = \sum_{p=0}^{N-1} A_{k,p} h_p \tag{B5}$$

$$X_k = \sum_{q=0}^{N-1} B_{k,q} x_q, \qquad k = 0, 1, 2, \cdots, M-1. \tag{B6}$$

Equation (B4) can be written

$$y_n = \sum_{k=0}^{M-1} C_{n,k} H_k X_k, \qquad n = 0, 1, \cdots, N-1. \tag{B7}$$

Substituting for H_k and X_k from (B5) and (B6), we get

$$y_n = \sum_{k=0}^{M-1} C_{n,k} \left\{ \sum_{p=0}^{N-1} A_{k,p} h_p \right\} \left\{ \sum_{q=0}^{N-1} B_{k,q} x_q \right\}$$

$$y_n = \sum_{p=0}^{N-1} \sum_{q=0}^{N-1} h_p x_q \left\{ \sum_{k=0}^{M-1} C_{n,k} A_{k,p} B_{k,q} \right\}. \tag{B8}$$

The CCP requires that

$$\sum_{k=0}^{M-1} C_{n,k} A_{k,p} B_{k,q} \begin{aligned} &= 1 \text{ if } p + q = n \bmod N \\ &= 0 \text{ otherwise.} \end{aligned} \tag{B9}$$

Equation (B9) is the necessary and sufficient condition for the CCP. It can be stated as follows. "The inner product of the pth column of A, the qth column of B, and the nth row of C should be 1 for $p + q = n \bmod N$ and zero otherwise."

For the square transform case ($M = N$), further restrictions can be placed on the A, B, and C matrices leading to the results of Agarwal and Burrus [2]. For this case, the transform matrices have the DFT structure and the computation of the transforms, in general, requires multiplications. But, if M is allowed to be greater than N, then more flexibility exists in choosing the A, B, and C matrices. As M is increased, one can obtain A, B, and C matrices with simpler coefficients. As an extreme case, one can take $M = N^2$, and in that case, each row of the A and B matrices and each column of the C matrix will have only one nonzero element. This case reduces to a direct computation of the convolution. Between the two extremes of the DFT structure ($M = N$) and the direct computation ($M = N^2$), various degrees of tradeoffs exist in the simplicity of the transformation matrices and the size of M. For very long sequences ($N \to \infty$) the DFT, using the FFT algorithm seems to be computationally optimal. We have chosen the algorithms of Appendix A so that M is small, but not always the minimum according to Winograd's theorem. The choice of a nonminimum M is made so that the transformation matrices are simple, meaning that their implementation requires only additions. This reduces the number of multiplications required for cyclic convolution to the given M-values.

References

[1] R. C. Agarwal and C. S. Burrus, "Fast one-dimensional digital convolution by multidimensional techniques," *IEEE Trans. Acoust., Speech, Signal Processing*, vol. ASSP-22, pp. 1-10, Feb. 1974.

[2] ——, "Fast convolution using Fermat number transforms with applications to digital filtering," *IEEE Trans. Acoust., Speech, Signal Processing*, vol. ASSP-22, pp. 87-99, Apr. 1974.

[3] ——, "Number theoretic transforms to implement fast digital convolution," *Proc. IEEE*, vol. 63, pp. 550-560, Apr. 1975.

[4] G. D. Bergland, "A fast Fourier transform algorithm using base 8 iterations," *Math. Comput.*, vol. 22, pp. 275-279, Apr. 1968.

[5] J. W. Cooley, P. A. W. Lewis, and P. D. Welch, "Historical notes on the fast Fourier transform," *IEEE Trans. Audio Electroacoust.*, vol. AU-15, pp. 76-79, June 1967.

[6] ——, "The fast Fourier transform: Programming considerations in the calculation of sine, cosine and Laplace transforms," *J. Sound Vib.*, vol. 22, pp. 315-337, July 1970.

[7] I. J. Good, "The interaction algorithm and practical Fourier analysis," *J. Royal Statist. Soc.*, ser. B. vol. 20, pp. 361-372, 1958; addendum, vol. 22, 1960, pp. 372-375, (MR 21 1674; MR 23 A4231).

[8] J. H. Griesmer, R. D. Jenks, and D. Y. Y. Yun, "SCRATCHPAD user's manual," IBM Res. Rep. RA 70, IBM Watson Res. Cen., Yorktown Heights, NY, June 1975; and SCRATCHPAD Technical Newsletter No. 1, Nov. 15, 1975.

[9] D. E. Knuth, "Seminumerical algorithms," in *The Art of Computer Programming*, vol. 2. Reading, MA: Addision-Wesley, 1971.

[10] T. Nagell, *Introduction to Number Theory*. New York: Wiley, 1951.

[11] P. J. Nicholson, "Algebraic theory of finite Fourier transforms," *J. Comp't. Syst. Sci.*, vol. 5, pp. 524-527, Oct. 1971.

[12] J. M. Pollard, "The fast Fourier transform in a finite field," *Math. Comput.*, vol. 25, no. 114, pp. 365-374, Apr. 1971.

[13] C. M. Rader, "Discrete convolutions via Mersenne transforms," *IEEE Trans. Comput.*, vol. C-21, pp. 1269-1273, Dec. 1972.

[14] I. S. Reed and T. K. Truong, "The use of finite fields to compute convolutions," *IEEE Trans. Inform. Theory*, vol. IT-21, pp. 208-213, Mar. 1975.

[15] ——, "Complex integer convolutions over a direct sum of Galois fields," *IEEE Trans. Inform. Theory*, vol. IT-21, pp. 657-661, Nov. 1975.

[16] R. C. Singleton, "An algorithm for computing the mixed radix fast Fourier transform," *IEEE Trans. Audio Electroacoust.*, vol. AU-17, pp. 93–103, June 1969.

[17] L. H. Thomas, "Using a computer to solve problems in physics," in *Applications of Digital Computers*. Boston, MA: Ginn and Co., 1963.

[18] S. Winograd, "Some bilinear forms whose multiplicative complexity depends on the field of constants," IBM Res. Rep. RC 5669, IBM Watson Research Cen., Yorktown Heights, NY, Oct. 10, 1975.

[19] ——, "On computing the discrete Fourier transform," *Proc. Nat. Acad. Sci. USA*, vol. 73, no. 4, pp. 1005–1006, Apr. 1976.

MATHEMATICS OF COMPUTATION, VOLUME 32, NUMBER 141
JANUARY 1978, PAGES 175–199

On Computing the Discrete Fourier Transform

By S. Winograd

Abstract. A new algorithm for computing the Discrete Fourier Transform is described. The algorithm is based on a recent result in complexity theory which enables us to derive efficient algorithms for convolution. These algorithms are then used to obtain the new Discrete Fourier Transform algorithm.

I. Introduction. A previous paper [1] investigated the minimum number of multiplications needed to obtain the coefficients of the product of two $(n-1)$st degree polynomials modulo an nth degree polynomial. In this paper we will use the results of [1] to obtain new algorithms for computing the Discrete Fourier Transform (DFT). These new algorithms use about the same number of additions as the algorithms proposed by Cooley and Tukey [2], but only about 20% of the number of multiplications which their algorithm requires.

In the second section we will summarize the results needed for the construction of the algorithms. The third section will describe the derivation of the algorithms for cyclic convolutions, the fourth section will use these algorithms to derive the algorithm for DFT's of a few tens to a few thousands of points. In the last section we will discuss algorithms for multidimensional DFT's as well as algorithms for computing the DFT of very large numbers.

II. Theoretical Background. Let

$$R_l(u) = \sum_{i=0}^{l} x_i u^i, \qquad S_m(u) = \sum_{i=0}^{m} y_i u^i$$

be two polynomials with indeterminate coefficients, and let $P(u) = u^n + \Sigma_{i=0}^{n-1} a_i u^i$ be a monic polynomial of degree n with coefficients in a field G. (In the applications we will use G as the field Q of the rationals, only in the last section we will use other fields.) Assume $P(u) = P_1(u) \cdot P_2(u)$ such that $P_1(u)$ and $P_2(u)$ are relatively prime, and let $n_1 = \deg(P_1)$ and $n_2 = \deg(P_2)$.

Using the Chinese Remainder Theorem, we obtain:

$$\begin{aligned} &R_l \cdot S_m \bmod P \\ &\quad = (Q_2 \cdot P_2(R_l \cdot S_m \bmod P_1) + Q_1 \cdot P_1 \cdot (R_l \cdot S_m \bmod P_2)) \bmod P, \end{aligned} \tag{1}$$

where Q_1 and Q_2 are polynomials such that

$$Q_1 P_1 + Q_2 \cdot P_2 = 1 \bmod P. \tag{2}$$

Received November 30, 1976; revised June 9, 1977.
AMS (MOS) subject classifications (1970). Primary 68A20.

Let $\widetilde{T}$ be the set of coefficients of $R_l \cdot S_m$ and let $\widetilde{T}_P$ be the set of coefficients of $R_l \cdot S_m$ mod P. It was shown in [3] that at least $l + m + 1$ multiplications are needed to compute $\widetilde{T}$ (multiplication by a fixed element $g \in G$ is not counted), and using the algorithm of [4] one can actually obtain an algorithm for computing $\widetilde{T}$ using $l + m + 1$ multiplications. Clearly, we can obtain $\widetilde{T}_P$ from $\widetilde{T}$ using only additions and multiplications by elements $g \in G$; thus, the number of multiplications, which are counted, needed to compute $\widetilde{T}_P$ is at most $l + m + 1$.

Another way of computing $\widetilde{T}_P$ when P has more than one irreducible factor is by the use of the identity (1), i.e., if $P = P_1 \cdot P_2$ such that P_1 and P_2 are relatively prime then we can use the algorithm for $\widetilde{T}_{P_1}$ to multiply (R mod P_1) $\cdot$ (S mod P_1) mod P_1, and the algorithm for $\widetilde{T}_{P_2}$ to multiply (R mod P_2) $\cdot$ (S mod P_2) mod P_2, and then obtain the algorithm for $\widetilde{T}_P$ using only additional additions and multiplications by elements of G. It was shown in [1] that for the case $l = m = n - 1$ the number of multiplications needed to compute $\widetilde{T}_P$ is $2n - k$, where k is the number of distinct irreducible factors of P. Moreover, every algorithm which computes $\widetilde{T}_P$ in $2n - k$ multiplications uses (1).

When P has only one irreducible factor, i.e., when P is a power of irreducible polynomial, we cannot use (1); but then we compute $\widetilde{T}_p$ by computing $\widetilde{T}$ first and then reducing modulo P.

There are two ways for computing $\widetilde{T}$ using only $l + m + 1$ multiplications. The first one uses the identity

$$(3) \qquad R_l(u) \cdot S_m(u) = R_l(u) \cdot S_m(u) \bmod \prod_{i=0}^{m+l} (u - \alpha_i),$$

where the α_i's are distinct elements of G. (We assume that G is large enough. Actually, we will use in this paper only G of characteristic 0.) The right-hand side of (3) can be computed using (1) in $m + l + 1$ multiplications. This algorithm is the same as the one described in [4].

A second algorithm uses the identity

$$(4) \qquad R_l(u) \cdot S_m(u) = R_l(u) \cdot S_m(u) \quad \bmod \prod_{i=1}^{m+l} (u - \beta_i) + x_l y_m \prod_{i=1}^{m+l} (u - \beta_i),$$

where the β_i's are distinct elements of G. It was shown in [1] that every algorithm for computing $\widetilde{T}$ in $m + l + 1$ multiplication uses either (3) or (4).

At times it is desirable to avoid the constants which the algorithms of (3) or (4) necessitate, even at the expense of additional multiplication. One way of accomplishing this is to choose a polynomial $P(u)$ of degree $l + m + 1$ with many distinct irreducible factors, but not necessarily only linear factors. Identity (3) is then modified to:

$$(5) \qquad R_l(u) \cdot S_m(u) = R_l(u) \cdot S_m(u) \quad \bmod P.$$

Similarly, we can modify identity (4).

Another theoretical development which will be needed in this paper is that of the dual or transpose of system. Let

$$\sum_{j=1}^{s}\sum_{i=1}^{r} a_{ijk}x_iy_j, \qquad k = 1, 2, \ldots, t, \tag{6}$$

be a system of bilinear forms, and assume that we have found an algorithm for computing this system using n multiplications without using the commutative law. That is,

$$\sum_{j=1}^{s}\sum_{i=1}^{r} a_{ijk}x_iy_j = \sum_{l=1}^{n} \gamma_{k,l}\left(\sum_{i=1}^{r}\alpha_{i,l}x_i\right)\left(\sum_{j=1}^{s}\beta_{j,l}y_j\right), \qquad k = 1, 2, \ldots, t. \tag{7}$$

Multiplying both sides of (7) by z_k and summing over k, we obtain

$$\sum_{k=1}^{t}\sum_{j=1}^{s}\sum_{i=1}^{r} a_{ijk}z_kx_iy_j = \sum_{l=1}^{n}\left(\sum_{k=1}^{t}\gamma_{k,l}z_k\right)\left(\sum_{i=1}^{r}\alpha_{i,l}x_i\right)\left(\sum_{j=1}^{s}\beta_{j,l}y_j\right). \tag{8}$$

Equating the coefficients of x_i, we obtain

$$\sum_{k=1}^{t}\sum_{j=1}^{r} a_{ijk}z_ky_j = \sum_{l=1}^{n}\alpha_{i,l}\left(\sum_{k=1}^{t}\gamma_{k,l}z_k\right)\left(\sum_{j=1}^{s}\beta_{j,l}y_j\right), \qquad i = 1, 2, \ldots, r. \tag{9}$$

The left-hand side of (9) is called a dual (or transpose) system, and the right-hand side of (9) provides an algorithm for computing it using n multiplications.

III. Cyclic Convolution. Consider the problem of computing the cyclic convolution of two sets of n points $(x_0, x_1, \ldots, x_{n-1})$ and $(y_0, y_1, \ldots, y_{n-1})$. This can be written as

$$\begin{pmatrix} x_0 & x_1 \cdots x_{n-2} & x_{n-1} \\ x_1 & x_2 \cdots x_{n-1} & x_0 \\ x_3 & x_4 \cdots x_0 & x_1 \\ \cdot & & \\ x_{n-1} & x_0 \cdots x_{n-3} & x_{n-2} \end{pmatrix}\begin{pmatrix} y_0 \\ y_1 \\ \cdot \\ \cdot \\ \cdot \\ y_{n-1}\end{pmatrix}. \tag{10}$$

It is readily verified that (10) is the system of coefficients of the polynomial

$$\begin{aligned}&(x_0 + x_1u + x_2u^2 + \cdots + x_{n-1}u^{n-1})\\ &\qquad (y_0 + y_{n-1}u + y_{n-2}u^2 + \cdots + y_1u^{n-1}) \bmod u^n - 1,\end{aligned} \tag{11}$$

and we can use the results described in the previous section to compute this system.

As an example we will take $n = 3$, that is, we consider the system

$$\begin{pmatrix} z_0 \\ z_1 \\ z_2 \end{pmatrix} = \begin{pmatrix} x_0 & x_1 & x_2 \\ x_1 & x_2 & x_0 \\ x_2 & x_0 & x_1 \end{pmatrix}\begin{pmatrix} y_0 \\ y_1 \\ y_2 \end{pmatrix}, \tag{12}$$

which is the system of coefficients of

$$(x_0 + x_1 u + x_2 u^2)(y_0 + y_2 u + y_1 u^2) \bmod (u^3 - 1). \tag{13}$$

Since $u^3 - 1 = (u - 1)(u^2 + u + 1)$, we have to compute

$$\begin{aligned}(x_0 + x_1 u + x_2 u^2)(y_0 + y_2 u + y_1 u^2) &\bmod (u - 1)\\ &= (x_0 + x_1 + x_2)(y_0 + y_1 + y_2)\end{aligned} \tag{14}$$

and

$$\begin{aligned}&(x_0 + x_1 u + x_2 u^2)(y_0 + y_2 u + y_1 u^2) \bmod (u^2 + u + 1)\\ &\quad = ((x_0 - x_2) + (x_1 - x_2)u) \cdot ((y_0 - y_1) + (y_2 - y_1)u) \bmod (u^2 + u + 1).\end{aligned} \tag{15}$$

The part of the computation (14) can be done in one multiplication. To compute (15) we first compute $\widetilde{T}$, that is, the coefficients of $((x_0 - x_2) + (x_1 - x_2)u) \cdot ((y_0 - y_1) + (y_2 - y_1)u)$. This is done using the identity:

$$\begin{aligned}&((x_0 - x_2) + (x_1 - x_2)u)((y_0 - y_1) + (y_2 - y_1)u)\\ &\quad = ((x_0 - x_2) + (x_1 - x_2)u)((y_0 - y_1) + (y_2 - y_1)u) \bmod u(u + 1)\\ &\quad + (x_1 - x_2)(y_2 - y_1)u(u + 1),\end{aligned} \tag{16}$$

which leads to the algorithm:

$$\begin{aligned}&m_1 = (x_0 - x_2)(y_0 - y_1), \qquad m_2 = (x_1 - x_2)(y_2 - y_1),\\ &m_3 = ((x_0 - x_2) - (x_1 - x_2))((y_0 - y_1) - (y_2 - y_1)) = (x_0 - x_1)(y_0 - y_2).\end{aligned}$$

$$\begin{aligned}&(x_0 - x_2)(y_0 - y_1) = m_1,\\ &(x_0 - x_2)(y_2 - y_1) + (x_1 - x_2)(y_0 - y_1) = m_1 + m_2 - m_3,\\ &(x_1 - x_2)(y_2 - y_1) = m_2.\end{aligned} \tag{17}$$

And consequently, the coefficients of (15) are

$$m_1 - m_2 \quad \text{and} \quad m_1 + m_2 - m_3 - m_2 = m_1 - m_3.$$

Defining $m_0 = (x_0 + x_1 + x_2)(y_0 + y_1 + y_2)$, we obtain

$$\begin{aligned}&(x_0 + x_1 u + x_2 u^2)(y_0 + y_2 u + y_1 u^2) \bmod (u - 1) = m_0,\\ &(x_0 + x_1 u + x_2 u^2)(y_0 + y_2 u + y_1 u^2) \bmod (u^2 + u + 1)\\ &\qquad = (m_1 - m_2) + (m_1 - m_3)u;\end{aligned} \tag{18}$$

and using (1), we obtain

$$(x_0 + x_1 u + x_2 u^2)(y_0 + y_2 u + y_1 u^2) \bmod (u^3 - 1)$$

$$= \frac{u^2 + u + 1}{3} m_0 + \left(\frac{-(u-1)(u+2)}{3}\right)((m_1 - m_2) + (m_1 - m_3)u) \quad \bmod (u^3 - 1)$$

(19)
$$= \left(\frac{m_0}{3} + \frac{m_1}{3} - \frac{2m_2}{3} + \frac{m_3}{3}\right) + \left(\frac{m_0}{3} + \frac{m_1}{3} + \frac{m_2}{3} - \frac{2m_3}{3}\right)u$$

$$+ \left(\frac{m_0}{3} - \frac{2m_1}{3} + \frac{m_2}{3} + \frac{m_3}{3}\right)u^2.$$

In many applications either the x_i's or the y_j's are known a priori, for example, they are the tap values; and therefore, computations involving only these variables can be done beforehand, and thus should not be counted. Assuming that the operations on the x_i's are not counted, we define

(20) $$m'_0 = \frac{x_0 + x_1 + x_2}{3}(y_0 + y_1 + y_2), \qquad m'_1 = \frac{x_0 - x_2}{3}(y_0 - y_1),$$

(20a) $$m'_2 = \frac{x_1 - x_2}{3}(y_2 - y_1), \qquad m'_3 = \frac{x_0 - x_1}{3}(y_0 - y_2);$$

and the three desired quantities are:

(20b) $$m'_0 + m'_1 - 2m'_2 + m'_3, \quad m'_0 + m'_1 + m'_2 - 2m'_3, \quad m'_0 - 2m'_1 + m'_2 + m'_3.$$

Another algorithm is obtained by noticing that the transpose of (12) is

(21) $$\begin{pmatrix} z_0 & z_2 & z_1 \\ z_1 & z_0 & z_2 \\ z_2 & z_1 & z_0 \end{pmatrix}\begin{pmatrix} y_0 \\ y_1 \\ y_2 \end{pmatrix} = \begin{pmatrix} z_0 & z_1 & z_2 \\ z_1 & z_2 & z_0 \\ z_2 & z_0 & z_1 \end{pmatrix}\begin{pmatrix} y_0 \\ y_2 \\ y_1 \end{pmatrix}.$$

Transposing algorithm (20), we obtain

(22a) $$m_0 = \frac{z_0 + z_1 + z_2}{3}(y_0 + y_1 + y_2), \qquad m_1 = \frac{z_0 + z_1 - 2z_2}{3}(y_0 - y_1),$$

$$m_2 = \frac{-2z_0 + z_1 + z_2}{3}(y_2 - y_1), \qquad m_3 = \frac{z_0 - 2z_1 + z_2}{3}(y_0 - y_2);$$

and the three quantities to be computed are:

(22b) $$m_0 + m_1 + m_3, \quad m_0 + m_2 - m_3, \quad m_0 - m_1 - m_2.$$

This method of obtaining simpler algorithms by transposing the system of bilinear forms is useful for other cyclic convolutions as well. Using the Chinese Remainder Theorem usually results in $Q_1 \cdot P_1$ and $Q_2 \cdot P_2$ coefficients other than 0, 1, -1, and transposing the algorithm results in moving these coefficients to what part which can be precomputed.

The matrix in (10) can be viewed as the "multiplication table" for the group z_n of addition modulo n. In case $n = n_1 \cdot n_2$ where n_1 and n_2 are relatively prime, then z_n is isomorphic to $z_{n_1} \times z_{n_2}$. Therefore, there exists a permutation of the rows and columns of the matrix of (10) such that the resulting matrix can be partitioned into blocks of $n_2 \times n_2$ cyclic matrices, and such that the blocks form an $n_1 \times n_1$ cyclic matrix.

For example, since $6 = 2 \times 3$ we have the isomorphism

$$\begin{array}{lll} 0 \longrightarrow (0,0), & 1 \longrightarrow (1,1), & 2 \longrightarrow (0,2), \\ 3 \longrightarrow (1,0), & 4 \longrightarrow (0,1), & 5 \longrightarrow (1,2); \end{array} \tag{23}$$

and therefore, if we have the cyclic convolution

$$\begin{pmatrix} \psi_0 \\ \psi_1 \\ \psi_2 \\ \psi_3 \\ \psi_4 \\ \psi_5 \end{pmatrix} = \begin{pmatrix} x_0 & x_1 & x_2 & x_3 & x_4 & x_5 \\ x_1 & x_2 & x_3 & x_4 & x_5 & x_0 \\ x_2 & x_3 & x_4 & x_5 & x_0 & x_1 \\ x_3 & x_4 & x_5 & x_0 & x_1 & x_2 \\ x_4 & x_5 & x_0 & x_1 & x_2 & x_3 \\ x_5 & x_0 & x_1 & x_2 & x_3 & x_4 \end{pmatrix} \begin{pmatrix} y_0 \\ y_1 \\ y_2 \\ y_3 \\ y_4 \\ y_5 \end{pmatrix} \tag{24}$$

and we arrange it in the order 0, 4, 2, 3, 1, 5 (that is, first those indices whose first coordinate is 0 and the second coordinate in ascending order, and then those indices whose first coordinate is 1 and the second coordinate in ascending order), we obtain

$$\begin{pmatrix} \psi_0 \\ \psi_4 \\ \psi_2 \\ \hline \psi_3 \\ \psi_1 \\ \psi_5 \end{pmatrix} = \left(\begin{array}{ccc|ccc} x_0 & x_4 & x_2 & x_3 & x_1 & x_5 \\ x_4 & x_2 & x_0 & x_1 & x_5 & x_3 \\ x_2 & x_0 & x_4 & x_5 & x_3 & x_1 \\ \hline x_3 & x_1 & x_5 & x_0 & x_4 & x_2 \\ x_1 & x_5 & x_3 & x_4 & x_2 & x_0 \\ x_5 & x_3 & x_1 & x_2 & x_0 & x_4 \end{array}\right) \begin{pmatrix} y_0 \\ y_4 \\ y_2 \\ \hline y_3 \\ y_1 \\ y_5 \end{pmatrix}, \tag{25}$$

which is the same as (24), yet exhibits the block structure.

This block structure can be used to derive an algorithm by composing two different algorithms. Using $u^2 - 1 = (u + 1)(u - 1)$, we immediately obtain:

$$\begin{pmatrix} x_0 & x_1 \\ x_1 & x_0 \end{pmatrix} \begin{pmatrix} y_0 \\ y_1 \end{pmatrix} = \begin{pmatrix} \frac{x_0 + x_1}{2}(y_0 + y_1) + \frac{x_0 - x_1}{2}(y_0 - y_1) \\ \frac{x_0 + x_1}{2}(y_0 + y_1) - \frac{x_0 - x_1}{2}(y_0 - y_1) \end{pmatrix} \tag{26}$$

for a cyclic convolution of two elements. If we define:

$$\Phi_0 = \begin{pmatrix} \psi_0 \\ \psi_4 \\ \psi_2 \end{pmatrix}, \quad \Phi_1 = \begin{pmatrix} \psi_3 \\ \psi_1 \\ \psi_5 \end{pmatrix}, \quad Y_0 = \begin{pmatrix} y_0 \\ y_4 \\ y_2 \end{pmatrix}, \quad Y_1 = \begin{pmatrix} y_3 \\ y_1 \\ y_5 \end{pmatrix},$$

(27)

$$X_0 = \begin{pmatrix} x_0 & x_4 & x_2 \\ x_4 & x_2 & x_0 \\ x_2 & x_0 & x_4 \end{pmatrix}, \quad X_1 = \begin{pmatrix} x_3 & x_1 & x_5 \\ x_1 & x_5 & x_3 \\ x_5 & x_3 & x_1 \end{pmatrix},$$

then we can write (25) as

(28)
$$\begin{pmatrix} \Phi_0 \\ \Phi_1 \end{pmatrix} = \begin{pmatrix} X_0 & X_1 \\ X_1 & X_0 \end{pmatrix} \begin{pmatrix} Y_0 \\ Y_1 \end{pmatrix};$$

and using algorithm (26), we get

(29)
$$M_1 = \frac{X_0 + X_1}{2}(Y_0 + Y_1), \quad M_2 = \frac{X_0 - X_1}{2}(Y_0 - Y_1),$$
$$\Phi_0 = M_1 + M_2, \quad \Phi_1 = M_1 - M_2.$$

Computing M_1 and M_2, we use algorithm (22). Thus we obtained an algorithm for (24) which uses eight multiplications and 34 additions. It should be noted that we could have factored 6 as 3×2 and obtained a different block structure, namely that of three point convolution of 2×2 blocks. In this case we would have obtained another algorithm for (24) using eight multiplications and 38 additions.

In Appendix A we give the algorithms derived for cyclic convolution of 2, 3, 4, 5 and 6 points. These algorithms are summarized in Table 1. The algorithm given for five point cyclic convolution does not use the minimum number of multiplications. Another algorithm could have been derived using only eight multiplications, but then the number of additions would have been much larger, and the constant coefficients would not have been 0, ± 1.

TABLE I

n	# mult.	# add.
2	2	4
3	4	11
4	5	15
5	10	31
6	8	34

IV. One Dimensional Fourier Transform. The Discrete Fourier Transform of n points

$$(30)\qquad A_k = \sum_{j=0}^{n-1} w^{kj} a_j, \qquad i = 0, 1, \dots, n-1, \quad w = e^{2\pi i/n},$$

can be written as $\mathbf{A} = W\mathbf{a}$ where $W_{i,j} = w^{ij}$. We will consider first the case that n is a prime. In this case the matrix $W|_{i,j\neq 0}$ can be viewed as the "multiplication table" for the group M_n of nonzero integers relatively prime to n with group operation of multiplication modulo n. As is well known, $M_{p^r} \cong z_{(p-1)p^{r-1}}$ for $p \neq 2$ a prime and $M_{2^r} \cong z_2 \times z_{2^{r-2}}$. That means that if n is a prime, we can rearrange the rows and columns of $W|_{i,j\neq 0}$ so the resulting matrix is cyclic. (The idea of rearranging the indices of the Discrete Fourier Transform of a prime number of points so as to obtain cyclic convolution was first suggested by C. M. Rader [5].) This is illustrated in (31) for the case $n = 7$, i.e., (31) is another way of writing the Discrete Fourier Transform for seven points.

$$(31)\qquad \begin{pmatrix} A_0 \\ A_1 \\ A_3 \\ A_2 \\ A_6 \\ A_4 \\ A_5 \end{pmatrix} = \begin{pmatrix} 1 & 1 & 1 & 1 & 1 & 1 & 1 \\ 1 & w^1 & w^3 & w^2 & w^6 & w^4 & w^5 \\ 1 & w^3 & w^2 & w^6 & w^4 & w^5 & w^1 \\ 1 & w^2 & w^6 & w^4 & w^5 & w^1 & w^3 \\ 1 & w^6 & w^4 & w^5 & w^1 & w^3 & w^2 \\ 1 & w^4 & w^5 & w^1 & w^3 & w^2 & w^6 \\ 1 & w^5 & w^1 & w^3 & w^2 & w^6 & w^4 \end{pmatrix} \begin{pmatrix} a_0 \\ a_1 \\ a_3 \\ a_2 \\ a_6 \\ a_4 \\ a_5 \end{pmatrix}, \quad w = e^{2\pi i/7}.$$

We can now use the algorithm for six point cyclic convolution developed in the previous section to compute the seven point Fourier Transform. Actually, for later use, it is better to compute first $A_i - A_0$, $i = 1, 2, \dots, 6$. (Note that we have not disturbed the symmetries and, therefore, can still use the algorithms developed in the previous section.) The resulting algorithm appears in Appendix B.

In case $n = p^r$ is a power of a prime number, the situation is very similar. We can permute the rows and columns of W so as to have copies of $M_{p^r}, M_{p^{r-1}}, \dots, M_{p^0}$. This permutation is best explained by means of an example. In (32) we illustrate the permutation of W for a nine point Fourier Transform. The algorithm for the nine points Fourier Transform is in Appendix B.

$$(32)\qquad \begin{pmatrix} A_0 \\ A_3 \\ A_6 \\ A_1 \\ A_2 \\ A_4 \\ A_8 \\ A_7 \\ A_5 \end{pmatrix} = \left(\begin{array}{ccc|ccc|ccc} 1 & 1 & 1 & 1 & 1 & 1 & 1 & 1 & 1 \\ 1 & 1 & 1 & w^3 & w^6 & w^3 & w^6 & w^3 & w^6 \\ 1 & 1 & 1 & w^6 & w^3 & w^6 & w^3 & w^6 & w^3 \\ \hline 1 & w^3 & w^6 & w^1 & w^2 & w^4 & w^8 & w^7 & w^5 \\ 1 & w^6 & w^3 & w^2 & w^4 & w^8 & w^7 & w^5 & w^1 \\ 1 & w^3 & w^6 & w^4 & w^8 & w^7 & w^5 & w^1 & w^2 \\ \hline 1 & w^6 & w^3 & w^8 & w^7 & w^5 & w^1 & w^2 & w^4 \\ 1 & w^3 & w^6 & w^7 & w^5 & w^1 & w^2 & w^4 & w^8 \\ 1 & w^6 & w^3 & w^5 & w^1 & w^2 & w^4 & w^8 & w^7 \end{array}\right) \begin{pmatrix} a_0 \\ a_3 \\ a_6 \\ a_1 \\ a_2 \\ a_4 \\ a_8 \\ a_7 \\ a_5 \end{pmatrix}.$$

An examination of the algorithms for the seven and nine point Fourier Transform reveals that the multiplicand which depends on the powers of w is either a real number or an imaginary number–never a general complex number. This is not a peculiarity of these two numbers, but a general property of these algorithms. In the case that $n = p^r$, $p \neq 2$, the group M_{p^r} is isomorphic to $Z_{(p-1)p^{r-1}}$, and the element -1 of M_{p^r} is mapped into $\frac{1}{2}(p-1)p^{r-1}$ under the isomorphism. But since $u^{(p-1)p^{r-1}} - 1 = (u^{\frac{1}{2}(p-1)p^{r-1}} - 1)(u^{\frac{1}{2}(p-1)p^{r-1}} + 1)$, the part of the algorithm for cyclic convolution which is based on computing modulo $u^{\frac{1}{2}(p-1)p^{r-1}} - 1$ depends on $w^j + w^{-j}$ which are real numbers; and the part of the algorithm which is based on $u^{\frac{1}{2}(p-1)p^{r-1}} + 1$ depends on $w^j - w^{-j}$ which are imaginary numbers. A similar argument establishes this fact for $n = 2^r$. Table 2 summarizes the algorithms for computing the Discrete Fourier Transform of 2, 3, 4, 5, 7, 8, 9, and 16 points. The actual algorithms are given in Appendix B. Since we will later have to consider multiplication by $w^0 = 1$ as a multiplication, this is also summarized in Table 2.

TABLE 2

n	# Mult.	# Mult. by w^0	# Add.
2	0	2	2
3	2	1	6
4	0	4	8
5	5	1	17
7	8	1	36
8	2	6	26
9	10	1	45
16	10	8	74

We now turn our attention to performing the Discrete Fourier Transform of n points where n is not a power of a prime. The idea of using the Chinese Remainder Theorem for "building up" an algorithm for computing the Discrete Fourier Transform of composite numbers originated with I. J. Good [6]. Since the way we "build up" the algorithm is somewhat different from Good's method, we will describe the whole process in detail.

Assume $n = n_1 \cdot n_2$ where n_1 and n_2 are relatively prime. By the Chinese Remainder Theorem we can represent every integer $0 \leqslant i \leqslant n$ by the pair (i_1, i_2) such that if i is represented by (i_1, i_2) and j by (j_1, j_2), then $i + j \bmod n$ is represented by $(i_1 + j_1 \bmod n, i_2 + j_2 \bmod n_2)$ and $i \cdot j \bmod n$ is represented by $(i_1 \cdot j_1 \bmod n, i_2 \cdot j_2 \bmod n_2)$.

Therefore, if we let w be the nth root of unity then:

$$w^{k \cdot j} = w^{k \cdot j \bmod n} = w^{(k_1, k_2)(j_1, j_2)} = w^{(k_1 \cdot j_1, k_2 \cdot j_2)} \tag{33}$$

$$= w^{(k_1 j_1, 0) + (0, k_2 j_2)} = w^{(k_1 j_1, 0)} w^{(0, k_2 j_2)} = (w^{(1,0)})^{k_1 j_1} \cdot (w^{(0,1)})^{k_2 j_2}.$$

That means that if we permute the rows and columns of the Discrete Fourier Transform matrix so as to arrange the indices to be in the lexicographical order of their representation, it can be partitioned in $n_1 \times n_1$ blocks each of dimensions $n_2 \times n_2$. The block in position r, s will be $(w^{(1,0)})^{r \cdot s} W_2$ when the u, v entry of W_2 is $(w^{(0,1)})^{u \cdot v}$. But $w^{(1,0)}$ is w_1^a where w_1 is the n_1th root of unity (note that the number represented by $(1, 0)$ is divisible by n_2), and $w^{(0,1)}$ is w_2^b where w_2 is the n_2th root of unity. Consequently, w_2 is the same as the Discrete Fourier Transform matrix for n_2 points except that w_2 is replaced by w_2^b. If we denote by W_1 the matrix of the Discrete Fourier Transform of n_1 points where w_1 is replaced by w_1^a, then the matrix of Discrete Fourier Transform of $n_1 \cdot n_2$ points has been transformed to the direct product of W_1 and W_2.

For example, take $n = 12 = 3 \cdot 4$. The correspondence according to the Chinese Remainder Theorem is

$$\begin{array}{llll} 0-(0,0) & 1-(1,1) & 2-(2,2) & 3-(0,3) \\ 4-(1,0) & 5-(2,1) & 6-(0,2) & 7-(1,3) \\ 8-(2,0) & 9-(0,1) & 10-(1,2) & 11-(2,3) \end{array}$$

and put in lexicographical order we get the rearrangement: 0, 9, 6, 3, 4, 1, 10, 7, 8, 5, 2, 11. Thus the Discrete Fourier Transform for 12 points can be written as:

(34)

$$\begin{pmatrix} A_0 \\ A_9 \\ A_6 \\ A_3 \\ A_4 \\ A_1 \\ A_{10} \\ A_7 \\ A_8 \\ A_5 \\ A_2 \\ A_{11} \end{pmatrix} = \begin{pmatrix}
1\cdot 1 & 1\cdot 1 & 1\cdot 1 & 1\cdot 1 & 1\cdot 1 & 1\cdot 1 & 1\cdot 1 & 1\cdot 1 & 1\cdot 1 & 1\cdot 1 & 1\cdot 1 & 1\cdot 1 \\
1\cdot 1 & 1\cdot -i & 1\cdot -1 & 1\cdot i & 1\cdot 1 & 1\cdot -i & 1\cdot -1 & 1\cdot i & 1\cdot 1 & 1\cdot -i & 1\cdot -1 & 1\cdot i \\
1\cdot 1 & 1\cdot -1 & 1\cdot 1 & 1\cdot -1 & 1\cdot 1 & 1\cdot -1 & 1\cdot 1 & 1\cdot -1 & 1\cdot 1 & 1\cdot -1 & 1\cdot 1 & 1\cdot -1 \\
1\cdot 1 & 1\cdot i & 1\cdot -1 & 1\cdot -i & 1\cdot 1 & 1\cdot i & 1\cdot -1 & 1\cdot -i & 1\cdot 1 & 1\cdot i & 1\cdot -1 & 1\cdot -i \\
1\cdot 1 & 1\cdot 1 & 1\cdot 1 & 1\cdot 1 & w\cdot 1 & w\cdot 1 & w\cdot 1 & w\cdot 1 & w^2\cdot 1 & w^2\cdot 1 & w^2\cdot 1 & w^2\cdot 1 \\
1\cdot 1 & 1\cdot -i & 1\cdot -1 & 1\cdot i & w\cdot 1 & w\cdot -i & w\cdot -1 & w\cdot i & w^2\cdot 1 & w^2\cdot -i & w^2\cdot -1 & w^2\cdot i \\
1\cdot 1 & 1\cdot -1 & 1\cdot 1 & 1\cdot -1 & w\cdot 1 & w\cdot -1 & w\cdot 1 & w\cdot -1 & w^2\cdot 1 & w^2\cdot -1 & w^2\cdot 1 & w^2\cdot -1 \\
1\cdot 1 & 1\cdot i & 1\cdot -1 & 1\cdot -i & w\cdot 1 & w\cdot i & w\cdot -1 & w\cdot -i & w^2\cdot 1 & w^2\cdot i & w^2\cdot -1 & w^2\cdot -i \\
1\cdot 1 & 1\cdot 1 & 1\cdot 1 & 1\cdot 1 & w^2\cdot 1 & w^2\cdot 1 & w^2\cdot 1 & w^2\cdot 1 & w\cdot 1 & w\cdot 1 & w\cdot 1 & w\cdot 1 \\
1\cdot 1 & 1\cdot -i & 1\cdot -1 & 1\cdot i & w^2\cdot 1 & w^2\cdot -i & w^2\cdot -1 & w^2\cdot i & w\cdot 1 & w\cdot -i & w\cdot -1 & w\cdot i \\
1\cdot 1 & 1\cdot -1 & 1\cdot 1 & 1\cdot -1 & w^2\cdot 1 & w^2\cdot -1 & w^2\cdot 1 & w^2\cdot -1 & w\cdot 1 & w\cdot -1 & w\cdot 1 & w\cdot -1 \\
1\cdot 1 & 1\cdot i & 1\cdot -1 & 1\cdot -i & w^2\cdot 1 & w^2\cdot i & w^2\cdot -1 & w^2\cdot -i & w\cdot 1 & w\cdot i & w\cdot -1 & w\cdot -i
\end{pmatrix} \begin{pmatrix} a_0 \\ a_9 \\ a_6 \\ a_3 \\ a_4 \\ a_1 \\ a_{10} \\ a_7 \\ a_8 \\ a_5 \\ a_2 \\ a_{11} \end{pmatrix},$$

where w is the cubic root of unity. Since $(1, 0)$ corresponds to four, we have $a = 1$, and since $(0, 1)$ corresponds to nine we have $b = 3$.

The decomposition of the 12 point Discrete Fourier Transform leads to an algorithm for its computation. If we define

(35)

$$\mathbf{a}_0 = \begin{pmatrix} a_0 \\ a_9 \\ a_6 \\ a_3 \end{pmatrix}, \quad \mathbf{a}_1 = \begin{pmatrix} a_4 \\ a_1 \\ a_{10} \\ a_7 \end{pmatrix}, \quad \mathbf{a}_2 = \begin{pmatrix} a_8 \\ a_5 \\ a_2 \\ a_{11} \end{pmatrix},$$

$$\mathbf{A}_0 = \begin{pmatrix} A_0 \\ A_9 \\ A_6 \\ A_3 \end{pmatrix}, \quad \mathbf{A}_1 = \begin{pmatrix} A_4 \\ A_1 \\ A_{10} \\ A_7 \end{pmatrix}, \quad \mathbf{A}_2 = \begin{pmatrix} A_8 \\ A_5 \\ A_2 \\ A_{11} \end{pmatrix},$$

then using Algorithm B2 we obtain:

$$M_0 = W_2 \cdot (a_0 + a_1 + a_2), \quad M_1 = \left(\cos\frac{2\pi}{3} - 1\right) W_2 \cdot (a_1 + a_2),$$

(36) $$M_2 = i \sin\frac{2\pi}{3} W_2 \cdot (a_1 - a_2),$$

$$A_0 = M_0, \quad A_1 = M_0 + M_1 + M_2, \quad A_2 = M_0 + M_1 - M_2,$$

where W_2 is the four point discrete Fourier Transform with i replaced by $-i$ (since $b = 3$). Therefore, we can use the Algorithm B3 to compute M_0, M_1, and M_2. In computing M_1, for example, we have to modify Algorithm B3 by first replacing i by $-i$ and second multiplying the constants in the multiplication steps by $(\cos 2\pi/3 - 1)$. These modifications are done initially when we derive the algorithm and, therefore, are not counted in analyzing the computation complexity of the algorithm. In the end of this section we will show how one can avoid the first modification.

It should be clear that the way we derived the algorithm for computing the 12 points Discrete Fourier Transform is quite general. If $n = n_1 \cdot n_2$ (where n_1, n_2 are relatively prime) and we have algorithms for computing the Discrete Fourier Transform of n_1 points using a_1 additions and m_1 multiplications, (including multiplication by 1) and of n_2 points using a_2 additions and m_2 multiplications, we can combine them to obtain an algorithm for computing the n points Discrete Fourier Transform using $m_1 \cdot m_2$ multiplications and $n_2 \cdot a_1 + m_1 \cdot a_2$ additions. Since we would have decomposed the n points discrete Fourier Transform using $n = n_2 \cdot n_1$ as well, we could have derived an algorithm using $m_1 \cdot m_2$ multiplications and $n_1 \cdot a_2 + m_2 \cdot a_1$ additions. In general, these two algorithms will differ in their number of additions. In Table 3 we summarize the number of multiplications and additions used in algorithms derived this way for various values of n. For the sake of comparison with FFT we also tabulate $2n \log_2 n$ and $3n \log_2 n$ (the formulas for the number of real multiplications and real additions, respectively, in FFT).

n	# Mult. Complex Data	# Add. Complex Data	$2n \log_2 n$	$3n \log_2 n$
30	72	384	295	442
48	108	636	537	805
60	144	888	709	1064
120	288	2076	1658	2487
168	432	3492	2484	3726
240	648	5016	3796	5693
420	1296	11352	7320	10980
504	1584	14642	9050	13574
840	2592	24804	16320	24480
1008	3564	34920	20115	30172
2520	9504	100188	56949	85423

TABLE 3

As we saw before, one of the modifications needed to compute the Discrete Fourier Transform of $n_1 \cdot n_2$ points is to replace w_1 by w_1^a and w_2 by w_2^b. One way of avoiding this modification is to use different permutations on the rows and columns of the matrix, that is, different permutations of the input and output data.

Let $b_0, b_1, \ldots, b_{n-1}$ be the reordering of the input data $a_0, a_1, \ldots, a_{n-1}$; and let $B_0, B_1, \ldots, B_{n-1}$ be the reordering of the output data $A_0, A_1, \ldots, A_{n-1}$. Choose r_1, r_2, s_1, s_2 such that

$$(37) \qquad r_1 \cdot s_1 \cdot n_1 = 1 \bmod n_2, \qquad r_2 \cdot s_2 \cdot n_2 = 1 \bmod n_1.$$

If we choose $b_{j_1 n_2 + j_2} = a_{r_1 j_1 n_1 + r_2 j_2 n_2 \bmod n}$ and $B_{k_1 n_2 + k_2} = A_{s_1 k_1 n_1 + s_2 k_2 n_2 \bmod n}$, then the resulting entry of $(k_1 n_2 + k_2, j_1 n_2 + j_2)$ of the Discrete Fourier Transform matrix is (remember that $w^n = 1$):

$$\begin{aligned}
W_{(k_1,k_2)(j_1,j_2)} &= w^{(s_1 k_1 n_1 + s_2 k_2 n_2)(r_1 j_1 n_1 + r_2 j_2 n_2)} \\
&= w^{(s_1 r_1 n_1) k_1 j_1 n_1 + (s_2 r_2 n_2) k_2 j_2 n_2} \\
(38) \qquad &= (w^{n_2})^{(s_2 r_2 n_2) k_2 j_2} \cdot (w^{n_1})^{(r_1 s_1 n_1) k_1 j_1} \\
&= w_1^{(s_2 r_2 n_2) k_2 j_2} \cdot w_2^{(r_1 s_1 n_1) k_1 j_1} \\
&= w_1^{k_2 j_2} \cdot w_2^{k_1 j_1} \qquad (w_1^{n_1} = 1, w_2^{n_2} = 1).
\end{aligned}$$

Therefore, this matrix is the direct product of the matrix for n_1 point Discrete Fourier Transform and n_2 point Discrete Fourier Transform.

One easy way of getting r_1, r_2, s_1 and s_2 is to choose $r_1 = r_2 = 1$ and s_1, s_2 according to the Chinese Remainder Theorem.

We will end this section with the remark that the algorithms developed here can be used in conjunction with FFT. The identity behind FFT states that computing the Discrete Fourier Transform of $n_1 \cdot n_2$ points (n_1, n_2 are not necessarily relatively prime) can be done by first performing the Discrete Fourier Transform of n_1 points n_2 times, then one performs $(n_1 - 1)n_2$ complex multiplications, and then one performs n_1 times the Discrete Fourier Transform of n_2 points. It is, of course, possible to use the algorithms developed here in the first and third stages of the FFT identity.

V. Multidimensional Fourier Transform. For the sake of concreteness we will consider only two dimensional Fourier Transform, even though it should be clear that the results apply to all dimensions. The $n_1 \times n_2$ points Discrete Fourier Transform is

$$(39) \quad A_{k,k'} = \sum_{j=0}^{n_1-1} \sum_{j'=0}^{n_2-1} w_1^{kj} w_2^{k'j'} a_{j,j'}, \qquad 0 \leqslant k \leqslant n_1 - 1, 0 \leqslant k' \leqslant n_2 - 1.$$

It is apparent from (39) that the matrix for the Discrete Fourier Transform of $n_1 \times n_2$ points is the direct product of the matrix of n_1 points by the matrix for n_2 points, and consequently the methods of the end of the last section are immediately applicable to multidimensional Discrete Fourier Transform. This can be stated even more strongly by noting that underlying the method for one dimensional Discrete Fourier Transform is the transformation to multidimensional transform.

The main results in this section consist of illustrating how the full strength of the investigation of a product of polynomials modulo a polynomial, *and their dependence on the field of scalars,* can be utilized. The discussion in the preceding paragraph indicates that the techniques to be described are applicable in the one dimensional case as well, but their exposition is simpler in the multidimensional case.

As was mentioned in Section II, the minimum number of multiplications needed to compute $\widetilde{T}_p$ is $2n - k$ where n is the degree of P and k is the number of distinct irreducible factors of P. By choosing a larger field of constants we can increase k and thus decrease the number of multiplications. For example, $\widetilde{T}_{u^4-1}$ requires five multiplications over the field Q of rationals, but only four over the field $Q(i)$. Similarly, $\widetilde{T}_{u^6-1}$ requires eight multiplications over Q, but only six over $Q(e^{2\pi i/3})$. Recalling that $\widetilde{T}_{u^4-1}$ is the cyclic convolution of four points and $\widetilde{T}_{u^6-1}$ is the cyclic convolution of six points, we see that if somehow we could take advantage of the larger fields we could reduce the complexity of cyclic convolutions and, therefore, of the Discrete Fourier Transform.

One way of utilizing algorithms over fields which are algebraic extensions of the rationals is to use them in the situation that the Discrete Fourier Transform is to be performed on more than one set of data. For the sake of concreteness assume that we have two independent sets of data: $\{a_0^{(1)}, a_1^{(2)}, \ldots, a_{N-1}^{(2)}\}$ and $\{a_0^{(2)}, a_1^{(2)}, \ldots, a_{N-1}^{(2)}\}$. We can "group" them together as $\{a_0, a_1, \ldots, a_{n-1}\}$ where $a_j = (a_j^{(1)}, a_j^{(2)})$. Assume that the field of constants we want to use is $Q(I)$, where $I^2 = -1$, i.e., the field of Gaussian rationals.

We can transform the vectors a_j into an algebra over $Q(I)$ by defining:

$$1.\quad (a_j^{(1)}, a_j^{(2)}) + (a_k^{(1)}, a_k^{(2)}) = (a_j^{(1)} + a_k^{(1)}, a_j^{(2)} + a_k^{(2)}),$$

$$I \cdot (a_j^{(1)}, a_j^{(2)}) = (-a_j^{(2)}, a_j^{(1)}),$$

$$(a_j^{(1)}, a_j^{(2)})(a_k^{(1)}, a_k^{(2)}) = (b^{(1)}, b^{(2)}), \tag{40}$$

$$b^{(1)} = a_j^{(1)}a_k^{(1)} - a_j^{(2)}a_k^{(2)} = a_j^{(1)}(a_k^{(1)} + a_k^{(2)}) - (a_j^{(1)} + a_j^{(2)})a_k^{(2)},$$

$$b^{(2)} = a_j^{(1)}a_k^{(2)} + a_j^{(2)}a_k^{(1)} = a_j^{(1)}(a_k^{(1)} + a_k^{(2)}) - (a_j^{(1)} - a_j^{(2)})a_k^{(1)}.$$

That is, we view the vector a_j as standing for $a_j^{(1)} + I \cdot a_j^{(2)}$ where $I^2 = -1$. This is of course possible whenever the number of independent sets of data is the same as the dimension of the extension field. We see that in this setting, multiplication by I is not counted (it amounts to interchanging the components of the vector and changing one of the signs), while multiplication of two elements of the algebra amounts to three multiplications of the components plus a certain number of additions.

The pairs of data could have been transformed into an algebra of $Q(\phi)$ where $\phi^2 + \phi + 1 = 0$, if instead of (40) we would have defined:

$$(a_j^{(1)}, a_j^{(2)}) + (a_k^{(1)}, a_k^{(2)}) = (a_j^{(1)} + a_k^{(1)}, a_j^{(2)} + a_k^{(2)}),$$

$$\phi \cdot (a_j^{(1)}, a_j^{(2)}) = (-a_j^{(2)}, a_j^{(1)} - a_j^{(2)}),$$

$$(a_j^{(1)}, a_j^{(2)}) \cdot (a_k^{(1)}, a_k^{(2)}) = (b^{(1)}, b^{(2)}), \tag{41}$$

$$b^{(1)} = (a_j^{(1)} + a_j^{(2)})a_k^{(1)} - a_j^{(2)}(a_k^{(1)} - a_k^{(2)}),$$

$$b^{(2)} = (a_j^{(1)}a_k^{(2)} + a_j^{(2)}(a_k^{(1)} - a_k^{(2)}).$$

That is if we view the vector a_j as standing for $a_j^{(1)} + \phi a_j^{(2)}$, where $\phi^2 = -\phi - 1$.

Having computed the Discrete Fourier Transform of the pair gives us the two desired Discrete Fourier Transforms.

As an example, consider performing the three dimensional Discrete Fourier Transform on $120 \times 120 \times 120$ data points, which are assumed to be complex. Using the method described in the beginning of this section would require 5,971,968 real multiplications and 97,203,456 real additions for each set of data. (For the sake of comparison with FFT, note that $2 \times 120^3 \times \log_2 120^3 = 71{,}610{,}641$ and $3 \times 120^3 \times \log_2 120^3 = 107{,}415{,}962$.) Since $120 = 8 \times 3 \times 5$, we could have reduced the number of multiplications if we could have done the four point cyclic convolution (which appears in the five point Fourier Transform) in four multiplications instead of five. Since $Q(I)$ splits $u^4 - 1$, we choose to view the pair of input data as an algebra over $Q(I)$. In Appendix C we give an algorithm for the five point Fourier Transform over $Q(I)$. Using this algorithm, we can perform the $120 \times 120 \times 120$ Fourier Transform in $2 \times 5{,}184{,}000$ real multiplications and $2 \times 96{,}840{,}800$ real additions. But since this yields the results of performing the Fourier Transform on two sets of data, we obtained 13% savings of the number of multiplications (and a slight reduction of the number of additions).

It should be clear that this construction is general. For example, computing the Discrete Fourier Transform of 252×252 points may be advantageously done over $Q(\phi)$; and the Fourier Transform of $140 \times 140 \times 140$ may be sped up by doing it over $Q(I, \phi)$.

Another, more subtle, way of utilizing the fact that $\widetilde{T}_p$ may require fewer multiplications when the field of constants is enlarged, is based on the construction in the beginning of [1].

The Chinese Remainder Theorem states that when P_1 and P_2 are relatively prime, the system $\widetilde{T}_{P_1 \cdot P_2}$ can be transformed, by appropriate change of variables, to the direct sum of $\widetilde{T}_{P_1}$ and $\widetilde{T}_{P_2}$. We will illustrate this by considering the four point cyclic convolution, i.e., $\widetilde{T}_{u^4-1}$. Since $u^4 - 1 = (u^2 - 1)(u^2 + 1)$, we obtain:

$$\begin{pmatrix} x_1 & x_2 & x_3 & x_4 \\ x_2 & x_3 & x_4 & x_1 \\ x_3 & x_4 & x_1 & x_2 \\ x_4 & x_1 & x_2 & x_3 \end{pmatrix}\begin{pmatrix} y_1 \\ y_2 \\ y_3 \\ y_4 \end{pmatrix} = \begin{pmatrix} 1 & 0 & 0 & 1 \\ 0 & 1 & 1 & 0 \\ 1 & 0 & 0 & -1 \\ 0 & 1 & -1 & 0 \end{pmatrix}$$

(42)

$$\cdot \begin{pmatrix} \frac{x_1+x_3}{2} & \frac{x_2+x_4}{2} & 0 & 0 \\ \frac{x_2+x_4}{2} & \frac{x_1+x_3}{2} & 0 & 0 \\ 0 & 0 & \frac{x_2-x_4}{2} & -\frac{x_1-x_3}{2} \\ 0 & 0 & \frac{x_1-x_3}{2} & \frac{x_2-x_4}{2} \end{pmatrix}\begin{pmatrix} y_1+y_3 \\ y_2+y_4 \\ y_1-y_3 \\ y_2-y_4 \end{pmatrix}.$$

This transformation can be carried directly into the appropriate Discrete Fourier Transform. Thus, the decomposition (42) translates into the following decomposition of the five point Discrete Fourier Transform

$$\begin{pmatrix} A_0 \\ A_1 \\ A_2 \\ A_3 \\ A_4 \end{pmatrix} = \begin{pmatrix} 1 & 0 & 0 & 0 & 0 \\ 1 & 1 & 0 & 0 & 1 \\ 1 & 0 & 1 & 1 & 0 \\ 1 & 1 & 0 & 0 & -1 \\ 1 & 0 & 1 & -1 & 0 \end{pmatrix}$$

$$\cdot \begin{pmatrix} 1 & 0 & 0 & 0 & 0 \\ 0 & \cos u - 1 & \cos 2u - 1 & 0 & 0 \\ 0 & \cos 2u - 1 & \cos u - 1 & 0 & 0 \\ 0 & 0 & 0 & i \sin 2u & -i \sin u \\ 0 & 0 & 0 & i \sin u & i \sin 2u \end{pmatrix}\begin{pmatrix} a_0 + a_1 + a_2 + a_3 + a_4 \\ a_1 + a_4 \\ a_2 + a_3 \\ a_1 - a_4 \\ a_2 - a_3 \end{pmatrix}.$$

If we consider the two dimensional Fourier Transform of 5×5 points it can be decomposed as

$$(\widetilde{T}_u \oplus \widetilde{T}_{u^2-1} \oplus \widetilde{T}_{u^2+1}) \otimes (\widetilde{T}_u \oplus \widetilde{T}_{u^2-1} \oplus \widetilde{T}_{u^2+1})$$

$$= \widetilde{T}_u \oplus 2 \cdot \widetilde{T}_{u^2-1} \oplus 2 \cdot \widetilde{T}_{u^2+1} \oplus (\widetilde{T}_{u^2-1} \otimes \widetilde{T}_{u^2+1})$$

$$\oplus\, 2 \cdot (\widetilde{T}_{u^2-1} \otimes \widetilde{T}_{u^2+1}) \oplus (\widetilde{T}_{u^2+1} \otimes \widetilde{T}_{u^2+1}).$$

In [1] we showed how to compute $\widetilde{T}_{u^2+1} \otimes \widetilde{T}_{u^2+1}$ using six multiplications, and therefore the total number of multiplications needed to perform the 5×5 two dimensional Fourier Transform is $1 + 4 + 6 + 4 + 12 + 6 = 33$ (instead of 36). Using this construction to obtain an algorithm for the $5 \times 5 \times 5$ Fourier Transform, and then incorporating it in computing the 120 × 120 × 120 points Fourier Transform we obtain an algorithm which uses 90,706,176 real additions and 4,810,752 real multiplications. (That is, 6.7% of the number of multiplications and 84% of the number of additions of FFT.)

It should be emphasized that the savings of the last construction occur at the expense of the length of the program. This construction calls for writing an algorithm to compute the 5 × 5 × 5 Discrete Fourier Transform.

Acknowledgment. The author wishes to thank Dr. Ramesh C. Agarwal of IBM Research for helping in simplifying the algorithm for DFT of nine points.

Appendix A

A1.
$$\begin{pmatrix}\psi_1\\ \psi_2\end{pmatrix} = \begin{pmatrix}x_1 & x_2\\ x_2 & x_1\end{pmatrix}\begin{pmatrix}y_1\\ y_2\end{pmatrix}$$

Algorithm:

$s_1 = y_1+y_2$ $\qquad$ $s_2 = y_1 - y_2$

$m_1 = \frac{x_1+x_2}{2} \cdot s_1$ $\qquad$ $m_2 = \frac{x_1- x_2}{2} \cdot s_2$

$s_3 = m_1+ m_2$ $\qquad$ $s_4 = m_1- m_2$

$\psi_1 = s_3$ $\qquad$ $\psi_4 = s_4$

A2.
$$\begin{pmatrix}\psi_1\\ \psi_2\\ \psi_3\end{pmatrix} = \begin{pmatrix}x_1 & x_2 & x_3\\ x_2 & x_3 & x_1\\ x_3 & x_1 & x_2\end{pmatrix}\begin{pmatrix}y_1\\ y_2\\ y_3\end{pmatrix}$$

Algorithm:

$s_1 = y_1+ y_2$ $\qquad$ $s_2 = y_1-y_2$ $\qquad$ $s_3 = y_2- y_3$

$s_4 = y_3- y_1$ $\qquad$ $s_5 = y_3+ s_1$

$m_1 = \frac{x_1+x_2+x_3}{3} \cdot s_5$ $\qquad$ $m_2 = \frac{2x_1-x_2-x_3}{3} \cdot s_2$

$m_3 = \frac{x_1+x_2-2x_3}{3} \cdot s_3$ $\qquad$ $m_4 = \frac{x_1-2x_2+x_3}{3} \cdot s_4$

$s_6 = m_2+ m_3$ $\qquad$ $s_7 = m_4- m_3$ $\qquad$ $s_8 = m_2+ m_4$

$s_9 = m_1+ s_6$ $\qquad$ $s_{10} = m_1+ s_7$ $\qquad$ $s_{11} = m_1- s_8$

$\psi_1 = s_9$ $\qquad$ $\psi_2 = s_{10}$ $\qquad$ $\psi_3 = s_{11}$

A3.

$$\begin{pmatrix} \psi_1 \\ \psi_2 \\ \psi_3 \\ \psi_4 \end{pmatrix} = \begin{pmatrix} x_1 & x_2 & x_3 & x_4 \\ x_2 & x_3 & x_4 & x_1 \\ x_3 & x_4 & x_1 & x_2 \\ x_4 & x_1 & x_2 & x_3 \end{pmatrix} \begin{pmatrix} y_1 \\ y_2 \\ y_3 \\ y_4 \end{pmatrix}$$

Algorithm:

$s_1 = y_1 + y_3 \qquad s_2 = y_1 - y_3 \qquad s_3 = y_4 + y_2 \qquad s_4 = y_4 - y_2$

$s_5 = s_1 + s_3 \qquad s_6 = s_1 - s_3 \qquad s_7 = s_2 + s_4$

$$m_1 = \frac{x_1+x_2+x_3+x_4}{4} \cdot s_5 \qquad m_2 = \frac{x_1-x_2+x_3-x_4}{4} \cdot s_6$$

$$m_3 = \frac{x_1+x_2-x_3-x_4}{2} \cdot s_2 \qquad m_4 = \frac{x_2-x_4}{2} \cdot s_7 \qquad m_5 = \frac{x_1-x_2-x_3+x_4}{2} \cdot s_4$$

$s_8 = m_1 + m_2 \qquad s_9 = m_1 - m_2 \qquad s_{10} = m_3 - m_4 \qquad s_{11} = m_4 + m_5$

$s_{12} = s_8 + s_{10} \qquad s_{13} = s_8 - s_{10} \qquad s_{14} = s_9 + s_{11} \qquad s_{15} = s_9 - s_{11}$

$\psi_1 = s_{12} \qquad \psi_2 = s_{14} \qquad \psi_3 = s_{13} \qquad \psi_4 = s_{15}$

A4.

$$\begin{pmatrix} \psi_1 \\ \psi_2 \\ \psi_3 \\ \psi_4 \\ \psi_5 \end{pmatrix} = \begin{pmatrix} x_1 & x_2 & x_3 & x_4 & x_5 \\ x_2 & x_3 & x_4 & x_5 & x_1 \\ x_3 & x_4 & x_5 & x_1 & x_2 \\ x_4 & x_5 & x_1 & x_2 & x_3 \\ x_5 & x_1 & x_2 & x_3 & x_4 \end{pmatrix} \begin{pmatrix} y_1 \\ y_2 \\ y_3 \\ y_4 \\ y_5 \end{pmatrix}$$

$s_1 = y_1 + y_4 \qquad s_2 = y_2 + y_3 \qquad s_3 = s_1 + s_2 \qquad s_4 = s_1 - s_2 \qquad s_5 = s_3 + y_5$

$s_6 = y_1 - y_5 \qquad s_7 = y_1 - y_2 \qquad s_8 = y_2 - y_5 \qquad s_9 = y_1 - y_3 \qquad s_{10} = y_2 - y_4$

$s_{11} = y_3 - y_5 \qquad s_{12} = y_3 - y_4 \qquad s_{13} = y_4 - y_5$

$$m_0 = \frac{x_1+x_2+x_3+x_4+x_5}{5} \cdot s_5 \qquad m_1 = \frac{x_1+x_2+x_3+x_4-4x_5}{5} \cdot s_6$$

$$m_2 = \frac{2x_1-3x_2+2x_3-3x_4+2x_5}{5} \cdot s_7 \qquad m_3 = \frac{-4x_1+x_2+x_3+x_4+x_5}{5} \cdot s_8$$

$$m_4 = \frac{2x_1+2x_2-3x_3-3x_4+2x_5}{5} \cdot s_9 \qquad m_5 = \frac{-x_1-x_2-x_3+2x_4-x_5}{5} \cdot s_4$$

$$m_6 = \frac{2x_1+2x_2+2x_3-3x_4-3x_5}{5} \cdot s_{10} \qquad m_7 = \frac{x_1-4x_2+x_3+x_4+x_5}{5} \cdot s_{11}$$

$$m_8 = \frac{-3x_1+2x_2+2x_3-3x_4+2x_5}{5} \cdot s_{12} \qquad m_9 = \frac{x_1+x_2-4x_3+x_4+x_5}{5} \cdot s_{13}$$

$s_{14} = m_4+m_5$ $\quad s_{15} = m_5-m_6$ $\quad s_{16} = m_0+m_1$ $\quad s_{17} = s_{16}+m_2$

$s_{18} = s_{17}+s_{14}$ $\quad s_{19} = m_0-m_2$ $\quad s_{20} = s_{19}+m_3$ $\quad s_{21} = s_{20}-s_{15}$

$s_{22} = m_0-s_{14}$ $\quad s_{23} = s_{22}+m_7$ $\quad s_{24} = s_{23}+m_8$ $\quad s_{25} = m_0+s_{15}$

$s_{26} = s_{25}-m_8$ $\quad s_{27} = s_{26}+m_9$ $\quad s_{28} = m_0-m_1$ $\quad s_{29} = s_{28}-m_3$

$s_{30} = s_{23}-m_7$ $\quad s_{31} = s_{30}-m_9$

$\psi_1 = s_{18}$ $\quad \psi_2 = s_{21}$ $\quad \psi_3 = s_{24}$ $\quad \psi_4 = s_{27}$ $\quad \psi_5 = s_{31}$

A5.
$$\begin{pmatrix}\psi_1\\ \psi_2\\ \psi_3\\ \psi_4\\ \psi_5\\ \psi_6\end{pmatrix} = \begin{pmatrix} x_1 & x_2 & x_3 & x_4 & x_5 & x_6\\ x_2 & x_3 & x_4 & x_5 & x_6 & x_1\\ x_3 & x_4 & x_5 & x_6 & x_1 & x_2\\ x_4 & x_5 & x_6 & x_1 & x_2 & x_3\\ x_5 & x_6 & x_1 & x_2 & x_3 & x_4\\ x_6 & x_1 & x_2 & x_3 & x_4 & x_5\end{pmatrix}\begin{pmatrix}y_1\\ y_2\\ y_3\\ y_4\\ y_5\\ y_6\end{pmatrix}$$

$s_1 = y_1+y_4$ $\quad s_2 = y_1-y_4$ $\quad s_3 = y_2+y_5$ $\quad s_4 = y_5-y_2$

$s_5 = y_3+y_6$ $\quad s_6 = y_3-y_6$ $\quad s_7 = s_1+s_3$ $\quad s_8 = s_7+s_5$

$s_9 = s_2+s_4$ $\quad s_{10} = s_9+s_6$ $\quad s_{11} = s_1-s_3$ $\quad s_{12} = s_3-s_5$

$s_{13} = s_5-s_1$ $\quad s_{14} = s_2-s_4$ $\quad s_{15} = s_4-s$ $\quad s_{16} = s_6-s_2$

$$m_1 = \frac{x_1+x_2+x_3+x_4+x_5+x_6}{6}\cdot s_8 \qquad m_2 = \frac{2x_1-x_2-x_3+2x_4-x_5-x_6}{6}\cdot s_{11}$$

$$m_3 = \frac{x_1\ x_2-2x_3+x_4+x_5-2x_6}{6}\cdot s_{12} \qquad m_4 = \frac{x_1-2x_2+x_3+x_4-2x_5+x_6}{6}\cdot s_{13}$$

$$m_5 = \frac{x_1-x_2+x_3-x_4+x_5-x_6}{6}\cdot s_{10} \qquad m_6 = \frac{2x_1+x_2-x_3-2x_4-x_5+x_6}{6}\cdot s_{14}$$

$$m_7 = \frac{x_1-x_2-2x_3-x_4+x_5+2x_6}{6}\cdot s_{15} \qquad m_8 = \frac{x_1+2x_2+x_3-x_4-2x_5-x_6}{6}\cdot s_{16}$$

$s_{17} = m_1+m_2$ $\quad s_{18} = s_{17}+m_3$ $\quad s_{19} = m_1-m_3$ $\quad s_{20} = s_{19}+m_4$

$s_{21} = m_1-m_2$ $\quad s_{22} = s_{21}-m_4$ $\quad s_{23} = m_5+m_6$ $\quad s_{24} = s_{23}+m_7$

$s_{25} = m_5-m_7$ $\quad s_{26} = s_{25}+m_8$ $\quad s_{27} = m_5-m_6$ $\quad s_{28} = s_{27}-m_8$

$s_{29} = s_{18}+s_{24}$ $\quad s_{30} = s_{18}-s_{24}$ $\quad s_{31} = s_{20}-s_{26}$ $\quad s_{32} = s_{20}+s_{26}$

$s_{33} = s_{22}+s_{28}$ $\quad s_{34} = s_{22}-s_{28}$

$\psi_1 = s_{29}$ $\quad \psi_2 = s_{31}$ $\quad \psi_3 = s_{33}$ $\quad \psi_4 = s_{30}$ $\quad \psi_5 = s_{32}$ $\quad \psi_6 = s_{34}$

Appendix B

B1.

$$\begin{pmatrix} A_0 \\ A_1 \end{pmatrix} = \begin{pmatrix} w^0 & w^0 \\ w^0 & -w^0 \end{pmatrix} \begin{pmatrix} a_0 \\ a_1 \end{pmatrix} \qquad w = e^{\frac{2\pi i}{2}} = -1$$

Algorithm:

$s_1 = a_0+a_1$ $\quad$ $s_2 = a_0-a_1$

$m_0 = 1\cdot s_1$ $\quad$ $m_1 = 1\cdot s_2$

$A_0 = m_0$ $\quad$ $A_1 = m_1$

B2.

$$\begin{pmatrix} A_0 \\ A_1 \\ A_2 \end{pmatrix} = \begin{pmatrix} w^0 & w^0 & w^0 \\ w^0 & w^1 & w^2 \\ w^0 & w^2 & w^1 \end{pmatrix} \begin{pmatrix} a_0 \\ a_1 \\ a_2 \end{pmatrix} \qquad w = e^{\frac{2\pi i}{3}}$$

Algorithm:

$s_1 = a_1+a_2$ $\quad$ $s_2 = a_1-a_2$ $\quad$ $s_3 = s_1+a_0$

$m_0 = 1\cdot s_3$ $\quad$ $m_1 = (\cos u-1)\cdot s_1$ $\quad$ $m_2 = i \sin u\cdot s_2$ $\quad$ $u = \frac{2\pi}{3}$

$s_4 = m_0+m_1$ $\quad$ $s_5 = s_4+m_2$ $\quad$ $s_6 = s_4-m_2$

$A_0 = m_0$ $\quad$ $A_1 = s_5$ $\quad$ $A_2 = s_6$

B3.

$$\begin{pmatrix} A_0 \\ A_1 \\ A_2 \\ A_3 \end{pmatrix} = \begin{pmatrix} w^0 & w^0 & w^0 & w^0 \\ w^0 & w^1 & -w^0 & -w^1 \\ w^0 & -w^0 & w^0 & -w^0 \\ w^0 & -w^1 & -w^0 & w^1 \end{pmatrix} \begin{pmatrix} a_0 \\ a_1 \\ a_2 \\ a_3 \end{pmatrix} \qquad w = e^{\frac{2\pi i}{4}}$$

$s_1 = a_0+a_2$ $\quad$ $s_2 = a_0-a_2$ $\quad$ $s_3 = a_1+a_3$ $\quad$ $s_4 = a_1-a_3$

$s_5 = s_1+s_3$ $\quad$ $s_6 = s_1-s_3$

$m_1 = 1\cdot s_5$ $\quad$ $m_2 = 1\cdot s_6$ $\quad$ $m_3 = 1\cdot s_2$ $\quad$ $m_4 = i \sin u\cdot s_4$ $\quad$ $u = \frac{2\pi}{4}$

$s_7 = m_3+m_4$ $\quad$ $s_8 = m_3-m_4$

$A_0 = m_1$ $\quad$ $A_1 = s_7$ $\quad$ $A_2 = m_2$ $\quad$ $A_3 = s_8$

B4.

$$\begin{pmatrix} A_0 \\ A_1 \\ A_2 \\ A_3 \\ A_4 \end{pmatrix} \begin{pmatrix} w^0 & w^0 & w^0 & w^0 & w^0 \\ w^0 & w^1 & w^2 & w^3 & w^4 \\ w^0 & w^2 & w^4 & w^1 & w^3 \\ w^0 & w^3 & w^1 & w^4 & w^2 \\ w^0 & w^4 & w^3 & w^2 & w^1 \end{pmatrix} \begin{pmatrix} a_0 \\ a_1 \\ a_2 \\ a_3 \\ a_4 \end{pmatrix} \qquad w = e^{\frac{2\pi i}{5}}$$

Algorithm:

$s_1 = a_1+a_4 \qquad s_2 = a_1-a_4 \qquad s_3 = a_3+a_2 \qquad s_4 = a_3-a_2$

$s_5 = s_1+s_3 \qquad s_6 = s_1-s_3 \qquad s_7 = s_2+s_4 \qquad s_8 = s_5+a_0$

$m_0 = 1\cdot s_8 \qquad m_1 = \left(\frac{\cos u+\cos 2u}{2} - 1\right)\cdot s_5 \qquad m_2 = \left(\frac{\cos u-\cos 2u}{2}\right)\cdot s_6 \qquad u = \frac{2\pi}{5}$

$m_3 = i(\sin u+\sin 2u)\cdot s_2 \qquad m_4 = i\ \sin 2u\cdot s_7 \qquad m_5 = i(\sin u-\sin 2u)\cdot s_4$

$s_9 = m_0+m_1 \qquad s_{10} = s_9+m_2 \qquad s_{11} = s_9-m_2 \qquad s_{12} = m_3-m_4$

$s_{13} = m_4+m_5 \qquad s_{14} = s_{10}+s_{12} \qquad s_{15} = s_{10}-s_{12} \qquad s_{16} = s_{11}+s_{13}$

$s_{17} = s_{11}-s_{13}$

$A_0 = m_0 \qquad A_1 = s_{14} \qquad A_2 = s_{16} \qquad A_3 = s_{17} \qquad A_4 = s_{15}$

B5.

$$\begin{pmatrix} A_0 \\ A_1 \\ A_2 \\ A_3 \\ A_4 \\ A_5 \\ A_6 \end{pmatrix} = \begin{pmatrix} w^0 & w^0 & w^0 & w^0 & w^0 & w^0 & w^0 \\ w^0 & w^1 & w^2 & w^3 & w^4 & w^5 & w^6 \\ w^0 & w^2 & w^4 & w^6 & w^1 & w^3 & w^5 \\ w^0 & w^3 & w^6 & w^2 & w^5 & w^1 & w^4 \\ w^0 & w^4 & w^1 & w^5 & w^2 & w^6 & w^3 \\ w^0 & w^5 & w^3 & w^1 & w^6 & w^4 & w^2 \\ w^0 & w^6 & w^5 & w^4 & w^3 & w^2 & w^1 \end{pmatrix} \begin{pmatrix} a_0 \\ a_1 \\ a_2 \\ a_3 \\ a_4 \\ a_5 \\ a_6 \end{pmatrix} \qquad w = e^{\frac{2\pi i}{7}}$$

Algorithm:

$s_1 = a_1+a_6 \qquad s_2 = a_1-a_6 \qquad s_3 = a_4+a_3 \qquad s_4 = a_4-a_3$

$s_5 = a_2-a_5 \qquad s_6 = a_2-a_5 \qquad s_7 = s_1+s_3 \qquad s_8 = s_7+s_5$

$s_9 = s_8+a_0 \qquad s_{10} = s_1-s_3 \qquad s_{11} = s_3-s_5 \qquad s_{12} = s_5-s_1$

$s_{13} = s_2+s_4 \qquad s_{14} = s_{13}+s_6 \qquad s_{15} = s_2-s_4 \qquad s_{16} = s_4-s_6$

$s_{17} = s_6-s_2$

$m_0 = 1\cdot s_9 \qquad m_1 = \left(\frac{\cos u+\cos 2u+\cos 3u}{3} -1\right)\cdot s_8 \qquad u = \frac{2\pi i}{7}$

$m_2 = \left(\frac{2\cos u-\cos 2u-\cos 3u}{3}\right)\cdot s_{10} \qquad m_3 = \left(\frac{\cos u-2\cos 2u+\cos 3u}{3}\right)\cdot s_{11}$

$m_4 = \left(\frac{\cos u+\cos 2u-2\cos 3u}{3}\right)\cdot s_{12} \qquad m_5 = i\left(\frac{\sin u+\sin 2u-\sin 3u}{3}\right)\cdot s_{14}$

$m_6 = i\left(\frac{2\sin u-\sin 2u+\sin 3u}{3}\right)\cdot s_{15}$ $\qquad m_7 = i\left(\frac{\sin u-2\sin 2u-\sin 3u}{3}\right)\cdot s_{16}$

$m_8 = i\left(\frac{\sin u+\sin 2u+2\sin 3u}{3}\right)\cdot s_{17}$

$s_{18} = m_0+m_1$ $\qquad s_{19} = s_{18}+m_2$ $\qquad s_{20} = s_{19}+m_3$ $\qquad s_{21} = s_{18}-m_2$

$s_{22} = s_{21}-m_4$ $\qquad s_{23} = s_{18}-m_3$ $\qquad s_{24} = s_{23}+m_4$ $\qquad s_{25} = m_5+m_6$

$s_{26} = s_{25}+m_7$ $\qquad s_{27} = m_5-m_6$ $\qquad s_{28} = s_{27}-m_8$ $\qquad s_{29} = m_5-m_7$

$s_{30} = s_{29}+m_8$ $\qquad s_{31} = s_{20}+s_{26}$ $\qquad s_{32} = s_{20}-s_{26}$ $\qquad s_{33} = s_{22}+s_{28}$

$s_{34} = s_{22}-s_{28}$ $\qquad s_{35} = s_{24}+s_{30}$ $\qquad s_{36} = s_{24}-s_{30}$

$A_0 = m_0$ $\qquad A_1 = s_{31}$ $\qquad A_2 = s_{33}$ $\qquad A_3 = s_{36}$

$A_4 = s_{35}$ $\qquad A_5 = s_{34}$ $\qquad A_6 = s_{32}$

B6.

$$\begin{pmatrix} A_0 \\ A_1 \\ A_2 \\ A_3 \\ A_4 \\ A_5 \\ A_6 \\ A_7 \end{pmatrix} = \begin{pmatrix} w^0 & w^0 & w^0 & w^0 & w^0 & w^0 & w^0 & w^0 \\ w^0 & w^1 & w^2 & w^3 & w^4 & w^5 & w^6 & w^7 \\ w^0 & w^2 & w^4 & w^6 & w^0 & w^2 & w^4 & w^6 \\ w^0 & w^3 & w^6 & w^1 & w^4 & w^7 & w^2 & w^5 \\ w^0 & w^4 & w^0 & w^4 & w^0 & w^4 & w^0 & w^4 \\ w^0 & w^5 & w^2 & w^7 & w^4 & w^1 & w^6 & w^3 \\ w^0 & w^6 & w^4 & w^2 & w^0 & w^6 & w^4 & w^2 \\ w^0 & w^7 & w^6 & w^5 & w^4 & w^3 & w^2 & w^1 \end{pmatrix} \begin{pmatrix} a_0 \\ a_1 \\ a_2 \\ a_3 \\ a_4 \\ a_5 \\ a_6 \\ a_7 \end{pmatrix} \qquad w = e^{\frac{2\pi i}{8}}$$

Algorithm.

$s_1 = a_0+a_4$ $\qquad s_2 = a_0-a_4$ $\qquad s_3 = a_2+a_6$ $\qquad s_4 = a_2-a_6$

$s_5 = a_1+a_5$ $\qquad s_6 = a_1-a_5$ $\qquad s_7 = a_3+a_7$ $\qquad s_8 = a_3-a_7$

$s_9 = s_1+s_3$ $\qquad s_{10} = s_1-s_3$ $\qquad s_{11} = s_5+s_7$ $\qquad s_{12} = s_5-s_7$

$s_{13} = s_9+s_{11}$ $\qquad s_{14} = s_9-s_{11}$ $\qquad s_{15} = s_6+s_8$ $\qquad s_{16} = s_6-s_8$

$m_1 = 1\cdot s_{13}$ $\qquad m_2 = 1\cdot s_{14}$ $\qquad m_3 = 1\cdot s_{10}$ $\qquad m_4 = i \sin 2u\cdot s_{12}$ $\qquad u = \frac{2\pi}{8}$

$m_5 = 1\cdot s_2$ $\qquad m_6 = i\cdot \sin 2u\cdot s_4$ $\qquad m_7 = i \sin u\cdot s_{15}$ $\qquad m_8 = \cos u\cdot s_{16}$

$s_{17} = m_3+m_4$ $\qquad s_{18} = m_3-m_4$ $\qquad s_{19} = m_5+m_8$ $\qquad s_{20} = m_5-m_8$

$s_{21} = m_6+m_7$ $\qquad s_{22} = m_6-m_7$ $\qquad s_{23} = s_{19}+s_{21}$ $\qquad s_{24} = s_{19}-s_{21}$

$s_{25} = s_{20}+s_{22}$ $\qquad s_{26} = s_{20}-s_{22}$

$A_0 = m_1 \qquad A_1 = s_{23} \qquad A_2 = s_{17} \qquad A_3 = s_{26}$

$A_4 = m_2 \qquad A_5 = s_{25} \qquad A_6 = s_{18} \qquad A_7 = s_{24}$

B7.

$$\begin{pmatrix} A_0 \\ A_1 \\ A_2 \\ A_3 \\ A_4 \\ A_5 \\ A_6 \\ A_7 \\ A_8 \end{pmatrix} = \begin{pmatrix} w^0 & w^0 & w^0 & w^0 & w^0 & w^0 & w^0 & w^0 & w^0 \\ w^0 & w^1 & w^2 & w^3 & w^4 & w^5 & w^6 & w^7 & w^8 \\ w^0 & w^2 & w^4 & w^6 & w^8 & w^1 & w^3 & w^5 & w^7 \\ w^0 & w^3 & w^6 & w^0 & w^3 & w^6 & w^0 & w^3 & w^6 \\ w^0 & w^4 & w^8 & w^3 & w^7 & w^2 & w^6 & w^1 & w^5 \\ w^0 & w^5 & w^1 & w^6 & w^2 & w^7 & w^3 & w^8 & w^4 \\ w^0 & w^6 & w^3 & w^0 & w^6 & w^3 & w^0 & w^6 & w^3 \\ w^0 & w^7 & w^5 & w^3 & w^1 & w^8 & w^6 & w^4 & w^2 \\ w^0 & w^8 & w^7 & w^6 & w^5 & w^4 & w^3 & w^2 & w^1 \end{pmatrix} \begin{pmatrix} a_0 \\ a_1 \\ a_2 \\ a_3 \\ a_4 \\ a_5 \\ a_6 \\ a_7 \\ a_8 \end{pmatrix}$$

Algorithm:

$s_1 = a_1+a_8 \qquad s_2 = a_1-a_8 \qquad s_3 = a_7+a_2 \qquad s_4 = a_7-a_2$

$s_5 = a_3+a_6 \qquad s_6 = a_3-a_6 \qquad s_7 = a_4+a_5 \qquad s_8 = a_4-a_5$

$s_9 = s_1+s_3 \qquad s_{10} = s_9+s_7 \qquad s_{11} = s_{10}+s_5 \qquad s_{12} = s_{11}+a_0$

$s_{13} = s_2+s_4 \qquad s_{14} = s_{13}+s_8 \qquad s_{15} = s_1-s_3 \qquad s_{16} = s_3-s_7$

$s_{17} = s_7-s_1 \qquad s_{18} = s_2-s_4 \qquad s_{19} = s_4-s_8 \qquad s_{20} = s_8-s_2$

$m_0 = 1\cdot s_{12} \qquad m_1 = (-\frac{1}{2})\cdot s_{10} \qquad m_2 = i\ \sin 3u\cdot s_{14} \qquad u = \frac{2\pi}{9}$

$m_3 = (\cos 3u-1)\cdot s_5 \qquad m_4 = i\sin 3u\cdot s_6 \qquad m_5 = (\frac{2\cos u-\cos 2u-\cos 4u}{3})\cdot s_{15}$

$m_6 = (\frac{\cos u+\cos 2u-2\cos 4u}{3})\cdot s_{16} \qquad m_7 = (\frac{\cos u-2\cos 2u+\cos 4u}{3})\cdot s_{17}$

$m_8 = i(\frac{2\sin u+\sin 2u-\sin 4u}{3})\cdot s_{18} \qquad m_9 = i(\frac{\sin u-\sin 2u-2\sin 4u}{3})\cdot s_{19}$

$m_{10} = i(\frac{\sin u+2\sin 2u+\sin 4u}{3})\cdot s_{20}$

$s_{21} = m_1+m_1 \qquad s_{22} = s_{20}+m_1 \qquad s_{23} = m_0+s_{22} \qquad s_{24} = s_{23}\ m_2$

$s_{25} = s_{23}-m_2 \qquad s_{26} = m_0+m_3 \qquad s_{27} = s_{26}+s_{21} \qquad s_{28} = s_{27}+m_5$

$s_{29} = s_{28}+m_6 \qquad s_{30} = s_{27}-m_6 \qquad s_{31} = s_{30}+m_7 \qquad s_{32} = s_{27}-m_5$

$s_{33} = s_{32}-m_7 \qquad s_{34} = m_4+m_8 \qquad s_{35} = s_{34}+m_9 \qquad s_{36} = m_4-m_9$

$s_{37} = s_{36}+m_{10} \qquad s_{38} = m_4-m_8 \qquad s_{39} = s_{38}-m_{10} \qquad s_{40} = s_{29}+s_{35}$

$s_{41} = s_{29}-s_{35} \qquad s_{42} = s_{31}+s_{37} \qquad s_{43} = s_{31}-s_{37} \qquad s_{44} = s_{33}+s_{39}$

$s_{45} = s_{33}-s_{39}$

$A_0 = m_0$ $\quad A_1 = s_{40}$ $\quad A_2 = s_{43}$ $\quad A_3 = s_{24}$ $\quad A_4 = s_{43}$

$A_5 = s_{45}$ $\quad A_6 = s_{23}$ $\quad A_7 = s_{42}$ $\quad A_8 = s_{41}$

B8. $\quad A_k = \sum_{j=0}^{15} w^{kj} a_j \qquad k=0,1,\ldots,15 \qquad w = e^{\frac{2\pi i}{16}}$.

Algorithm.

$s_1 = a_0+a_8$ $\quad s_2 = a_0-a_8$ $\quad s_3 = a_4+a_{12}$ $\quad s_4 = a_4-a_{12}$

$s_5 = a_2+a_{10}$ $\quad s_6 = a_2-a_{10}$ $\quad s_7 = a_6+a_{14}$ $\quad s_8 = a_6-a_{14}$

$s_9 = a_1+a_9$ $\quad s_{10} = a_1-a_9$ $\quad s_{11} = a_5+a_{13}$ $\quad s_{12} = a_5-a_{13}$

$s_{13} = a_3+a_{11}$ $\quad s_{14} = a_3-a_{11}$ $\quad s_{15} = a_7+a_{15}$ $\quad s_{16} = a_7-a_{15}$

$s_{17} = s_1+s_3$ $\quad s_{18} = s_1-s_3$ $\quad s_{19} = s_5+s_7$ $\quad s_{20} = s_5-s_7$

$s_{21} = s_9+s_{11}$ $\quad s_{22} = s_9-s_{11}$ $\quad s_{23} = s_{13}+s_{15}$ $\quad s_{24} = s_{13}-s_{15}$

$s_{25} = s_{17}+s_{19}$ $\quad s_{26} = s_{17}-s_{19}$ $\quad s_{27} = s_{21}+s_{23}$ $\quad s_{28} = s_{21}-s_{23}$

$s_{29} = s_{25}+s_{27}$ $\quad s_{30} = s_{25}-s_{27}$ $\quad s_{31} = s_{22}+s_{24}$ $\quad s_{32} = s_{22}-s_{24}$

$s_{33} = s_6+s_8$ $\quad s_{34} = s_6-s_8$ $\quad s_{35} = s_{10}+s_{16}$ $\quad s_{36} = s_{10}-s_{16}$

[illegible] $= s_{12}+s_{14}$ $\quad s_{38} = s_{12}-s_{14}$ $\quad s_{39} = s_{35}+s_{37}$ $\quad s_{40} = s_{36}+s_{38}$

$m_1 =$ [illegible] $\quad m_2 = 1\cdot s_{30}$ $\quad m_3 = 1\cdot s_{26}$ $\quad m_4 = i\ \sin 4u\cdot s_{28}$ $\quad u = \frac{2\pi}{16}$

$m_5 = 1\cdot$ [illegible] $= i\ \sin 4u\cdot s_{20}$ $\quad m_7 = i\ \sin 2u\cdot s_{31}$ $\quad m_8 = \cos 2u\cdot s_{32}$

$m_9 = 1\cdot$ [illegible] $_2$ [illegible] $\sin 4u\cdot$ [illegible] $_4$ $\quad m_{11} = i\ \sin 2u\cdot$ [illegible] $_{33}$ $\quad m_{12} = \cos 2u\cdot s_{34}$

$m_{13} = i\ \sin 3u\cdot$ [illegible] $= i(\sin u-\sin 3u)\cdot s_{35}$ $\quad m_{15} = i(\sin u + \sin 3u)\cdot s_{37}$

$m_{16} = \cos 3u\cdot s_{40}$ [illegible] $\cos u + \cos 3u)\cdot s_{36}$ $\quad m_{18} = (\cos 3u - \cos u)\cdot s_{38}$

$s_{41} = m_3+ m_4$ $\quad s_{42}$ [illegible] $\quad s_{43} = m_5 + m_7$ $\quad s_{44} = m_5 - m_7$

$s_{45} = m_6 + m_8$ $\quad s_{46} = m_6$ [illegible] $\quad s_{47} = s_{43} + s_{45}$ $\quad s_{48} = s_{43}- s_{45}$

$s_{49} = s_{44} + s_{46}$ $\quad s_{50} = s_{44} - s_{46}$ $\quad s_{51} = m_9 + m_{12}$ $\quad s_{52} = m_9 - m_{12}$

$s_{53} = m_{10} + m_{11}$ $\quad s_{54} = m_{10}- m_{11}$ $\quad s_{55} = m_{13}+ m_{14}$ $\quad s_{56} = m_{13} - m_{15}$

$s_{57} = m_{17}- m_{16}$ $\quad s_{58} = m_{18}- m_{16}$ $\quad s_{59} = s_{51}+ s_{55}$ $\quad s_{60} = s_{51} - s_{55}$

$s_{61} = s_{52}+ s_{56}$ $\quad s_{62} = s_{52}- s_{56}$ $\quad s_{63} = s_{53}+ s_{57}$ $\quad s_{64} = s_{53} - s_{57}$

$s_{65} = s_{54}+ s_{58}$ $\quad s_{66} = s_{54}- s_{58}$ $\quad s_{67} = s_{59}+ s_{63}$ $\quad s_{68} = s_{59}- s_{63}$

$s_{69} = s_{60} + s_{64}$ $s_{70} = s_{60} - s_{64}$ $s_{71} = s_{61} + s_{65}$ $s_{72} = s_{61} - s_{65}$

$s_{73} = s_{62} + s_{66}$ $s_{74} = s_{62} - s_{66}$

$A_0 = m_1$ $A_1 = s_{67}$ $A_2 = s_{47}$ $A_3 = s_{72}$ $A_4 = s_{41}$ $A_5 = s_{71}$

$A_6 = s_{48}$ $A_7 = s_{68}$ $A_8 = m_2$ $A_9 = s_{69}$ $A_{10} = s_{49}$ $A_{11} = s_{74}$

$A_{12} = s_{42}$ $A_{13} = s_{73}$ $A_{14} = s_{50}$ $A_{15} = s_{70}$

Appendix C

C1.

$$\begin{pmatrix} \psi_1 \\ \psi_2 \\ \psi_3 \\ \psi_4 \end{pmatrix} = \begin{pmatrix} x_1 & x_2 & x_3 & x_4 \\ x_2 & x_3 & x_4 & x_1 \\ x_3 & x_4 & x_1 & x_2 \\ x_4 & x_1 & x_2 & x_3 \end{pmatrix} \begin{pmatrix} y_1 \\ y_2 \\ y_3 \\ y_4 \end{pmatrix}$$

Algorithm:

$s_1 = y_1 + y_3$ $s_2 = y_1 - y_3$ $s_3 = y_4 + y_2$ $s_4 = y_4 - y_2$

$s_5 = s_1 + s_3$ $s_6 = s_1 - s_3$ $s_7 = s_2 + I \cdot s_4$ $s_8 = s_2 - I \cdot s_4$

$m_1 = \frac{x_1+x_2+x_3+x_4}{4} \cdot s_5$ $m_2 = \frac{x_1-x_2+x_3-x_4}{4} \cdot s_6$

$m_3 = \frac{(x_1-x_3) + I(x_2-x_4)}{4} \cdot s_7$ $m_4 = \frac{(x_1-x_3) - I(x_2-x_4)}{4} \cdot s_8$

$s_9 = m_1 + m_2$ $s_{10} = m_1 - m_2$ $s_{11} = m_4 + m_3$ $s_{12} = m_4 - m_3$

$s_{13} = s_9 + s_{11}$ $s_{14} = s_9 - s_{11}$ $s_{15} = s_{10} + I \cdot s_{12}$ $s_{16} = s_{10} - I \cdot s_{12}$

$\psi_1 = s_{13}$ $\psi_2 = s_{15}$ $\psi_3 = s_{14}$ $\psi_4 = s_{16}$

C2.

$$\begin{pmatrix} A_0 \\ A_1 \\ A_2 \\ A_3 \\ A_4 \end{pmatrix} = \begin{pmatrix} w^0 & w^0 & w^0 & w^0 & w^0 \\ w^0 & w^1 & w^2 & w^3 & w^4 \\ w^0 & w^2 & w^4 & w^1 & w^3 \\ w^0 & w^3 & w^1 & w^4 & w^2 \\ w^0 & w^4 & w^3 & w^2 & w^1 \end{pmatrix} \begin{pmatrix} a_0 \\ a_1 \\ a_2 \\ a_3 \\ a_4 \end{pmatrix} .$$

Algorithm:

$s_1 = a_1 + a_4$ $\quad s_2 = a_1 - a_4$ $\quad s_3 = a_3 + a_2$ $\quad s_4 = a_3 - a_2$

$s_5 = s_1 + s_3$ $\quad s_6 = s_1 - s_3$ $\quad s_7 = s_2 + I \cdot s_4$ $\quad s_8 = s_2 - I \cdot s_4$

$s_9 = a_0 + s_5$

$m_0 = 1 \cdot s_9$ $\quad m_1 = (\frac{\cos u + \cos 2u}{2} - 1) \cdot s_5 \quad u = \frac{2\pi}{5}$

$m_2 = \frac{\cos u - \cos 2u}{2} \cdot s_6$ $\quad m_3 = \frac{i(\sin u + I \sin 2u)}{2} \cdot s_7$

$m_5 = \frac{i(\sin u - I \sin 2u)}{2} \cdot s_8$

$s_{10} = m_0 + m_1$ $\quad s_{11} = s_{10} + m_2$ $\quad s_{12} = s_{10} - m_2$ $\quad s_{13} = m_4 + m_3$

$s_{14} = m_4 - m_3$ $\quad s_{15} = s_{11} + s_{13}$ $\quad s_{16} = s_{11} - s_{13}$ $\quad s_{17} = s_{12} + I \cdot s_{14}$

$s_{18} = s_{12} - I \cdot s_{14}$

$A_0 = m_0$ $\quad A_1 = s_{15}$ $\quad A_2 = s_{17}$ $\quad A_3 = s_{18}$ $\quad A_4 = s_{16}$

Mathematical Science Department
IBM Thomas J. Watson Research Center
Yorktown Heights, New York 10598

1. S. WINOGRAD, "Some bilinear forms whose multiplicative complexity depends on the field of constants," to be published in *Mathematical Systems Theory*, Vol. 10.

2. J. W. COOLEY & J. W. TUKEY, "An algorithm for the machine calculation of complex Fourier series," *Math. Comp.*, v. 19, 1965, pp. 297–301.

3. C. M. FIDUCCIA & Y. ZALCSTEIN, *Algebras Having Linear Multiplicative Complexities,* Technical Report 46, Dept. of Computer Science, State University of New York, Stony Brook, August 1975.

4. A. L. TOOM, "The complexity of a scheme of functional elements simulating the multiplication of integers," *Dokl. Akad. Nauk SSSR*, v. 150, 1963, pp. 496–498 = *Soviet Math. Dokl.*, v. 4, 1963, pp. 714–716.

5. C. M. RADER, "Discrete Fourier transforms when the number of data samples is prime," *Proc. IEEE,* v. 5, no. 6, June 1968, pp. 1107–1108.

6. I. J. GOOD, "The interaction of algorithm and practical Fourier series," *J. Roy. Statist. Soc. Ser. B*, v. 20, 1958, pp. 361–372; Addendum, v. 22, 1960, pp. 372–375.

Reprinted from IEEE TRANSACTIONS
ON *COMPUTERS*
Volume C-20, Number 3, March, 1971
pp. 310–317

The Relationship Between Two Fast Fourier Transforms

I. J. GOOD

Abstract—The purpose of this note is to show as clearly as possible the mathematical relationship between the two basic fast methods used for the calculation of discrete Fourier transforms and to generalize one of the methods a little further. This method applies to all those linear transformations whose matrices are expressible as direct products.

Index Terms—Algorithms, circulices, direct product of matrices, discrete Fourier transforms, fast Fourier transforms, frequency analysis, Hadamard transform, harmonic analysis, multidimensional linear transformation, sparse matrices.

A fast Fourier transform (FFT) is a fast method for the calculation of a discrete Fourier transform (DFT) whose definition we shall give below in order to make the paper adequately self-contained. For applications of FFTs and DFTs see, for example, [1]–[4]. See also the bibliographies [5], [6], [4]. Here we shall be concerned only with mathematical and computational aspects. The main purpose of this note is to show as clearly as possible the relationship between the two basic algorithms for FFTs [7], [8]. Other methods are refinements of these two. Each method has its own advantages; one method can be applied to a wider variety of moduli, the other is more appropriate for a wide class of multidimensional transforms, not necessarily Fourier.

A one-dimensional mod t DFT of a vector (a sequence of t numbers $a_0, a_1, \cdots, a_{t-1}$ assumed to be real in this paper) is another vector of t complex numbers $a_0^*, a_1^*, \cdots, a_{t-1}^*$ defined by the equations

$$a_s^* = \sum_{s=0}^{t-1} a_r \omega^{rs} \qquad (s = 0, 1, \cdots, t-1) \tag{1}$$

where $\omega = \exp(2\pi i/t)$, a tth root of unity. It has elegant analogs of the properties of an ordinary Fourier transform like an inversion formula, a formula for the transform of a convolution, and even a Poisson summation formula (for example, [9]–[11]). A *multidimensional* mod $(t_1, t_2, \cdots, t_n)$ DFT of the n-dimensional array of $t_1 t_2 \cdots t_n$ numbers $a_{\boldsymbol{r}}$, where $\boldsymbol{r}$ is the vector $(r_1, r_2, \cdots, r_n)$ $(r_j = 0, 1, \cdots, t_j - 1;$ $j = 1, 2, \cdots, n)$, is the n-dimensional array of complex numbers $a_{\boldsymbol{s}}^*$ defined by the equations

$$a_{\boldsymbol{s}}^* = \sum_{\boldsymbol{r}} a_{\boldsymbol{r}} \omega_1^{r_1 s_1} \cdots \omega_n^{r_n s_n}$$

$$(r_j, s_j = 0, 1, \cdots, t_j - 1; j = 1, \cdots, n) \tag{2}$$

where $\omega_j = \exp(2\pi i/t_j)$ $(j = 1, \cdots, n)$. The arrays $a_{\boldsymbol{r}}$ and $a_{\boldsymbol{s}}^*$ can also be regarded as vectors if a rule is given for ordering the $\tau = t_1 t_2 \cdots t_n$ elements of each as sequences or "strings." When all the t_j's are equal to t, we can refer to the n-dimensional mod t DFT. The components of a multidimensional mod 2 DFT (of an array of real number $a_{\boldsymbol{r}}$) are all real because in this case all the ω_j's are equal to -1. The mod 2 multidimensional DFT is also an example of a Hadamard transform [12]. The matrix of a Hadamard transform is orthogonal and each of its elements is $+1$ or -1. It is also known as an "anallagmatic pavement" [13].

A DFT, unidimensional or multidimensional, is, of course, a linear transformation of a vector (with t or $t_1 t_2 \cdots t_n$ components), a multiplication of a vector by a matrix, in which the matrix of the transformation is (ω^{rs}) or $(\omega_1^{r_1 s_1} \cdots \omega_n^{r_n s_n})$. In the latter matrix some convention is required for the ordering of the rows and columns. The simplest convention is to order the $t_1 t_2 \cdots t_n$ vectors $\boldsymbol{r} = (r_1, \cdots, r_n)$ lexicographically as if they were words in a dictionary, and similarly for the "columns" $\boldsymbol{s} = (s_1, \cdots, s_n)$. Then the components of $a_{\boldsymbol{r}}$ must be similarly ordered in order to convert it from an array into a vector.

The first FFT to be described here depends on the fact that a one-dimensional DFT modulo τ, where $\tau = t_1 t_2 \cdots t_n$, and $t_1, t_2, \cdots, t_n$ are mutually prime in pairs, can be expressed as a mod $(t_1, t_2, \cdots, t_n)$ multidimensional DFT. For the sake of completeness the procedure is quoted from Good [7].[1] We first set up (1–1) correspondences between the scalar r and the vector $\boldsymbol{r}$ and also between s and $\boldsymbol{s}$ in the following manner.

We begin by recalling the Chinese Remainder Theorem, known by Sun-Tsu in the first century A.D. [14], [15]. It states that if an integer is known modulo n integers (that is, its remainder is known when it is divided by each of these n integers), these integers being mutually prime in pairs (that is, each pair has highest common factor equal to 1), then it can be uniquely determined modulo the product. The solution can be expressed succinctly [9], [7] as follows. If $t_1, \cdots, t_n$ are mutually prime in pairs and $\tau = t_1 \cdots t_n$, and if the remainders are $s_1, \cdots, s_n$, or in the standard "congruence" notation, if

$$s \equiv s_1 (\text{mod } t_1), \cdots, s \equiv s_n (\text{mod } t_n),$$

then

$$s \equiv \frac{\tau s_1}{t_1} \left(\frac{\tau}{t_1}\right)_{t_1}^{-1} + \cdots + \frac{\tau s_n}{t_n} \left(\frac{\tau}{t_n}\right)_{t_n}^{-1}$$

$$(\text{mod } \tau; 0 \le s < \tau; 0 \le s_1 < t_1, \text{etc.}) \tag{3}$$

where $(\tau/t_1)_{t_1}^{-1}$, for example, means the modulo t_1 reciprocal of the integer τ/t_1. This reciprocal exists and is unique modulo t_1 because the integers τ/t_1 and t_1 are mutually

Manuscript received December 18, 1969; revised August 21, 1970.
The author is with the Department of Statistics and Statistical Laboratory, Virginia Polytechnic Institute and State University, Blacksburg, Va. 24061.

[1] There are misprints in [7]. Four incorrect entries in the matrix at the foot of p. 363 have been found. On p. 364, line 10, $\delta_{r_{\nu-1}}$ should be $\delta_{r_{\nu+1}}$. The lower limit of the product at the foot of p. 365 should be 1. On p. 368, line 13, an asterisk was omitted.

prime. Hence the first term of this expression for s is unique modulo τ, and similarly the other terms are unique. The reader can verify at once that the value of s does satisfy the n given congruences. Moreover, each of the τ possible sets of n congruences has at least one solution given by (3). These solutions must be unique, since there are only τ residues modulo τ and none of them can correspond to more than one set of n congruences. Formula (3) is analogous to Lagrange's interpolation formula. We use it to set up a (1–1) correspondence between s and $\mathbf{s}=(s_1, \cdots, s_n)$ which I shall call the *Sino correspondence* or the *Sino representation* of s. For example, if $t_1=3$, $t_2=8$, $\tau=24$, then $s\equiv 16s_1+9s_2$ (mod 24). Once we fix $t_1, \cdots, t_n$ the coefficients (here 16 and 9) are known once for all. On the other hand, the (1–1) relationship between r and $\mathbf{r}$ is given by

$$r \equiv \frac{\tau}{t_1} r_1 + \cdots + \frac{\tau}{t_n} r_n \qquad (\text{mod } \tau, 0 \le r < \tau; 0 \le r_1 < t_1, \text{etc.}) \quad (4)$$

which I shall call the *Ruritanian correspondence*. Again, if $t_1=3$, $t_2=8$, $\tau=24$, this gives $r\equiv 8r_1+3r_2$ (mod 24).

I now assert that the equations

$$\hat{a}_s^* = \sum_{r=0}^{\tau-1} \hat{a}_r \omega^{rs} \qquad (\omega = e^{2\pi i/\tau})$$

and

$$a_{\mathbf{s}}^* = \sum_{r_1=0}^{t_1-1} \cdots \sum_{r_n=0}^{t_n-1} a_{\mathbf{r}} \omega_1^{r_1 s_1} \cdots \omega_n^{r_n s_n}$$

are equivalent, where $\hat{a}_r = a_{\mathbf{r}}$ and $\hat{a}_s^* = a_{\mathbf{s}}^*$ by definition, and where the correspondences are Sino and Ruritanian, respectively. To see this, note that

$$r \equiv (\tau/t_\nu) r_\nu \,(\text{mod } t_\nu), \; s \equiv \frac{\tau s_\nu}{t_\nu}\left(\frac{\tau}{t_\nu}\right)_{t_\nu}^{-1} (\text{mod } t_\nu)$$

so that $rs \equiv r_\nu s_\nu \tau/t_\nu$ (mod t_ν) and hence, by Sun-Tsu's theorem in the form given above,

$$rs \equiv \frac{\tau r_1 s_1}{t_1} \cdot \frac{\tau}{t_1}\left(\frac{\tau}{t_1}\right)_{t_1}^{-1} + \cdots \equiv \sum_\nu r_\nu s_\nu \tau/t_\nu.$$

Moreover,

$$\omega_\nu = \omega^{\tau/t_\nu}$$

so

$$\hat{a}_s^* = \sum_r \hat{a}_r \omega_1^{r_1 s_1} \cdots \omega_n^{r_n s_n} = a_{\mathbf{s}}^*,$$

as claimed.

A unidimensional mod $t_1 t_2 \cdots t_n$ DFT can therefore be computed with about the same speed as a multidimensional mod $(t_1, \cdots, t_n)$ DFT where $t_1, \cdots, t_n$ are mutually prime in pairs. Accordingly we now consider the calculation of a multidimensional DFT, or more generally, a linear transformation from $a_{\mathbf{r}}$ to $a_{\mathbf{s}}^*$ of the form

$$a_{\mathbf{s}}^* = \sum_{r_1=0}^{t_1-1} \cdots \sum_{r_n=0}^{t_n-1} a_{\mathbf{r}} K_1(r_1, s_1) \cdots K_n(r_n, s_n) \qquad (5)$$

where we have replaced $\omega_1^{r_1 s_1}$ by $K_1(r_1, s_1)$, etc., and where we do not insist for the present on any constraints on $t_1, \cdots, t_n$. (The following discussion of (5) and (6) was given in [4].) We shall describe (5) as an *n-dimensional K-transform*. Although each of the functions $K_1, K_2, \cdots, K_n$ can be regarded as a square matrix, the right side of (5) is, of course, not as it stands in the form of the product of a vector by several matrices, although it can be regarded as the product of a vector by one large matrix whose rows are indexed by $\mathbf{r}$ and columns by $\mathbf{s}$. This large matrix will later be seen to be the "direct product" of the smaller matrices, and also to be the product of large matrices which correspond in turn to K_1, $K_2, \cdots, K_n$. For the moment we shall point out a simple procedure for summing (5) without matrix notation.

The summation can be performed by first summing over r_1, then over r_2, etc. The first summation is

$$\sum_{r_1} a_{\mathbf{r}} K_1(r_1, s_1)$$

and the result is of the form $L_1(r_2, r_3, \cdots, r_n, s_1)$, that is, a function of $r_2, r_3, \cdots, r_n, s_1$. The next stage involves the sum

$$\sum_{r_2} L_1(r_2, r_3, \cdots, r_n, s_1) K_2(r_2, s_2)$$

and is of the form $L_2(r_3, \cdots, r_n, s_1, s_2)$, and so on. Thus at each stage we need store only an n-dimensional array, and altogether we need do only $(t_1+t_2+\cdots+t_n)\tau$ scalar multiplications. This is the basis of the method of [7] for forming a multidimensional DFT, except that it was there expressed in matrix notation. The method saves arithmetic as compared with the multiplication of the vector by the single large matrix mentioned above which would require τ^2 scalar multiplications. For example, with $\tau=5.7.8.9=2520$, the number of scalar multiplications is less than an eightieth of what would be required ($\tau^2=2520^2$) if $a_{\mathbf{r}}$ were multiplied by the $\tau\times\tau$ matrix $K_1(r_1, s_1)\cdots K_n(r_n, s_n)$.

We shall later discuss the inverse of the multidimensional K-transform, but first we consider a generalization of the above simple summation procedure which leads up to the second method for calculating an FFT.

We consider the analogous multiple sum

$$\sum_{r_1=0}^{t_1-1} \cdots \sum_{r_n=0}^{t_n-1} a_{\mathbf{r}} K_1(r_1, s_1) K_2(r_2, s_1, s_2) \cdots K_n(r_n, s_1, s_2, \cdots, s_n) \quad (6)$$

and we find that the above argument goes through virtually unchanged for this more general summation. The summation of $a_{\mathbf{r}} K_1$ with respect to r_1 leads to an expression of the form $L_1(r_2, \cdots, r_n, s_1)$; then the summation of $L_1 K_2$ with respect to r_2 leads to an expression of the form $L_2(r_3, \cdots, r_n, s_1, s_2), \cdots$, and finally the summation of $L_{n-1} K_n$ with respect to r_n to an expression of the form $L_n(s_1, s_2, \cdots, s_n)$. At each stage only an n-dimensional array needs to be stored, but the dimensions of the array vary from stage to stage unless the t_ν's are all equal.

The method used by Cooley and Tukey [8] for computing a one-dimensional DFT is in effect to reduce it to the form (6). Their method applies to a modulus $\tau = t_1 t_2 \cdots t_n$, but it was first spelled out for the case $\tau=2^n$, that is $t_1=t_2=\cdots$

$=t_n=2$. The general case was explained in more detail by Bingham, Godfrey, and Tukey [16]. In this method the correspondence between the one-dimensional DFT and a multiple sum depends on setting up (1–1) correspondences between r and $\boldsymbol{r}$ and between s and $\boldsymbol{s}$ different from the Sino–Ruritanian correspondences or representations. Instead r and s are each represented in mixed radix form, but each with a different kind of reversal, as follows.

$$\begin{aligned} r &= t_1t_2\cdots t_{n-1}r_1 + t_1t_2\cdots t_{n-2}r_2 + \cdots + t_1r_{n-1} + r_n \\ s &= t_2t_3\cdots t_ns_n + t_3t_4\cdots t_ns_{n-1} + \cdots + t_ns_2 + s_1. \end{aligned} \tag{7}$$

If we think of t_1 as analogous to 10, t_1t_2 to 100 and so on, we see that the representation of r is mixed radical, but with the *components* of $\boldsymbol{r}$ in the reverse order to the *digits* of r. We could call this the *reversed-digit* (mixed-radix) representation. On the other hand, the representation of s reverses the order of the radices $t_1, t_2, \cdots, t_n$ but leaves the components of $\boldsymbol{s}$ in the same order as the digits of s. Thus there are two different kinds of reversal so the representation of r and s together could be described as the *reversed-radical* or *reactionary* representation or correspondence. There are variants of the method, depending on various mixed-radix representations of integers, which, of course, are not mixed when $t_1=t_2=\cdots=t_n$.

The single summation

$$\sum_r a_r\omega^{rs} \qquad (\omega = e^{2\pi i/\tau})$$

can be regarded as a multiple summation over the digits of r in its mixed-radix representation. We also substitute the mixed-radix representation of s and obtain

$$\sum_{\boldsymbol{r}} a_r\omega^{(t_1t_2\cdots t_{n-1}r_1+t_1t_2\cdots t_{n-2}r_2+\cdots+t_1r_{n-1}+r_n)(t_2\cdots t_ns_n+\cdots+t_ns_2+s_1)} = \sum_{\boldsymbol{r}} a_rK_1(r_1, s_1)K_2(r_2, s_1, s_2)\cdots K_n(r_n, s_1, s_2, \cdots, s_n) \tag{8}$$

where (since $\omega^\tau=1$)

$$\begin{aligned} K_1(r_1, s_1) &= \omega^{t_1t_2\cdots t_{n-1}r_1s_1} \\ K_2(r_2, s_1, s_2) &= \omega^{t_1t_2\cdots t_{n-2}r_2(t_ns_2+s_1)} \\ &\cdots\cdots\cdots\cdots\cdots \\ K_n(r_n, s_1, \cdots, s_n) &= \omega^{r_n(t_2\cdots t_ns_n+\cdots+t_ns_2+s_1)}. \end{aligned} \tag{9}$$

If the elements of a_r are now permuted in accordance with the mixed-radical representation of r, the right side of (8) is of the form of (6) and can be computed by the simple procedure described above. Note that it would not be strictly correct to write a_r in place of $a_{\boldsymbol{r}}$ in the summation because we would then be using the symbol a to denote two different functions.

For this Cooley–Tukey or mixed-radix method the transformation from a unidimensional DFT to the multiple sum does not require that $t_1, t_2, \cdots, t_n$ be mutually prime in pairs. In this respect it has an advantage over the Sino–Ruritanian method, being more widely applicable when we are concerned with DFT's. It is perhaps used most often with the t_ν's all equal to 2, which is the simplest case to program for a binary computer. On the other hand, the multiple sum (5) is simpler than (6) and therefore more likely to occur (as a multidimensional K-transform) in contexts other than the DFT, especially with $K_1, \cdots, K_n$ all the same function. Some such contexts are mentioned, for example, by Andrews and Caspari [17]. Also, even for the DFT itself, if very fast special-purpose equipment is to be built, using fixed values of $t_1, t_2, \cdots, t_n$, the Sino–Ruritanian representation of r and s might turn out to be more convenient than the mixed-radical representation.

In order to discuss the algorithms in matrix notation it is necessary to make use of *direct products* (=*Kronecker products*) of matrices. (See, for example, [18]). The direct product $B\times C$ of two matrices $B=(b_{ij})$ $(i, j=0, 1, \cdots)$ and $C=(c_{\mu\nu})$ $(\mu, \nu=0, 1, \cdots)$, which need not be either square nor of related shapes and sizes, is defined as the block matrix

$$D = \begin{bmatrix} b_{00}C & b_{01}C\cdots \\ b_{10}C & b_{11}C\cdots \\ \cdots & \cdots \end{bmatrix}. \tag{10}$$

this being of course an abbreviated notation for an ordinary larger matrix whose elements are scalars. This definition is equivalent to saying that the elements of $B\times C$ are

$$d_{\boldsymbol{r},\boldsymbol{s}} = b_{r_1,s_1}c_{r_2,s_2} \tag{11}$$

where $\boldsymbol{r}=(r_1, r_2)$ and $\boldsymbol{s}=(s_1, s_2)$ are ordered lexicographically. For example, if B and C are both 2×2,

$$\begin{array}{c|cccc} \boldsymbol{s} & 00 & 01 & 10 & 11 \\ \boldsymbol{r} & & & & \\ 00 & b_{00}c_{00}, & b_{00}c_{01}, & b_{01}c_{00}, & b_{01}c_{01} \\ 01 & b_{00}c_{10}, & b_{00}c_{11}, & b_{01}c_{10}, & b_{01}c_{11} \\ 10 & b_{10}c_{00}, & b_{10}c_{01}, & b_{11}c_{00}, & b_{11}c_{01} \\ 11 & b_{10}c_{10}, & b_{10}c_{11}, & b_{11}c_{10}, & b_{11}c_{11} \end{array} = B\times C. \tag{12}$$

Thus, for example, $b_{11}c_{00}$ corresponds to $r_1=1$, $s_1=1$, $r_2=0$, $s_2=0$. (The reader should now hold in mind Savage's metamathematical principle: he should sit bolt upright in a hard chair with a pencil in his hand [19]). The general definition for the direct product of several matrices is the obvious inductive one, which turns out to be associative, so that parentheses are not required. Moreover, formula (11) generalizes in the natural manner. The direct product of n matrices,

$$M^{(1)}\times M^{(2)}\times\cdots\times M^{(n)}$$

where $M^{(\nu)}=(m_{i,j})$ $(i=0, 1, \cdots, p_\nu-1;\ j=0, 1, \cdots, q_\nu-1)$ has its $(\boldsymbol{r}, \boldsymbol{s})$ element

$$\prod_{\nu=1}^{N} m^{(\nu)}_{r_\nu,s_\nu} \tag{13}$$

wherein the ordering of the rows is lexicographic as is that of the columns. This again is true even if the matrices are not square and are of various shapes and sizes.

If the ordinary matrix product $M^{(\nu)}N^{(\nu)}$ can be formed, that is, if the number of columns of $M^{(\nu)}$ is the same as the number of rows of $N^{(\nu)}$ $(\nu=1, 2, \cdots)$, then

$$\begin{aligned} &(M^{(1)}\times M^{(2)}\times\cdots)(N^{(1)}\times N^{(2)}\times\cdots) \\ &\qquad = (M^{(1)}N^{(1)})\times(M^{(2)}N^{(2)})\times\cdots. \end{aligned} \tag{14}$$

In particular, if M is $t \times t$ and is nonsingular, then

$$(M \times M \times \cdots)(M^{-1} \times M^{-1} \times \cdots) = I_t \times I_t \times \cdots = I_{t^n}$$

where I_u is the $u \times u$ identity matrix. More briefly

$$(M^{[n]})^{-1} = (M^{-1})^{[n]} \tag{15}$$

where the symbol $[n]$ indicates the "direct nth power." Now the n-dimensional K-transform, as defined by (5), has as its matrix the direct product of the matrices $K_1, K_2, \cdots, K_n$, where, of course, the matrix K_1 has $K_1(r_1, s_1)$ as its elements, and so on. This is true provided that we order the rows and columns lexicographically. This explains the relevance of direct products of matrices to the previous discussion.

Now it is an interesting fact that a direct product of n matrices can also be expressed as an ordinary product of *sparse* matrices in accordance with the following theorem. (A sparse matrix is one consisting mainly of zeros.)

$$M = M^{(1)} \times M^{(2)} \times \cdots \times M^{(n)} = A^{(1)}A^{(2)} \cdots A^{(n)} \tag{16}$$

where

$$A^{(\nu)} = (a^{(\nu)}_{\mathbf{r},\mathbf{s}})$$

where

$$a^{(\nu)}_{\mathbf{r},\mathbf{s}} = m^{(\nu)}_{r_1, s_n} \delta^{s_1}_{r_2} \delta^{s_2}_{r_3} \cdots \delta^{s_{n-1}}_{r_n} \tag{17}$$

$$\begin{aligned}(0 \le r_1 < p_\nu, 0 \le r_2 < p_{\nu+1}, \cdots, 0 \le r_{n-\nu+1} < p_n,\\ 0 \le r_{n-\nu+2} < q_1, \cdots, 0 \le r_n < q_{\nu-1}\\ 0 \le s_1 < p_{\nu+1}, 0 \le s_2 < p_{\nu+2}, \cdots, 0 \le s_{n-\nu} < p_n,\\ 0 \le s_{n-\nu+1} < q_1, \cdots, 0 \le s_n < q_\nu).\end{aligned}$$

(The "Kronecker delta" is defined by $\delta^b_a = 1$ if $a = b$, otherwise $\delta^b_a = 0$.) Note that the nonzero elements of $A^{(\nu)}$ are, apart from replications, the same as those of $M^{(\nu)}$, but its *size* depends on all the t's.

Thus $A^{(1)}$ is a $(p_1 p_2 \cdots p_n) \times (p_2 p_3 \cdots p_n q_1)$ matrix having at most $p_1 p_2 \cdots p_n q_1$ nonzero elements. $A^{(2)}$ is a $(p_2 p_3 \cdots p_n q_1) \times (p_3 p_4 \cdots p_n q_1 q_2)$ matrix having at most $p_2 p_3 \cdots p_n q_1 q_2$ nonzero elements. $A^{(\nu)}$ is a $(p_\nu \cdots p_n q_1 \cdots q_{\nu-1}) \times (p_{\nu+1} \cdots p_n q_1 \cdots q_\nu)$ matrix having most $p_\nu p_{\nu+1} \cdots p_n q_1 q_2 \cdots q_\nu$ nonzero elements. $A^{(n)}$ is a $(p_n q_1 q_2 \cdots q_{n-1}) \times (q_1 q_2 \cdots q_n)$ matrix having at most $p_n q_1 q_2 \cdots q_n$ nonzero elements. $A^{(1)} \cdots A^{(n)}$ is a $(p_1 p_2 \cdots p_n) \times (q_1 q_2 \cdots q_n)$ matrix, namely M.

The product $\mathbf{a}'M$, where the components of $\mathbf{a}$ are $a_{\mathbf{r}}(0 \le r_1 < p_1, \cdots, 0 \le r_n < p_n)$ can thus be effected as

$$(\cdots((\mathbf{a}'A^{(1)})A^{(2)}) \cdots A^{(n)}). \tag{18}$$

As an example, take $n = 2$, $p_1 = 1$, $q_1 = 2$, $p_2 = 3$, $q_2 = 4$, and note that

$$\begin{array}{c|cccccc} p_2q_1 & 00 & 01 & 10 & 11 & 20 & 21 \\ p_1p_2 & & & & & & \\ 00 & a_{00} & a_{01} & 0 & 0 & 0 & 0 \\ 01 & 0 & 0 & a_{00} & a_{01} & 0 & 0 \\ 02 & 0 & 0 & 0 & 0 & a_{00} & a_{01} \end{array} \quad \begin{array}{cccccccc|c} 00 & 01 & 02 & 03 & 10 & 12 & 13 & 14 & q_1q_2 \\ & & & & & & & & p_2q_1 \\ b_{00} & b_{01} & b_{02} & b_{03} & 0 & 0 & 0 & 0 & 00 \\ 0 & 0 & 0 & 0 & b_{00} & b_{01} & b_{02} & b_{03} & 01 \\ b_{10} & b_{11} & b_{12} & b_{13} & 0 & 0 & 0 & 0 & 10 \\ 0 & 0 & 0 & 0 & b_{10} & b_{11} & b_{12} & b_{13} & 11 \\ b_{20} & b_{21} & b_{22} & b_{23} & 0 & 0 & 0 & 0 & 20 \\ 0 & 0 & 0 & 0 & b_{20} & b_{21} & b_{22} & b_{23} & 21 \end{array}$$

$$= \begin{bmatrix} a_{00}b_{00} & a_{00}b_{01} & a_{00}b_{02} & a_{00}b_{03} & a_{01}b_{00} & a_{01}b_{01} & a_{01}b_{02} & a_{01}b_{03} \\ a_{00}b_{10} & a_{00}b_{11} & a_{00}b_{12} & a_{00}b_{13} & a_{01}b_{10} & a_{01}b_{11} & a_{01}b_{12} & a_{01}b_{13} \\ a_{00}b_{20} & a_{00}b_{21} & a_{00}b_{22} & a_{00}b_{23} & a_{01}b_{20} & a_{01}b_{21} & a_{01}b_{22} & a_{01}b_{23} \end{bmatrix}$$

$$= A \times B.$$

In this example, if we were to multiply a horizontal 3-vector or a bolt-upright 8-vector by $A \times B$ it would involve $3 \times 8 \times 8 = 192$ scalar products, whereas multiplying first by $A^{(1)}$ and then by $A^{(2)}$, or vice versa, requires only $4 \times 8 + 3 \times 2 = 38$ scalar products. Generally, the reduction is from

$$p_1 p_2 \cdots p_n q_1 q_2 \cdots q_n$$

to

$$\sum_\nu p_\nu p_{\nu+1} \cdots p_n q_1 q_2 \cdots q_\nu \tag{19}$$

scalar multiplications.

The identification of this procedure with that in (5), when $p_\nu = q_\nu = t_\nu$ $(\nu = 1, 2, \cdots, n)$, is exemplified by the following algebra for the case $t_1 = 2$, $t_2 = 3$.

$$[a_{00}, a_{01}, a_{02}, a_{10}, a_{11}, a_{12}] \begin{array}{cccccc|c} 00 & 01 & 10 & 11 & 20 & 21 & t_2t_1 \\ & & & & & & t_1t_2 \\ c_{00} & c_{01} & 0 & 0 & 0 & 0 & 00 \\ 0 & 0 & c_{00} & c_{01} & 0 & 0 & 01 \\ 0 & 0 & 0 & 0 & c_{00} & c_{01} & 02 \\ c_{10} & c_{11} & 0 & 0 & 0 & 0 & 10 \\ 0 & 0 & c_{10} & c_{11} & 0 & 0 & 11 \\ 0 & 0 & 0 & 0 & c_{10} & c_{11} & 12 \end{array}$$

$$= [a_{00}c_{00} + a_{10}c_{10}, a_{00}c_{01} + a_{10}c_{11}, a_{01}c_{00} + a_{11}c_{10}, a_{01}c_{01} + a_{11}c_{11}, a_{02}c_{00} + a_{12}c_{10}, a_{02}c_{01} + a_{12}c_{11}]$$

whose elements are $\sum_{r_1} a_{r_1,r_2} c_{r_1,s_1} = l_{r_2,s_1}$ just as after the first summation for (5). Further,

$$\begin{array}{r} \\ \\ [l_{00}, l_{01}, l_{10}, l_{11}, l_{20}, l_{21}] \end{array} \begin{array}{c} \begin{array}{cccccc} 00 & 01 & 02 & 10 & 11 & 12 \end{array} \; t_1 t_2 \\ \end{array} \left[\begin{array}{cccccc} d_{00} & d_{01} & d_{02} & 0 & 0 & 0 \\ 0 & 0 & 0 & d_{00} & d_{01} & d_{02} \\ d_{10} & d_{11} & d_{12} & 0 & 0 & 0 \\ 0 & 0 & 0 & d_{10} & d_{11} & d_{12} \\ d_{20} & d_{21} & d_{22} & 0 & 0 & 0 \\ 0 & 0 & 0 & d_{20} & d_{21} & d_{22} \end{array}\right] \begin{array}{c} t_2 t_1 \\ 00 \\ 01 \\ 10 \\ 11 \\ 20 \\ 21 \end{array}$$

$$= [l_{00}d_{00} + l_{10}d_{10} + l_{20}d_{20}, l_{00}d_{01} + l_{10}d_{11} + l_{20}d_{21}, \cdots, l_{01}d_{02} + l_{11}d_{12} + l_{21}d_{22}]$$

whose elements are $\sum_{r_2} l_{r_2,s_1} d_{r_2,s_2}$, as in the second summation for (5). The first term in this vector is $(a_{00}c_{00} + a_{10}c_{10})d_{00} + (a_{01}c_{00} + a_{11}c_{10})d_{10} + (a_{02}c_{00} + a_{12}c_{10})d_{20} = \sum_{r_1,r_2} a_{r_1,r_2} c_{r_1,0} d_{r_2,0}$, and similarly the other five terms also agree with (5).

The theorem given by (16) was proven for the case $p_1 = \cdots = p_n = q_1 = \cdots = q_n$ in [7], and the proof extends to the general case.

The inverse of a multidimensional K-transform exists if the matrices $M^{(v)}$, defined by $M^{(v)} = (K_v(r_v, s_v))$, are all square and nonsingular. This can be easily proved by verifying (29), which is given in the Summary and Discussion. We here express the proof in terms of direct products of matrices. If we write $\boldsymbol{a}$ and $\boldsymbol{a}^*$ for the column vectors whose components are $a_{\boldsymbol{r}}$ and $a_{\boldsymbol{s}}^*$, these components being arranged in the lexicographic order of $\boldsymbol{r}$ and $\boldsymbol{s}$, then (5) can be written

$$\boldsymbol{a}^{*\prime} = \boldsymbol{a}'(M^{(1)} \times M^{(2)} \times \cdots \times M^{(n)}) = \boldsymbol{a}'M. \tag{20}$$

Now $M^{(v)}(M^{(v)})^{-1} = I_{t_v}$, and we see readily from (14) that

$$\boldsymbol{a}' = \boldsymbol{a}^*\{(M^{(1)})^{-1} \times \cdots \times (M^{(n)})^{-1}\}. \tag{21}$$

It may be noted that the determinant of M

$$|M| = \Delta_1^{\tau/t_1} \cdots \Delta_n^{\tau/t_n} \tag{22}$$

where $\Delta_v = |M^{(v)}|$. This follows inductively from the case $n = 2$ which is given by MacDuffee [18] who references Hensel, Netto, and von Sterneck. In the case where $M^{(v)}$ is independent of v and has order t and determinant Δ, we have

$$|M| = \Delta^{nt^{n-1}}. \tag{23}$$

If we wish to express the inverse multidimensional K-transform by means of ordinary matrix multiplications, we can apply (16) to the inverses of the $M^{(v)}$'s. Alternatively, we can denote the cofactors in $M^{(v)}$ by $M_{i,j}^{(v)}$ $(i, j = 0, 1, \cdots, t_{v-1})$ and write $B^{(v)} = b_{\boldsymbol{r},\boldsymbol{s}}^{(v)}$ where

$$b_{\boldsymbol{r},\boldsymbol{s}}^{(v)} = M_{s_1,r_n}^{(v)} \delta_{s_2}^{r_1} \delta_{s_3}^{r_2} \cdots \delta_{s_n}^{r_{n-1}} \tag{24}$$

$(s_1, r_n = 0, 1, \cdots, t_v - 1; s_2, r_1 = 0, 1, \cdots, t_{v+1} - 1; \cdots; s_{n-v+1}, r_{n-v} = 0, 1, \cdots, t_n - 1; \cdots; s_n, r_{n-1} = 0, 1, \cdots, t_{v-1} - 1)$.

Then

$$A^{(v)}B^{(v)} = \Delta_v I_\tau. \tag{25}$$

This is true even if $A^{(v)}$ is singular, in which case the right side of (25) is zero. If $A^{(v)}$ is nonsingular, it follows from (25) that $(A^{(v)})^{-1} = B^{(v)}/\Delta_v$. A proof of (25) is

$$\begin{aligned} \sum_{\boldsymbol{q}} m_{r_1,q_n}^{(v)} \delta_{r_2}^{q_1} \delta_{r_3}^{q_2} \cdots \delta_{r_n}^{q_{n-1}} \cdot M_{s_1,q_1}^{(v)} \delta_{s_2}^{q_1} \delta_{s_3}^{q_2} \cdots \delta_{s_n}^{q_{n-1}} &= \sum_{q_n} m_{r_1,q_n}^{(v)} M_{s_1,q_n}^{(v)} \delta_{r_2}^{s_2} \delta_{r_3}^{s_3} \cdots \delta_{r_n}^{s_n} \\ &= \Delta_v \delta_{r_1}^{s_1} \delta_{r_2}^{s_2} \cdots \delta_{r_n}^{s_n}. \end{aligned}$$

An example of (25) is

$$\begin{bmatrix} m_{00} & m_{01} & 0 & 0 \\ 0 & 0 & m_{00} & m_{01} \\ m_{10} & m_{11} & 0 & 0 \\ 0 & 0 & m_{10} & m_{11} \end{bmatrix} \begin{bmatrix} M_{00} & 0 & M_{10} & 0 \\ M_{01} & 0 & M_{11} & 0 \\ 0 & M_{00} & 0 & M_{10} \\ 0 & M_{01} & 0 & M_{11} \end{bmatrix} = \begin{bmatrix} \Delta & 0 & 0 & 0 \\ 0 & \Delta & 0 & 0 \\ 0 & 0 & \Delta & 0 \\ 0 & 0 & 0 & \Delta \end{bmatrix}$$

where

$$\Delta = m_{00}m_{11} - m_{01}m_{10}, M_{00} = m_{11}, M_{01} = -m_{10}, M_{10} = -m_{01}, M_{11} = m_{00}.$$

Note that $A^{(v)}$ and its inverse are equally sparse.

The determinants of $A^{(v)}$ can be neatly obtained from (22) by supposing that $\Delta_1 = \cdots = \Delta_{v-1} = \Delta_{v+1} = \cdots \Delta_n = 1$. It follows from this, together with the omitted argument for determining the sign, that

$$|A^{(v)}| = \Delta_v^u (-1)^{u(u-1)/2} \qquad (u = \tau/t_v) \tag{26}$$

and therefore, from (25), or from a familiar property of adjugates [20, p. 165]

$$|B^{(v)}| = \Delta_v^{\tau(t_v - 1)/t_v} (-1)^{u(u-1)/2}. \tag{27}$$

Of course if all the $M^{(v)}$'s are orthogonal (or else unitary), then their inverses are equal to their (conjugate) transposes, and the multidimensional K-transform and its inverse involve nearly the same calculation. The matrix of the multidimensional DFT becomes unitary if divided by $\sqrt{\tau}$ (or if each $M^{(\nu)}$ is divided by $\sqrt{t_\nu}$), just as the multidimensional Fourier transform is unitary if a suitable normalizing factor is used, such as $1/\sqrt{2\pi}$ per integral. When each $M^{(\nu)}$ is square, with $p_\nu = q_\nu = t_\nu$ as before, then another way of factorizing $M^{(1)} \times \cdots \times M^{(n)}$ into n sparse matrices is [7]

$$M^{(1)} \times \cdots \times M^{(n)} = C^{(1)}C^{(2)} \cdots C^{(n)} \tag{28}$$

where

$$C^{(1)} = M^{(1)} \times I_{t_2} \times \cdots \times I_{t_n}$$
$$C^{(2)} = I_{t_1} \times M^{(2)} \times \cdots \times I_{t_n}$$
$$C^{(n)} = I_{t_1} \times I_{t_2} \times \cdots \times M^{(n)}.$$

This follows from (14) and might sometimes be more convenient to use than (16) although the patterns of $C^{(1)}, \cdots, C^{(n)}$ are less simple than those of $A^{(1)}, \cdots, A^{(n)}$.

Summary and Discussion

Yates [21] introduced a simple adding and subtracting algorithm for the calculation of the interactions in 2^n-factorial experiments. The fact that Yates' algorithm could be expressed as a multidimensional mod 2 discrete Fourier transform (DFT) was pointed out by Good [22]. Good [7] showed that the algorithm can be generalized to apply to any multidimensional DFT and even more generally. This depends on the first three of the following facts.

1) A multidimensional mod t DFT is a linear transformation and corresponds to multiplication by a matrix (of order t^n where n is the dimensionality) which can be expressed as the nth direct power of a matrix of order t.

2) The nth direct power of a matrix M is equal to the nth ordinary power of a matrix A defined as

$$A = \{M_{r_1, s_n} \delta_{r_2}^{s_1} \delta_{r_3}^{s_2} \cdots \delta_{r_n}^{s_{n-1}}\}.$$

3) Although A is t^n by t^n it is sparse and contains only t^{n+1} nonzero elements, so that multiplication by M involves only nt^{n+1} ordinary multiplications instead of t^{2n}. This, and the remarks under facts 1 and 2, extended readily to the case of unequal moduli $t_1, t_2, \cdots, t_n$ and even more generally. We thus have a fast algorithm for the computation of an n-dimensional K-transform, defined by (5), of which the n-dimensional DFT is a special case. The remarks following (5) give the algorithm in nonmatrix language.

When $K_1, \cdots, K_n$ in (5) are all the same function, the matrices $A^{(1)}, \cdots, A^{(n)}$ are all equal and this matrix formulation is then at its simplest. In equation (6) the K's cannot be identical unless the equation reduces to (5).

A mod 2 multidimensional DFT is an example of a Hadamard transform which is discussed and applied by Pratt, Kane, and Andrews [12]. Another example of a K-transform is the Andrews–Caspari transform [17] in which the orthogonal matrix of the transform has elements

$$m_{r,s} = \alpha^n \beta^{[r,s]}(-1)^{r \cdot s} \quad (t_\nu = 2;\ \nu = 1, 2, \cdots, n)$$

where α is real and positive,

$$\beta = \sqrt{1 - \alpha^2}/\alpha, \quad \text{and} \quad [r, s] = \sum_{\mu=0}^{n-1} (r_\mu \oplus s_\mu)$$

where $\oplus$ denotes mod 2 addition (EXCLUSIVE OR), but the summation is ordinary summation which happens to reduce to counting.

4) If, in (5), which defines a multidimensional K-transform, the functions $K_1(r_1, s_1), K_2(r_2, s_2), \cdots$, when regarded as elements of matrices, are square and nonsingular, and if the inverse matrices have elements $J_1(r_1, s_1)$, $J_2(r_2, s_2)$, etc., then the inverse transform can be written

$$a_r = \sum_s a_s^* J_1(s_1, r_1) \cdots J_n(s_n, r_n). \tag{29}$$

This can be readily proved by using Kronecker deltas and is another way of expressing (21).

5) When one wishes to carry out a multiplication by a large matrix it can be worthwhile to look for a factorization into several sparse matrices so as to cut down on the work.

6) New coding methods, generalizing the DFT, can be invented ad lib by taking several sparse square matrices of equal orders and using their product as the matrix of the transformation or coding. In order that unique decoding should be possible, it is necessary that the sparse matrices should all be nonsingular and convenient if their inverses are sparse. Apart from the methods discussed in this paper, the matrices could, for example, be taken as permutation matrices, these being the simplest and sparsest possible nonsingular matrices. They might be too sparse for some purposes since they merely permute the elements of the vector on which they operate and so have no tendency to flatten the variation from one component to another. Such flattening is apt to be convenient for communication systems, but the permutation by itself gives some robustness against bursts of noise. It might also turn out to be useful to use functions $K_1, K_2, \cdots, K_n$ as in (6) instead of (5), although it is presumably more difficult to find elegant schemes of this kind.

7) Unidimensional DFTs can be expressed as multidimensional ones by either the Sino–Ruritanian method or the "reactionary" method, the latter being more widely applicable for this purpose. Once in the multidimensional form, the streamlining described above is applicable. The paper [7] provoked the influential paper by Cooley and Tukey in which the reactionary method was described. Their original acknowledgments to my paper were somewhat too generous since they gave the impression that the two methods were almost identical. Their methods more closely resembled the methods of Runge [23] and Danielson and Lanczos [24], whose papers had been generally overlooked by oceanographers, X-ray crystallographers, and many others who make frequent use of harmonic and spectral analysis. These authors wrote before the age of electronic

computers and were in this respect ahead of their time, although a fast Fourier transform would have been of use for more than 100 years. In the design of special-purpose equipment, both the Sino–Ruritanian and reactionary representations should be kept in mind.

8) One application of the DFT is for the inversion of a circulix, that is, a $t \times t$ matrix (c_{s-r}) where $s-r$ is taken modulo t [25]. A 1 000 000 × 1 000 000 circulix could easily be inverted since it requires only about 100 000 000 operations instead of some quintillion if the matrix were random. The eigenvalues of a circulix are given by the DFT of its top row, and the inverse DFT of their reciprocals gives the inverse circulix. Similarly a recursively blocked circulix, defined recursively as either a circulix or a block matrix whose elements are recursively blocked circulices, or equally defined as $(a_{\mathbf{s}-\mathbf{r}})$ where $\mathbf{r}$ and $\mathbf{s}$ are ordered lexicographically and $\mathbf{s}-\mathbf{r}$ is taken modulo $(t_1, t_2, \cdots, t_n)$, has eigenvalues given by the multidimensional DFT of its top row, and so can be inverted in a similar manner. (I have avoided the name "block circulix" tout court, because a block circulant is defined by Muir [20, p. 485] in a somewhat different sense, apart, of course, from its being a determinant and not a matrix.) The eigenvectors of a recursively blocked circulix are the columns of the matrix $(\omega_1^{r_1 s_1}\omega_2^{r_2 s_2} \cdots)$.

9) Let p be a prime number and g one its primitive roots; that is $1, g, g^2, \cdots, g^{p-2}$ are all distinct modulo p. Let α_r $(r = 0, 1, \cdots, p-2)$ be a vector of integers and let us define a *number-theoretic Fourier transform:*

$$\alpha_s^* = \sum_{r=0}^{p-2} \alpha_r g^{rs} \pmod p. \tag{30}$$

Then it is easily proved that the inverse transform is

$$\alpha_r = -\sum_{s=0}^{p-2} \alpha_s^* g^{-rs} \pmod p \tag{31}$$

and the transform of the convolution

$$\sum_{q=0}^{p-2} \alpha_q \beta_{r-q}$$

is $\alpha_s^*\beta_s^*$. A $(p-1)\times(p-1)$ circulix of integers has eigenvalues equal to the transform of its top row, and it can be inverted modulo p by using modulo p reciprocals of the eigenvalues, almost as in fact 8. Recursively blocked circulices of integers, modulo p, have their eigenvalues equal to multidimensional transforms of their top rows, these transforms being defined in an obvious way, modulo a *fixed* prime p. These multidimensional transforms could be used for a coding scheme for vectors of integers modulo p. The calculation can be done by the simple method following (5), or by the equivalent matrix methods, except that the arithmetic must of course be modulo p.

If p is replaced by a composite number m we can define

$$\alpha_s^* = \sum_{\substack{(r,m)=1 \\ r<m}} \alpha_r g^{rs} \pmod m,$$

but this seems uninteresting because it does not have an elegant inverse even when m has a primitive root (whose powers take all values prime to m). But (30) can be generalized in a different direction, as we now indicate.

10) We can generalize (30) and its multidimensional form to a finite (Galois) field F. A finite field contains a prime power p^ν of elements and always has a primitive root g for which $1, g, g^2, \cdots, g^{p^\nu-2}$ are distinct elements of the field [26]. If α_r $(r=0, 1, \cdots, p^\nu-2)$ is a vector of field elements we can define

$$\alpha_s^* = \sum_{r=0}^{p^\nu-2} \alpha_r g^{rs} \quad (s = 0, 1, \cdots, p^\nu - 2) \tag{32}$$

as its (unidimensional) *Fourier–Galois tranform.* Similarly, for an array $\alpha_{\mathbf{r}}(\mathbf{r}=(r_1, \cdots, r_n))$, we can define a multidimensional transform

$$\alpha_{\mathbf{s}}^* = \sum_{\mathbf{r}} \alpha_{\mathbf{r}} g_1^{r_1 s_1} \cdots g_n^{r_n s_n} \tag{33}$$

where $g_1, g_2, \cdots, g_n$ are primitive elements of F, not necessarily distinct. This can also be written in the form

$$\alpha_{\mathbf{s}}^* = \sum_{\mathbf{r}} \alpha_{\mathbf{r}} g^{c_1 r_1 s_1 + \cdots + c_n r_n s_n} \tag{34}$$

where $c_1, \cdots, c_n$ are certain constants prime to $p^\nu - 1$. The inversion formula is

$$\alpha_{\mathbf{r}} = (-1)^n \sum_{\mathbf{s}} \alpha_{\mathbf{s}}^* g_1^{-r_1 s_1} \cdots g_n^{-r_n s_n} \tag{35}$$

and there is a convolution formula

$$\sum_{\mathbf{r}} \left(\sum_{\mathbf{q}} \alpha_{\mathbf{q}} \beta_{\mathbf{r}-\mathbf{q}} \right) g_1^{r_1 s_1} \cdots g_n^{r_n s_n} = \alpha_{\mathbf{s}}^* \beta_{\mathbf{s}}^*. \tag{36}$$

A recursively blocked circulix of group elements has, as eigenvalues, the multidimensional Galois–Fourier transform of its top row, and, as eigenvectors, the columns of the matrix $(g_1^{r_1 s_1}) \times \cdots \times (g_n^{r_n s_n})$. The transform could be used for coding "sentences" of "words" where each word, being an element of F, could be expressed by means of ν modulo p integers, namely the coefficients of the word when the group is represented by means of polynomials modulo both p and a fixed irreducible polynomial [27]. (Finite fields have often been used in other ways in coding theory and in the design of statistical experiments.)

For the multidimensional transform we can again apply the simple algorithm following (5), or the equivalent matrix methods, the arithmetic being in the field. I can see no elegant transformation from unidimensional to multidimensional Fourier–Galois transforms.

References

[1] *IEEE Trans. Audio Electroacoust.*, vol. AU-15, June 1967.
[2] *IEEE Trans. Audio and Electroacoust.*, vol. AU-17, June 1969.
[3] I. J. Good, "Polynomial algebra: An application of the fast Fourier transform," *Nature*, vol. 222, 1969, p. 1302.
[4] ——, "The discrete Fourier transform: Work by I. J. Good," June, 1969, pp. 1–6, mimeographed.
[5] R. C. Singleton, "A short bibliography on the fast Fourier transform," in [2], pp. 166–169.
[6] I. J. Good, "The fast Fourier transform and discrete Fourier transform: Bibliography," June 1969, pp. 1–5, mimeographed.
[7] ——, "The interaction algorithm and practical Fourier analysis,"

J. Roy. Statist. Soc. Ser. B, vol. 20, 1958, pp. 361–372; vol. 22, 1960, pp. 372–375.
[8] J. W. Cooley and J. W. Tukey, "An algorithm for the machine calculation of complex Fourier series," *Math. Computation*, vol. 19, 1965, pp. 297–301.
[9] I. J. Good, "Random motion on a finite Abelian group," *Proc. Cambridge Phil. Soc.*, vol. 47, 1951, pp. 756–762.
[10] ——, "The serial test and other tests for randomness," *Proc. Cambridge Phil. Soc.*, vol. 49, 1953, pp. 276–284.
[11] ——, "Analogues of Poisson's summation formula," *Amer. Math. Mon.*, vol. 69, 1962, pp. 259–266.
[12] W. K. Pratt, J. Kane, and H. C. Andrews, "Hadamard transform image coding," *Proc. IEEE*, vol. 57, Jan. 1969, pp. 58–68.
[13] W. W. R. Ball, *Mathematical Recreations and Essays*. London: Macmillan, 1940. (Revised by H. M. S. Coxeter.)
[14] G. H. Hardy and E. M. Wright, *An Introduction to the Theory of Numbers*. Oxford: Clarendon Press, 1938.
[15] P. Bachmann, *Niedere Zahlentheorie*, pt. 1. Leipzig: Teubner, 1902, p. 83.
[16] C. Bingham, M. D. Godfrey, and J. W. Tukey, "Modern techniques of power spectral estimation," in [1], pp. 56–66.
[17] H. C. Andrews and K. L. Caspari, "A generalized technique for spectral analysis," *IEEE Trans. Computers*, vol. C-19, Jan. 1970, pp. 16–25.
[18] C. C. MacDuffee, *The Theory of Matrices*. New York: Chelsea, 1956, p. 82.
[19] L. J. Savage, *The Foundations of Statistics*. New York: Wiley, 1954, p. *viii*.
[20] T. Muir, *A Treatise on the Theory of Determinants*. New York: Dover, 1960.
[21] F. Yates, *The Design and Analysis of Factorial Experiments*. Harpenden: Imperial Bureau of Soil Science, 1937.
[22] I. J. Good, review of M. H. Quenouille's "Design and analysis of experiment," in *Ann. Eugen.*, vol. 18, 1953, p. 263.
[23] C. Runge, "Über die Zerlegung empirisch gegebener periodischer Funktionen in Sinnuswellen," *Z. Math. Physik*, vol. 48, 1903, pp. 443–456; vol. 52, 1905, p. 117.
[24] G. C. Danielson and C. Lanczos, "Some improvements in practical Fourier analysis and their application to X-ray scattering from liquids," *J. Franklin Inst.*, vol. 233, 1942, pp. 365–380 and 435–452.
[25] I. J. Good, "On the inversion of circulant matrices," *Biometrika*, vol. 37, 1950, pp. 185–186.
[26] G. Birkhoff and S. MacLane, *A Survey of Modern Algebra*. New York: Macmillan, 1944, pp. 428–431.
[27] R. D. Carmichael, *Introduction to the Theory of Groups of Finite Order*. Boston: Ginn, 1937, p. 255.

MATHEMATICS OF COMPUTATION, VOLUME 25, NUMBER 114, APRIL, 1971

The Fast Fourier Transform in a Finite Field

By J. M. Pollard

Abstract. A transform analogous to the discrete Fourier transform may be defined in a finite field, and may be calculated efficiently by the 'fast Fourier transform' algorithm. The transform may be applied to the problem of calculating convolutions of long integer sequences by means of integer arithmetic.

1. Introduction and Basic Properties. Let GF(p^n), or F for short, denote the Galois Field (Finite Field) of p^n elements, where p is a prime and n a positive integer. Let d be a divisor of $p^n - 1$ (possibly $d = p^n - 1$), and r be a member of F of order d in the multiplicative group, F^* say, of the nonzero elements of F (which certainly exists, since this group is cyclic of order $p^n - 1$, [1, p. 125]). Then one can define the transform of a sequence (a_i) $(0 \leqq i \leqq d - 1)$ of members of F to be the sequence (A_i) where

$$A_i = \sum_{j=0}^{d-1} a_j r^{ij}. \tag{1}$$

The transformed sequence (A_i) depends on the choice of r, which will be considered fixed throughout.

The inverse transform to (1), (that is, an equivalent set of equations giving the (a_i) in the terms of the (A_i)) is

$$a_j = -d' \cdot \sum_{i=0}^{d-1} A_i r^{-ij}, \tag{2}$$

where d' is the integer for which

$$d'd = p^n - 1. \tag{3}$$

The transform (1) and its inverse (2) may be calculated by the 'fast Fourier transform' (FFT) algorithm ([2], [3]), of which many versions have been described in detail (see [4] for other references). These authors were concerned with the calculation of the 'discrete Fourier transform' of sequences of complex numbers (see Section 7), but these algorithms are equally applicable to the present case; the roots of unity

$$e(k/d) = e^{2\pi ik/d}, \qquad (k \text{ an integer}),$$

are replaced by r^k throughout, and the operations of complex addition and multiplication become the corresponding operations in F. The transform (1), like the 'conventional' FFT, is simplest to perform when the integer d is highly composite (the product of many small factors, e.g. $d = 2^m$), but more complicated versions cope with any

Received February 24, 1970, revised November 18, 1970.

AMS 1970 *subject classifications.* Primary 42A68, 12C15, 10A40, 10L99.

Key words and phrases. Finite field, fast Fourier transform, residue arithmetic, primitive polynomials, Mersenne primes.

d ([5] and Section 8). In any case, the total number of operations (in F or in the complex field) is reduced to $O(d \cdot \log d)$ from the $O(d^2)$ operations required to calculate these transforms in the most obvious way. In the case of (1), this estimate assumes that a suitable r has already been found, and the powers r^i $(0 \leqq i \leqq d-1)$ computed (see Section 5). The principal reason for interest in the transform (1) lies in the following 'convolution property'. Suppose that three pairs of sequences, (a_i) and (A_i), (b_i) and (B_i), (c_i) and (C_i) $(0 \leqq i \leqq d-1)$ form transform pairs (i.e. are related by (1)) and that

$$C_i = A_i B_i, \qquad (0 \leqq i \leqq d-1). \tag{4}$$

Then,

$$c_i = \sum_{\substack{j=0 \\ j+k \equiv i \pmod d}}^{d-1} \sum_{k=0}^{d-1} a_j b_k, \qquad (0 \leqq i \leqq d-1). \tag{5}$$

If we extend the definition of (b_i) to all i by making the sequence periodic with period d, (5) may be written

$$c_i = \sum_{j=0}^{d-1} a_j b_{i-j}. \tag{5A}$$

Thus, the calculation of the 'cyclic convolution' of the sequences (a_i) and (b_i), as defined by (5), may be obtained by transforming the sequences, multiplying the results term-by-term as in (4), and performing the inverse transform (2). The corresponding process for sequences of complex numbers is well known (see [4]).

Another simple property of (1) is that

$$\sum_{j=0}^{d-1} A_j^2 = d \cdot \sum_{i=0}^{d-1} a_i a_{d-i}, \tag{6}$$

and, similarly, that

$$\sum_{i=0}^{d-1} a_i^2 = (-d') \cdot \sum_{j=0}^{d-1} A_j A_{d-j}. \tag{7}$$

The properties expressed in the equations (1) to (7) will be proved in the next section. These properties are analogous to well-known properties of the discrete Fourier transform; in fact, these properties hold, more generally, in any field F, finite or not, provided that r has finite order d in the multiplicative group F^* of F.

2. Proof of the Basic Properties. We prove first that for any integer k,

$$\begin{aligned} \sum_{i=0}^{d-1} r^{ik} &= d, \quad \text{if } k \equiv 0 \pmod d, \\ &= 0, \quad \text{otherwise.} \end{aligned} \tag{8}$$

(8) is trivially true when $k \equiv 0 \pmod d$ since each term on the left is unity; when $k \not\equiv 0 \pmod d$ we use the relation

$$\left(\sum_{i=0}^{d-1} r^{ik}\right)(1 - r^k) = 1 - r^{dk} = 0$$

from which (8) follows since $1 - r^k \neq 0$.

To show that (1) implies (2), we simplify the right side of (2) using (1), (3) and (8), as

$$-d'\cdot\sum_{i=0}^{d-1} r^{-ij}\cdot\sum_{l=0}^{d-1} a_l r^{il} = (-d')\cdot\sum_{l=0}^{d-1} a_l\left(\sum_{i=0}^{d-1} r^{i(l-j)}\right)$$
$$= -d'\cdot a_j\cdot d = a_j.$$

The proof that (2) implies (1), and the proof of the other properties expressed in (4)–(7) are equally simple deductions from (8).

3. Applications of the Transform.

(i) *Multiplication of Polynomials over* GF(p^n). Let the polynomials to be multiplied be

(9) $$f(x) = \sum_{i=0}^{m_1} a_i x^i,\qquad g(x) = \sum_{i=0}^{m_2} b_i x^i,\quad ((a_i), (b_i) \in \mathrm{GF}(p^n)),$$

where $m_1 \geqq m_2$. The product is

$$h(x) = \sum_{i=0}^{m_1+m_2} e_i x^i,$$

where

(10) $$e_k = \sum_{0\leqq i\leqq m_1;0\leqq j\leqq m_2;i+j=k} a_i b_j,\qquad (0 \leqq k \leqq m_1 + m_2).$$

Assume first that $(p^n - 1)$ is highly composite and $m_1 + m_2 < p^n - 1$. Choose a divisor d of $(p^n - 1)$ for which $d > m_1 + m_2$ and extend the definitions of (a_i), (b_i) by setting

$$a_i = 0,\quad (m_1 < i \leqq (d-1)),\qquad b_i = 0,\quad (m_2 < i \leqq (d-1)).$$

The 'convolution algorithm' of Section 1 with $F = \mathrm{GF}(p)$ applied to these extended sequences gives the required (e_k); more precisely the (c_k) so obtained satisfy the equations,

$$c_k = e_k,\quad (0 \leqq k \leqq m_1 + m_2),\qquad c_k = 0,\quad (m_1 + m_2 < k \leqq d - 1).$$

If $m_1 + m_2 \geqq p^n - 1$ or $p^n - 1$ is not highly composite it may be feasible to operate in an extension field, $F = \mathrm{GF}(p^{ln})$, say, (l a positive integer), of GF(p^n); this requires that $p^{ln} - 1$ be highly composite and $m_1 + m_2 < p^{ln} - 1$. The conditions can certainly be met for the important practical case $p = 2$, $n = 1$ (e.g. [6], [15]). Otherwise, see Section 3(ii) and Section 8.

(ii) *Multiplication of Integer Polynomials* (*or Convolution of Integer Sequences*). We require to calculate the (e_k) defined by (10) where the (a_i), (b_i) are now ordinary integers. An upper bound is needed for the members of these sequences, so suppose that

(11) $$|a_i| \leqq M_1,\quad (0 \leqq i \leqq m_1),\qquad |b_i| \leqq M_2,\quad (0 \leqq i \leqq m_2),$$

so that, by (10)

$$|e_k| \leqq m_2 M_1 M_2 \qquad (= L,\ \text{say}).$$

Let p be a prime and d a highly composite integer such that $d \mid (p - 1)$, $d > (m_1 + m_2)$ and $p > 2L$; then

(12) $$|e_k| \leqq L < p/2.$$

Applying the process of Section 3(i) with $F = \mathrm{GF}(p)$, we obtain a sequence (c_i) with

$$c_i \equiv e_i \pmod p, \qquad (0 \leqq k \leqq m_1 + m_2),$$

so that, by (12), the least absolute residue (mod p) of c_k $(0 \leqq k \leqq m_1 + m_2)$ is e_k.

It is possible to replace the one possibly very large prime p required by a number of smaller primes. Choose pairs

$$(p_i, d_i), \qquad (1 \leqq i \leqq t, \text{ say})$$

of primes p_i and highly composite integers d_i with

$$d_i > m_1 + m_2, \qquad (1 \leqq i \leqq t),$$

$$p_i \equiv 1 \pmod{d_i}, \qquad (1 \leqq i \leqq t)$$

and

$$P = p_1 p_2 \cdots p_t > 2L.$$

By applying the previous algorithm with each pair (p_i, d_i) in turn, one obtains the residues of each e_k to each modulus p_i; from these, one can find the least absolute residues of the (e_k) (mod $P = p_1 \cdots p_t$) by the 'Chinese Remainder Theorem' ([8, p. 94], [9, p. 27]) and, since by (12) $|e_k| < P/2$, these are again the required integer values.

(iii) *The Multiplication of Very Large Integers*. Suppose the integers are expressed to the base u as

$$f = \sum_{i=0}^{m_1} a_i u^i, \qquad g = \sum_{i=0}^{m_2} b_i u^i, \quad (0 \leqq a_i < u, 0 \leqq b_i < u).$$

Their product is

$$h = \sum_{i=0}^{m_1+m_2} e_i u^i,$$

where the (e_i), defined by (10), do not necessarily satisfy $0 \leqq e_i < u$. So, after obtaining the (e_i) by the method of Section 3(ii), it remains to perform some carrying to obtain h in standard form; but with a suitable choice of parameters, this is a relatively trivial operation. This algorithm is applicable to Lucas' method for testing the primality of Mersenne numbers, $M_p = 2^p - 1$, p prime ([8, p. 16 and p. 223], [10]).

As an example, I give some details of a method of multiplying two integers of 10,000 bits (binary digits) in a digital computer of word length 24 bits. I take $u = 2^{21}$ so that m_1 and m_2 are less than $2^9 = 512$. In Section 3(ii), I take $t = 3$, $d_1 = d_2 = d_3 = 2^{10} = 1024$ and so require three primes p_1, p_2, p_3 satisfying

$$2^{21} < p_1 < p_2 < p_3 < 2^{23}, \quad p_i \equiv 1 \pmod{2^{10}};$$

for example $p_1 = 6946817$, $p_2 = 7340033$, $p_3 = 7667713$ which in fact satisfy $p_i \equiv 1 \pmod{2^{16}}$. The condition $p_1 p_2 p_3 > 2L$ is seen to be easily satisfied.

Using these values, the algorithm has been programmed in machine language for the ICL 4120 computer. The multiplication of two different integers takes 45 seconds while a straightforward 'long multiplication' algorithm, using all 24 bits of each word, takes 95 seconds. For the problem of squaring a single integer (the main part of one step of Lucas' algorithm for a prime p around 10,000), the times are 30 and 48 seconds, respectively. In this case, we perform two transforms to each prime modulus instead of three, while 'long multiplication' requires just half as many operations as before. In absolute terms, these times are slow by modern standards [10].

It is probable that the highest speeds would be obtained in a special purpose computer, taking advantage of the possibility of extensive parallel computation in the transforms. Such computers have been built for the conventional transform [4]. But, of course, it is possible to use parallel computation for the long multiplication process also.

(iv) *Division of Polynomials Over* GF(p). Let (a_i) $(0 \leqq i \leqq d-1)$ be a sequence in the finite field $F = \mathrm{GF}(p^n)$, where $d \mid p^n - 1$, and let $f(x) = \sum_{i=0}^{d-1} a_i x^i$.

Then, Eq. (1) could be written

$$A_j = f(r^j), \qquad (0 \leqq j \leqq d-1).$$

So, what I have termed a fast Fourier transform in F might be regarded simply as 'fast evaluation of a polynomial over F'.

If $d = p^n - 1$, so that every nonzero element of F is of the form r^j $(0 \leqq j \leqq d-1)$, it follows that the condition that the transformed sequence (A_j) contains no zeros is just that $f(x)$ has no roots in F, except perhaps for $x = 0$; while if $d < p^n - 1$ then this condition is sufficient but not necessary.

This remark is used in the following division algorithm. Let

$$f(x) = \sum_{i=0}^{m_1} a_i x^i \quad \text{and} \quad h(x) = \sum_{i=0}^{m_1+m_2} e_i x^i, \qquad (m_1 > 1,\ m_2 \geqq 1),$$

be polynomials over $F = \mathrm{GF}(p)$, $f(x)$ being given to be irreducible. We wish to find whether

$$f(x) \mid h(x), \tag{13}$$

and if so, to find the quotient, say

$$g(x) = \sum_{i=0}^{m_2} b_i x^i. \tag{14}$$

Choose integers $n \geqq 1$ and $d \geqq 1$ (and highly composite) such that

$$d > m_1 + m_2, \qquad d \mid p^n - 1, \qquad m_1 \nmid n,$$

extend the sequences (a_i), (e_i) to d terms (i.e., $0 \leqq i \leqq d-1$) by the addition to zeros (as in Section 3(i)) and apply the transform (1) in the field $F' = \mathrm{GF}(p^n)$ to obtain sequences (A_i), (E_i), $(0 \leqq i \leqq d-1)$. Since $m_1 \nmid n$, the field F' does not contain $\mathrm{GF}(p^{m_1})$, the splitting field of $f(x)$ (see [1, Chapter IV]), so F' contains no roots of $f(x)$ and so $A_i \neq 0$, $(0 \leqq i \leqq d-1)$.

Define (B_i) by

$$B_i = E_i / A_i, \qquad (0 \leqq i \leqq d-1) \tag{15}$$

and perform the inverse transform (2) in F' to obtain (b_i), $(0 \leqq i \leqq d-1)$. If

$$b_i = 0, \qquad (m_2 < i \leqq d-1), \tag{16}$$

then (13) holds, (14) gives the quotient and $b_i \in \mathrm{GF}(p)$ $(0 \leqq i \leqq m_2)$; otherwise (13) is false.

Proof. Suppose first that (13) holds, with quotient (14); extend the sequence (b_i) to d terms by the addition of zeros. By the multiplication algorithm of Section 3(i), the transform (B_i) satisfies

$$A_i B_i = E_i, \qquad (0 \leqq i \leqq d-1).$$

Hence the division algorithm leads to a sequence (b_i) with the properties stated.

Now suppose that (16) holds; by defining $g(x)$ by (14), it follows from (15) that

$$f(x) \cdot g(x) = h(x), \qquad (\text{over } F'). \tag{17}$$

Suppose that some of (b_i) $(0 \leqq i \leqq m_2)$ are not in F. If b_K is the first such b, then by taking coefficients of x^{K+m_1} in (17)

$$b_K a_{m_1} + b_{K+1} a_{m_1-1} + \cdots + b_{K+m} a_{m_1-m} = e_{K+m_1}, \qquad (m = \min(m_1, m_2 - K))$$

in which every element except b_K is in F and $a_{m_1} \neq 0$. This is a contradiction so $b_i \in F$ $(0 \leqq i \leqq m_2)$ as asserted.

4. The Transform in the Ring of Integers Modulo m. A version of the transform (1) can also be given in a particular ring which is not a field, namely the ring R_m of integers modulo a composite integer m. Suppose first that m is prime power, say $m = p^n$, $p > 2$, and that d is a divisor of $(p - 1)$. The multiplicative group, R^* say, of the members of R_m prime to m is still cyclic and of order $p^{n-1}(p - 1)$ divisible by d, and so it contains a number r of order d. The transform of a sequence of members of R_m can still be defined by (1) and the results of Section 1 still hold. To prove this it suffices to show that (8) still holds (as a congruence mod m).

In the nontrivial case $K \not\equiv 0 \pmod d$, we have now

$$\left(\sum_{i=0}^{d-1} r^{iK}\right)(1 - r^K) \equiv 1 - r^{dK} \equiv 0, \qquad (\text{mod } m = p^n),$$

and need to show that $r^K \not\equiv 1 \pmod p$.

The numbers r^i $(0 \leqq i \leqq d-1)$ form a subgroup of R^* of order d, and the members of R_m congruent to 1 (mod p) form a subgroup of order p^{n-1}; the orders of these subgroups being coprime, they can have only the unit element in common, which proves the assertion.

More generally, we can take $m = p_1^{a_1} \cdots p_t^{a_t}$, where $d \mid (p_i - 1)$ $(1 \leqq i \leqq t)$ and where r has order d (mod $p_i^{a_i}$) for each i $(1 \leqq i \leqq t)$ and hence also order d (mod m).

5. Practical Methods of Performing the Transform.

(i) *The Case* $F = \mathrm{GF}(p)$. The obvious method is to represent the elements of F by the integers $0, 1, \cdots, (p - 1)$, and perform addition and multiplication followed by reduction (mod p). Alternatively, if p is not too large, one can replace the multiplications (mod p) by additions (mod $(p - 1)$) and 'table look-up' operations by means of previously constructed tables of powers of a primitive root w (mod p), [11, p. 63]. This is well known in connection with 'residue-system' computers [9].

A suitable choice for r (of order d in F^*) is $r \equiv w^{d'}$ (mod p), where, as in Section 1, $d'd = p - 1$. One method to find a primitive root w is to test successively the positive integers by the following trivial theorem, which applies also to the case $F = \mathrm{GF}(p^n)$.

THEOREM. *The element* w *in* $F = \mathrm{GF}(p^n)$ *generates* F^*, *the multiplicative group of* F, *if and only if for each prime factor* q *of* $p^n - 1$ (*with* $q < p^n - 1$) *we have* $w^{((p^n-1)/q)} \neq 1$.

The application of the theorem can be made by a device discussed in [12, Section 3].

(ii) *The Case* $F = \mathrm{GF}(p^n)$, *where* $n > 1$. The following table 'look-up' method [7, pp. 37–50] is probably best, when possible, that is, when p^n is not too large. We take a generator w of F^* and represent the nonzero elements of F by their exponents when written as powers of w, with a special symbol for the zero element (say, $0 = w^*$). Multiplication in F is merely addition of exponents (mod $(p^n - 1)$); for the addition of nonzero elements w^a, w^b ($a \leq b$) we write

$$w^a + w^b = w^a(1 + w^{b-a}),$$

and then consult a table giving the solution x of

$$w^x = 1 + w^y,$$

for given y (where possibly $x = *$).

The use of a table of p^n entries can be avoided by the following less efficient method. Let $F[x]$ be the ring of polynomials in a single variable x over the field $\mathrm{GF}(p)$ and let $h(x) \in F[x]$ be irreducible and of degree n. The members of $F = \mathrm{GF}(p^n)$ can be represented by the residue classes (mod $h(x)$) of $F[x]$, and so can be written as

$$f(x) = \sum_{i=0}^{h-1} a_i x^i \qquad (a_i \in \mathrm{GF}(p)). \tag{18}$$

Addition in F is now straightforward; but multiplication necessitates forming the product in $F[x]$ of two polynomials of form (18) and then division by $h(x)$ to give a remainder of this form again.

It is possible to choose $h(x)$ in such a way that the element $w(x) = x$ is itself a generator of F^*, and so take

$$r(x) = \{w(x)\}^{d'} = x^{d'},$$

as the required element of order d. The suitable choices for $h(x)$, the 'primitive polynomials over F', are the irreducible factors in $F[x]$ of $\Phi_{p^n-1}(x)$, where $\Phi_m(x)$ denotes the cyclotomic polynomial

$$\Phi_m(x) = \prod_{k \mid m} (x^k - 1)^{\mu(m/k)},$$

[1, p. 132]. These polynomials are of interest in connection with 'shift register sequences', [6], [13]. An efficient test for a given $h(x) \in P[x]$, of degree n but not known to be irreducible is provided by the following theorem, which generalises that of [13].

THEOREM. *Let* $h(x)$ *be a polynomial of degree* n (≥ 1) *in* $F[x]$. *Then*

$$h(x) \mid \Phi_{p^n-1}(x), \quad (in\ F[x])$$

if and only if

(a) $x^{p^n-1} \equiv 1 \pmod{h(x)}$, *in* $F[x]$, *and*

(b) *for each prime factor* q *of* $p^n - 1$ *with* $q < p^n - 1$ (*if any exist*), *the polynomials* $h(x)$, $x^m - 1$, *where* $m = (p^n - 1)/q$, *are relatively prime in* $F[x]$.

Proof. The necessity of the conditions (a) and (b) is trivial. Suppose these conditions hold and that $h_1(x)$ is an irreducible factor of $h(x)$ in $F[x]$. Then $h_1(x) \mid \Phi_{p^n-1}(x)$

which implies that $h_1(x)$ has degree n, and so $h(x) = k \cdot h_1(x)$ $(k \in F,\ k \neq 0)$ and $h(x) \mid \Phi_{p^n-1}(x)$.

To apply (a), calculate the residue (mod $h(x)$) of $x^{(p^n-1)} - 1$ by the device mentioned in Section 5(i). For (b), first find the residue of $x^m - 1$ by the same means, then the greatest common divisor of this polynomial and $h(x)$ by Euclid's algorithm (for polynomials over a field, analogous to [12, Section 2]); the result should be a nonzero constant, i.e. member of F.

(iii) *Division in* GF(p^n), $n \geqq 1$. The division algorithm of Section 3(iv) requires the operation of dividing elements of $F = \mathrm{GF}(p^n)$, (Eq. (15)). By writing

$$a/b = a \cdot b^{-1}, \qquad (a, b \in F,\ b \neq 0),$$

the problem becomes that of computing inverses in F, discussed in [12], (this is for GF(p), but extends to any GF(p^n)); one method is to use

$$b^{-1} = b^{p^n-2} \qquad (b \in F,\ b \neq 0).$$

(iv) *Program Layout.* As with the conventional fast Fourier transform, it is simplest to divide the calculation of the transform (1) (or (2)) into two separate stages, a permutation and a calculation stage, which may be performed in either order [17]. But for most applications, including those of Section 3, it is sufficient to obtain the transformed sequence in 'digit-reversed' order and the permutation stage can therefore be omitted from the transforms, as has been done in convolution algorithms for real numbers.

6. Generalisation to k Dimensions. The transform (1) which has been considered up to this point, whether in GF(p) or GF(p^n) with $n > 1$, is to be regarded as 1-dimensional in that it acts on a 1-dimensional sequence (a_i) of elements of F. A generalisation to higher dimensions is possible.

Let $F = \mathrm{GF}(p^n)$ $(n \geqq 1)$ be a finite field and $d_1, d_2, \cdots, d_k$ any series of divisors of $p^n - 1$. Let w be a generator of F^* and put

$$d_i' d_i = p^n - 1 \qquad (1 \leqq i \leqq k),$$

so that the elements $w^{d_i'}$ have order d_i in F^*. Then the transform of the k-dimensional matrix $(a_{i_1,\cdots,i_k})$ $(0 \leqq i_l \leqq d_l - 1$ for $l = 1, \cdots, k)$ may be defined to be

$$A_{i_1,\cdots,i_k} = \sum_{j_1=0}^{d_1-1} \cdots \sum_{j_k=0}^{d_k-1} a_{j_1,\cdots,j_k} w^{(d_1' i_1 j_1 + \cdots + d_k' i_k j_k)} . \tag{19}$$

The inverse transform is

$$a_{j_1,\cdots,j_k} = \prod_{i=1}^{k} (-d_i') \cdot \sum_{i_1=0}^{d_1-1} \cdots \sum_{i_k=0}^{d_k-1} A_{i_1,\cdots,i_k} w^{-(d_1' i_1 j_1 + \cdots + d_k' i_k j_k)}, \tag{20}$$

$$(0 \leqq j_l \leqq d_l - 1 \text{ for } l = 1, \cdots, k).$$

The properties expressed by Eqs. (4), (5), (6), (7) have simple generalisations and permit the extension of the algorithms of Section 3 to 'k-dimensional convolutions', or multiplication of polynomials in k variables. Thus, the equations analogous to (4) and (5A) are

$$C_{i_1,\cdots,i_k} = A_{i_1,\cdots,i_k} B_{i_1,\cdots,i_k} \qquad (0 \leqq j_l \leqq d_l - 1 \text{ for } l = 1, \cdots, k) \tag{21}$$

and

$$c_{i_1,\ldots,i_k} = \sum_{j_1=0}^{d_1-1} \cdots \sum_{j_k=0}^{d_k-1} a_{j_1,\ldots,j_k} b_{i_1-j_1,\ldots,i_k-j_k}. \tag{22}$$

7. Comparison with the Conventional Discrete Fourier Transform. The k-dimensional form of this transform may be written

$$A_{i_1,\ldots,i_k} = \sum_{j_1=0}^{d_1-1} \cdots \sum_{j_k=0}^{d_k-1} a_{j_1,\ldots,j_k} e\left(\frac{i_1 j_1}{d_1} + \cdots + \frac{i_k j_k}{d_k}\right), \tag{23}$$

where $e(x) = e^{2\pi i x}$, and the (d_l) $(1 \leqq l \leqq k)$ are now any series of positive integers. The inverse transform is

$$a_{j_1,\ldots,j_k} = \left(\prod_{l=1}^{k} d_l\right)^{-1} \cdot \sum_{i_1=0}^{d_1-1} \cdots \sum_{i_k=0}^{d_k-1} A_{i_1,\ldots,i_k} \cdot e\left(-\frac{i_1 j_1}{d_1} - \cdots - \frac{i_k j_k}{d_k}\right). \tag{24}$$

The convolution property is again expressed by Eqs. (21), (22), where these equations are now, of course, ordinary equalities between complex numbers. Suppose this transform is applied to the problem of calculating convolutions of integer sequences exactly. In general, this is most simply done by conventional 'real arithmetic', in which the real and imaginary parts of the complex quantities are stored and handled to only a finite accuracy. This accuracy must be such that when the results of the convolution process are rounded to the nearest integer the true sums are obtained. However, in special cases this difficulty does not exist. Consider first the case

$$d_1 = \cdots = d_k = 2. \tag{25}$$

The exponential factors in (23) and (24) now both become

$$(-1)^{i_1 j_1 + \cdots + i_k j_k}.$$

The corresponding factors in (19) and (20) also reduce to this expression since the unique member of $F = \mathrm{GF}(p^n)$ of order 2 in F^* is (-1) (the choice (25) implies that p is odd to satisfy $d_i = 2 \mid (p^n - 1)$).

The transforms now require only additions and subtractions for their calculation, and so may be calculated faster and more simply than the general case. This is true for (23) even when the data is not integral (see [14]). But when it is integral then (23) and (24) are obtained by integer arithmetic and provide a simple means of calculating the convolution (22) exactly; so that the advantages of the finite field transform (19) vanish in this case (see [15, p. 29–32]). It is perhaps worth pointing out that the conventional transform (23) with integral data could in principle always be calculated by integer operations; suppose for simplicity that $d_1 = \cdots = d_k = d$, and put $\theta = e(1/d) = e^{2\pi i/d}$. θ is an algebraic number of degree $\phi = \phi(d)$ ([16, Chapter IV]), whose minimal polynomial is $\Phi_d(x)$ (see Section 5(ii)). The quantities to be manipulated during the evaluation of (23) and (24) are integer polynomials in θ, which may be written in the form

$$z = \sum_{i=0}^{\phi-1} b_i \theta^i.$$

Multiplication of numbers in this form, however, consists of polynomial multiplication, followed by division by the monic polynomial $\Phi_d(\theta)$. This is a relatively complicated process and renders the algorithm impracticable for all but small values of d. (Compare with Section 5(ii), where the polynomials have degree only n; here their order may be near p^n.)

8. Calculation of (1) When d is not Highly Composite. The device given in [5] for the discrete Fourier transform can be applied also to the case $F = \mathrm{GF}(p)$ of the transform (1). Considering for simplicity the 1-dimensional case (with the notation of Section 1), we must make the condition $2d \mid p - 1$, instead of $d \mid p - 1$ as previously. Then we can find $s \in F$ such that $s^2 = r$ and write (1) as

$$A_i = \sum_{j=0}^{d-1} a_j s^{2ij} = s^{-i^2} \cdot \sum_{j=0}^{d-1} a_j s^{-j^2} \cdot s^{(i+j)^2}.$$

This sum has the form of a noncyclic convolution on the sequences

$$a_k s^{-k^2}, \quad (0 \leqq k \leqq d - 1), \qquad s^{k^2}, \quad (0 \leqq k \leqq 2d - 1).$$

By representing the elements of F by the integers $0, 1, \cdots, (p - 1)$, this convolution can be performed by the method of Section 3(ii).

Mathematics Laboratory
Plessey Telecommunications Research Ltd.
Taplow Court, Taplow, Nr. Maidenhead
Berkshire, England.

1. I. T. Adamson, *Introduction to Field Theory*, Oliver & Boyd, London; Interscience, New York, 1964. MR **33** #7325.

2. J. W. Cooley & J. W. Tukey, "An algorithm for the machine calculation of complex Fourier series," *Math. Comp.*, v. 19, 1965, pp. 297–301. MR **31** #2843.

3. W. M. Gentleman, "Matrix multiplication and fast Fourier transformations," *Bell System Tech J.*, v. 47, 1968, pp. 1099–1102.

4. G. D. Bergland, "A guided tour of the fast Fourier transform," *IEEE Spectrum*, v. 6, no. 7, 1969, pp. 41–53.

5. L. I. Bluestein, *A Linear Filtering Approach to the Computation of the Discrete Fourier Transform*, Northeast Electronics Research and Engineering Meeting Record, v. 10, 1968, pp. 218–219.

6. S. W. Golomb, *Shift Register Sequences*, Holden-Day, San Francisco, Calif., 1967. MR **39** #3906.

7. R. F. Churchouse & J. C. Herz (Editors), *Computers in Mathematical Research*, North-Holland, Amsterdam, 1968. MR **38** #1972.

8. G. H. Hardy & E. M. Wright, *An Introduction to the Theory of Numbers*, Clarendon Press, Oxford, 1938.

9. N. S. Szabo & R. I. Tanaba, *Residue Arithmetic and its Applications to Computer Technology*, McGraw-Hill, New York, 1967.

10. D. B. Gillies, "Three new Mersenne primes and a statistical theory," *Math. Comp.*, v. 18, 1964, pp. 93–97. MR **28** #2990.

11. H. Davenport, *The Higher Arithmetic. An Introduction to the Theory of Numbers*, Hutchinson, London; Longmens, Green, New York, 1952. MR **14**, 352.

12. G. E. Collins, "Computing multiplicative inverses in GF(p)," *Math. Comp.*, v. 23, 1969, pp. 197–200. MR **39** #3676.

13. E. P. Rodemich & H. Rumsey, "Primitive trinomials of high degree," *Math. Comp.*, v. 22, 1968, pp. 863–865. MR **39** #177.

14. W. K. Pratt, J. Kane & H. C. Andrews, "Hadamard transform image coding," *Proc. IEEE*, v. 57, 1969, pp. 58–68.

15. H. B. Mann (Editor), *Error Correcting Codes*, Proc. Sympos. Math. Res. Center (University of Wisconsin, Madison, Wis., 1968), Wiley, New York, 1968. MR **38** #3071.

16. H. Pollard, *The Theory of Algebraic Numbers*, Carus Math. Monographs, no. 9, Math. Assoc. Amer.; Distributed by Wiley, New York, 1950. MR **12**, 243.

17. W. T. Cochran et al., "What is the fast Fourier transform?" *Proc. IEEE*, v. 55, 1967, pp. 1664–1674.

Fast Convolution Using Fermat Number Transforms with Applications to Digital Filtering

RAMESH C. AGARWAL AND CHARLES S. BURRUS, MEMBER, IEEE

Abstract—**The structure of transforms having the convolution property is developed. A particular transform is proposed that is defined on a finite ring of integers with arithmetic carried out modulo Fermat numbers. This Fermat number transform (FNT) is ideally suited to digital computation, requiring on the order of $N \log N$ additions, subtractions and bit shifts, but no multiplications. In addition to being efficient, the Fermat number transform implementation of convolution is exact, i.e., there is no roundoff error. There is a restriction on sequence length imposed by word length but multidimensional techniques are discussed which overcome this limitation. Results of an implementation on the IBM 370/155 are presented and compared with the fast Fourier transform (FFT) showing a substantial improvement in efficiency and accuracy.**

I. INTRODUCTION

THE convolution property of certain transforms can be used to efficiently compute the cyclic convolution of two long discrete sequences. One such transform is the discrete Fourier transform (DFT) with the fast Fourier transform (FFT) implementation [1] which operates on sequences in the complex number field. To compute the DFT of a sequence of length N requires on the order of $(N/2) \log_2 (N/2)$ complex multiplications, which often take most of the computation time. In addition, the FFT implementation of cyclic convolution introduces significant amounts of roundoff error [2], thus deteriorating the signal-to-noise ratio. When working with digital machines, the data are available only with some finite precision and, therefore, without loss of generality, the data can be considered to be integers with some upper bound. In this digital domain, the complex number field of the continuous domain can be replaced with a finite field or, more generally, a finite ring of integers with addition and multiplication modulo some integer F. In this ring, using transforms similar to the DFT, the cyclic convolution can be performed very efficiently and without any roundoff error. One such transform presented here is a number theoretic transform which operates in the ring of integers modulo a Fermat number. Computation of this transform of length N, analogous to an FFT, requires on the order of $N \log_2 N$ additions, bit shifts, and subtractions, but no multiplication. This is considerably simpler than the FFT.

Knuth [3] has proposed the use of transforms in finite fields. Pollard [4] discussed transforms having the cyclic convolution property in a finite field and also gives the conditions for having transforms with the cyclic convolution property in a finite ring of integers. Good [5] also mentioned the use of transforms in a finite ring of integers. Schonhage and Strassen [6] defined transforms having the cyclic convolution property modulo a Fermat number and discussed their application to fast multiplication of very large integers. Knuth [7] elaborated on the work of Schonhage and Strassen. Nicholson [8] presented an algebraic theory of finite Fourier transforms in any ring and established fast FFT-type algorithms to compute these transforms. Rader [9], [10] proposed finite transforms in rings of integers modulo both Mersenne and Fermat numbers. He first proposed the application to digital signal processing, showed that the transforms could be calculated using only additions and bit shifting, showed the word-length constraint, and suggested two-dimensional transforms as a possible relaxation of that constraint. Agarwal and Burrus [11] defined Fermat transforms and also proposed their application for fast digital convolution.

The purpose of this paper is to develop the general structure of transforms with the convolution property and then present the number theoretic transform modulo a Fermat number as a practical scheme for use with digital computation. The properties are then discussed for the implementation of digital filters and the results of a software implementation on an IBM 370/155 are given. It is believed that the presentation here is different from previous works and gives more insight into these transforms. One result of this approach allows the maximum length of sequences that can be convolved to be increased by a factor of 2 over what had previously been reported in [10].

The choice of terminology in a new area is always difficult. In this paper the general name, number-theoretic transform, is used for any transform using number-theoretic concepts in its definition. The two particular transforms with arithmetic modulo Mersenne and Fermat numbers are called Mersenne number transforms (MNT's) and Fermat number transforms (FNT's). The particular MNT's and FNT's with a basis function $\alpha = 2$ are called Rader transforms (RT's).

These transforms are truly digital transforms, taking into account the quantization in amplitude and the finite precision of digital signals. They bear the same relation to digital signal as the DFT does to discrete-time or sampled data signals and the Fourier or Laplace transforms do to continuous-time signals. In the same manner that the relation of discrete-time signals to continuous-

Manuscript received July 16, 1973; revised November 19, 1973. This work was supported by NSF Grant GK-23697.

The authors are with the Department of Electrical Engineering, Rice University, Houston, Tex. 77001.

time signals through sampling involves a possible folding or aliasing in the frequency domain, the relation of calculations with the DFT to calculations with the number theoretic transforms involves a possible folding of the amplitude that must be taken into account.

In Section II, the structures of transforms having the cyclic and the noncyclic convolution properties are developed. It is shown that what we call the DFT structure is the only structure which supports the cyclic convolution property and any transform having the DFT structure will have the cyclic convolution property. In Section III, the necessary and sufficient conditions for the existence of transforms in a ring of integers modulo of an integer F are established. In Section IV, FNT's are defined which, computationally, appear to be the best transforms for implementing convolution. In Section V, reference is given to a two-dimensional transform implementation for long one-dimensional convolution. In Section VI, a hardware implementation to compute the FNT is suggested. Comparison of the FNT with the FFT implementation of the DFT, to implement convolution, is made in Section VII. In Section VIII, a software implementation of the FNT on an IBM 370/155 is presented and timing comparisons are made with the FFT. For cyclic convolution lengths up to 256, FNT implementations are faster by a factor of 3 to 5 over the best FFT implementations which make use of the symmetry of the DFT for real data. For computers with slower multiplication, the advantages look even better. Finally, several generalizations and applications are presented.

II. THE STRUCTURE OF TRANSFORMS HAVING THE CONVOLUTION PROPERTY

In this section the structure of transforms having the cyclic convolution property, i.e., the transform of cyclic convolution of two sequences is equal to the product of their transforms, is established. The class of transforms considered is the set of linear nonsingular (invertible) transforms which map a sequence of length N to another sequence of length N. It is shown that what we call the DFT structure is the only structure which supports the cyclic convolution property and, moreover, any transform having the DFT structure will have the cyclic convolution property.

Let x_n and h_n, $n = 0,1,\cdots,N-1$, be two sequences which are to be cyclically convolved as given by

$$y_n = x_n * h_n = \sum_{k=0}^{N-1} x_k h_{n-k} \qquad n = 0,1,\cdots N-1. \tag{1}$$

This definition assumes the sequences are periodically extended with period N or the indices are evaluated modulo N. Here we will describe sequences as vectors of length N. Since only linear invertible transforms are considered, they can be represented by an $N \times N$ nonsingular matrix T whose elements are $t_{k,m}$, $k,m = 0,1,\cdots, N-1$. Let capital letters denote the transformed sequences, i.e.,

$$X = Tx$$

$$H = Th$$

$$Y = Ty.$$

Consider the conditions on $t_{k,m}$'s so that

$$Y = X \otimes H \tag{3}$$

where $\otimes$ denotes term by term multiplication of the vectors.

Equations (2) and (1) are combined to give

$$Y_k = \sum_{n=0}^{N-1} t_{k,n} y_n = \sum_{n=0}^{N-1} t_{k,n} \sum_{m=0}^{N-1} x_m h_{n-m}$$

$$= \sum_{m=0}^{N-1} \sum_{l=0}^{N-1} x_m h_m t_{k,m+l}$$

$$X_k = \sum_{m=0}^{N-1} t_{k,m} x_m$$

$$H_k = \sum_{l=0}^{N-1} t_{k,l} h_l \qquad k = 0,1,\cdots,N-1. \tag{4}$$

The individual terms in (3) can be rewritten as

$$Y_k = X_k H_k$$

which, using (4), becomes

$$\sum_{m=0}^{N-1} \sum_{l=0}^{N-1} x_m h_l t_{k,m+l} = \sum_{m=0}^{N-1} \sum_{l=0}^{N-1} x_m h_l t_{k,m} t_{k,l}. \tag{5}$$

Matching the multiples of $x_m h_l$ on both sides of (5) gives

$$t_{k,m+l} = t_{k,m} t_{k,l} \qquad k,l,m = 0,1,\cdots,N-1. \tag{6}$$

Repeated applications of (6) gives

$$t_{k,m} = t_{k,1}{}^m \qquad k,m = 0,1,\cdots,N-1. \tag{7}$$

And, since the convolution is cyclic, the indices in (6) are added modulo N. This gives

$$t_{k,m}{}^N = 1 \qquad k,m = 0,1,\cdots,N-1. \tag{8}$$

Therefore, the $t_{k,m}$'s are Nth roots of unity. For T to be nonsingular, all the $t_{k,1}$'s should be distinct and, since there are only N distinct Nth roots of unity, the $t_{k,1}$'s must be these N distinct roots. Without loss of generality, $t_{1,1}$ can be taken as a root of order N, i.e., N is the least positive integer such that $t_{1,1}{}^N = 1$. Therefore, all $t_{k,1}$'s can be written as some powers of $t_{1,1}$. Again without loss of generality, the rows of T can be arranged so that

$$t_{k,1} = t_{1,1}{}^k. \tag{9}$$

Combining (7) and (9) and denoting $t_{1,1}$ by α gives the following

$$t_{k,m} = \alpha^{km} \qquad k,m = 0,1,\cdots,N-1. \tag{10}$$

The structure thus established causes the transform to be orthogonal. The elements $\bar{t}_{k,m}$ of T^{-1} are given by

$$\bar{t}_{k,m} = N^{-1}\alpha^{-km} \qquad k,m = 0,1,\cdots,N-1. \tag{11}$$

To prove this, consider

$$TT^{-1} = I$$

or

$$N^{-1} \sum_{n=0}^{N-1} \alpha^{kn} \alpha^{-ln} = \delta_{k,l} \quad (12)$$

where $\delta_{k,l}$ is the Kronecker delta function. Taking $k - l = p$, (12) can be rewritten as

$$N^{-1} \sum_{n=0}^{N-1} \alpha^{pn} = \begin{cases} 1 & \text{if } p = 0 (\text{mod } N) \\ 0 & \text{otherwise.} \end{cases} \quad (13)$$

To prove the orthogonality, we have to prove (13). If $p = 0(\text{mod } N)$, then $\alpha^p = 1$ and the first part is thus established. If $p \neq 0(\text{mod } N)$, $\alpha^p \neq 1$ or $(\alpha^p - 1) \neq 0$. Multiplying (13) by $(\alpha^p - 1)$, we obtain

$$N^{-1}(\alpha^p - 1) \sum_{n=0}^{N-1} \alpha^{pn} = N^{-1}(\alpha^{pN} - 1) = 0.$$

Thus (13) is established.

An examination of the preceding development reveals that the existence of an $N \times N$ transform having the cyclic convolution property depends only on the existence of an α that is a root of unity of order N, and the existence of N^{-1}. The structures of the transform matrix T and its inverse are given by (10) and (11) respectively, and these are the only structures which support the cyclic convolution property. This structure is called the DFT structure. Because of the DFT structure, any transform having the cyclic convolution property also has a fast computational algorithm similar to the FFT if N is highly composite [8]. In the complex number field the DFT, with $\alpha = e^{-j2\pi/N}$, is the only transform having the cyclic convolution property. But, in a finite field or, more generally, a ring we might also have transforms with the cyclic convolution property if there exists a root of unity of order N and if N^{-1} exists. In the next section we establish results for the special case of a finite ring of integers modulo of an integer, which seem to be most appropriate for practical digital computation. Appendix A lists some of the properties of these transforms.

Thus far we have discussed only the transforms having the cyclic convolution property. Noncyclic convolution can be implemented using cyclic convolution by padding the sequences with zeros [1]. If only noncyclic convolution is desired, a more general class of transforms is possible. Consider two sequences of lengths L and M, respectively, that are noncyclically convolved giving an output sequence of length $N = L + M - 1$. First we pad the sequences with zeros to make them of length N each. For the $N \times N$ transform T, the desired restrictions on $t_{k,m}$'s are still given by (6), but now the indices are not added modulo N; this modifies (6) to be

$$t_{k,m+l} = t_{k,m} t_{k,l} \qquad k = 0,1,\cdots,N-1 \quad m = 0,1,\cdots,M-1 \quad l = 0,1,\cdots,L-1. \quad (14)$$

Repeated applications of (14) gives

$$t_{k,m} = k_{k,1}{}^m \qquad k,m = 0,1,\cdots,N-1. \quad (15)$$

This is the same as (7), but now we do not have (8) which implies cyclic convolution. Because of the absence of (8), $t_{k,1}$'s are not restricted to Nth roots of unity. The only restriction is the nonsingularity constraint which requires all $t_{k,1}$'s to be distinct. A similar procedure is discussed by Knuth [3] in connection with the multiplication of large integers which is almost analogous to noncyclic convolution. When the $t_{k,1}$'s are not restricted to Nth roots of unity, in general there is no FFT-type algorithm to compute the transform or the inverse transform. As a matter of fact, the inverse transform does not have a simple structure. Because of these limitations we restrict ourselves to transforms having the cyclic convolution property.

III. TRANSFORMS IN THE RING OF INTEGERS MODULO AN INTEGER

In any practical situation, or when working with digital machines, the data are available only with some finite precision, and therefore, without loss of generality, the data can be considered to be integers with some upper bound. To compute convolution in this digital domain, operations in the complex number field of the continuous domain can be imitated with a finite field or, more generally, a finite ring of integers under additions and multiplications modulo some integer F; and an integer α of order N replaces $e^{-j2\pi/N}$ used in a DFT. In this ring, when two integer sequences $x(n)$ and $h(n)$ are convolved, the output integer sequence $y(n)$ is congruent to the convolution of $x(n)$ and $h(n)$ modulo F. In the ring of integers modulo F, conventional integers can be unambiguously represented if their absolute value is less than $F/2$. If the input integer sequences $x(n)$ and $h(n)$ are so scaled that $|y(n)|$ never exceeds $F/2$, we would get the same results by implementing convolution in the ring of integers modulo F as that obtained with normal arithmetic. This is similar to the overflow constraint in fixed-point digital machines.

In most digital filtering applications, $h(n)$ represents the impulse response and is known a priori; also the maximum magnitude of the input signal is usually known. In this situation, we can bound the peak output magnitude by,

$$|y(n)| \leq |x(n)|_{\max} \sum_{n=0}^{N-1} |h(n)|.$$

When we relate continuous-time signals to discrete-time signals through sampling, we obtain an aliasing effect in the frequency domain; all the frequencies are obtained modulo the sampling frequency. A similar effect occurs when the amplitude is discretized and convolution is performed in the integer domain modulo an integer F. We obtain aliasing in amplitude which can be avoided as noted before.

In order to have a computational advantage in implementing cyclic convolution, we will derive a transform having the cyclic convolution property in this ring. In addition, we would like this transform to have a fast computational algorithm. First, we will present results about the existence of such transforms which were shown to depend on the existence of an α of order N and existence N^{-1} in the ring. The number theory necessary to understand the derivations is very basic and can be found in any elementary book on number theory [14].

Let

$$F = p_1^{r_1} p_2^{r_2} \cdots p_l^{r_l} \tag{16}$$

be the prime factorization of F. Since α is of order N, we have

$$\alpha^N = 1 (\text{mod } F) \tag{17}$$

which implies

$$\alpha^N = 1 (\text{mod } p_i^{r_i}) \qquad i = 1,2,\cdots,l. \tag{18}$$

If the transform matrix T is to be nonsingular, no two rows of T can be the same, which implies

$$\alpha^p \not\equiv \alpha^p (\text{mod } p_i^{r_i}) \qquad i = 1,\cdots,l, p \neq q$$

$$0 \leq p,q \leq N - 1. \tag{19}$$

Equations (18) and (19) imply that α should be of order N with respect to each prime power factor $p_i^{r_i}$ of F, i.e., N is the least positive integer such that

$$\alpha^N = 1 (\text{mod } p_i^{r_i}) \qquad i = 1,2,\cdots,l. \tag{20}$$

If α is of order less than N with respect to any moduli $p_i^{r_i}$, then T would be singular with respect to that modulus. Also, for N^{-1} to exist, N should be relatively prime to F, i.e., p_i should not be a factor of N. These two considerations give Theorem 1, which is proven in Appendix B.

Theorem 1

An $N \times N$ transform T having the cyclic convolution property in the ring of integers modulo an integer F exists if and only if N divides $0(F) \triangleq gcd(p_1 - 1, p_2 - 1, \cdots, p_l - 1)$.

IV. FERMAT NUMBER TRANSFORMS

For number theoretic transforms to be more attractive in comparison to the FFT for implementing convolution, they should be computationally efficient. There are three requirements that will be considered. First, N should be highly composite (preferably a power of 2) for a fast FFT type algorithm to exist. Second, since complex multiplications take most of the computational effort in calculating the FFT, it is important that the multiplication by powers of α be a simple operation. This is possible if the powers of α have very few bit binary representations; preferably also a power of two, where multiplication by a power of α reduces to a word shift. Third, in order to facilitate arithmetic modulo F, F should also have a very few bit binary representation.

Now we shall make a systematic investigation of good choices for F, for which the maximum transform length N is not too small. As noted above, we would like F to have a very few bit binary representation. The first possibility is 2^k; it has a prime factor 2 and therefore according to Theorem 1 of Section III, the maximum possible transform length is 1. For $2^k - 1$, let k be a composite PQ, where P is prime. Then $2^P - 1$ divides $2^{PQ} - 1$ and the maximum possible length of the transform will be governed by the length possible for $2^P - 1$. Therefore, only the primes P need to be considered interesting. Numbers of this form are known as Mersenne numbers and Rader [10] has discussed convolution using Mersenne numbers in detail. For MNT's [10], it can be shown that transforms of length at least $2P$ exist and the corresponding $\alpha = -2$. Mersenne number transforms are not of as much interest because $2P$ is not highly composite and, therefore, we do not have fast FFT-type computational algorithms to compute the transforms. For $2^k + 1$, say k is odd. Then 3 divides $2^k + 1$ and the largest possible transform length is 2. Thus k is even. Let k be $s2^t$, where s is an odd integer. Then $2^{2^t} + 1$ divides $2^{s2^t} + 1$ and the length of the possible transform will be governed by the length possible for $2^{2^t} + 1$. Therefore, integers of the form $2^{2^t} + 1$ are of interest. These numbers are known as Fermat numbers and will be discussed in detail in this paper. Fermat numbers seem to be optimum in the sense of having transforms whose length is interesting while the word size is moderate. Numbers of the form $2^{s2^t} + 1$ are also of limited interest and are discussed in Section IX. A systematic investigation of those F which require more than two bit representation is difficult. Our preliminary investigation in that direction does not seem to be very encouraging.

In the light of the above discussion, it is proposed to use $F_t \triangleq 2^b + 1, b = 2^t$, t being a positive integer, as good choices for F. F_t is known as the tth Fermat number. Arithmetic modulo F_t can be performed using b-bit arithmetic, as will be discussed in Section VI. The number of bits used for the signal and the filter coefficients decide the value of b to be used so as not to cause error in the output due to overflow. The value of b required is slightly more than double the number of bits used for the signal representation. One simple bound on the output was presented in the last section. Computationally, Fermat numbers seem to be the best choices for F, therefore, as far as is practicable, they should be used. If the value of b obtained from the overflow consideration is not a power os 2, it should be increased to the next power of 2 in order that Fermat numbers can be used. Some other possibilities are discussed in Section IX. Since the arithmetic is done modulo a Fermat number, we call the associated transforms FNT's. Below we define FNT's and their inverse transforms for sequences of length $N = 2^m$:

$$X(k) = \sum_{n=0}^{N-1} x(n) \alpha^{nk} (\text{mod } F_t) \qquad k = 0,1,\cdots,N-1 \tag{21}$$

$$x(n) = (2^{-m}) \sum_{k=0}^{N-1} X(k) \alpha^{-nk} (\text{mod } F_t)$$

$$F_t = 2^b + 1, \qquad b = 2^t \tag{22}$$

α is an integer of order N, i.e., N is the least positive integer such that

$$\alpha^N = 1 (\text{mod } F_t).$$

Note that $\alpha^{-1} = \alpha^{N-1}$ and $2^{-m} = -2^{b-m} (\text{mod } F_t)$.

As discussed in Sections II and III, we can perform cyclic convolution of two integer sequences of length N, using the FNT, if certain conditions are satisfied. Now we discuss the possible values of N for which an FNT exists, given a choice of F_t. As implied by Theorem 1 of Section III, for transforms (mod F_t) of length N to exist, N should divide $0(F_t)$.

Fermat numbers up to F_4 are all primes, therefore for these cases $0(F_t) = 2^b$, and we can have an FNT for any length $N = 2^m$, $m \le b$. For these Fermat primes the integer 3 is an α of order $N = 2^b$, giving the largest possible transform length. Of course there are 2^{b-1} other integers also which are of order 2^b. They can be obtained by taking odd powers of 3. If an integer α is of order N, then $\alpha^p (\text{mod } F_t)$ will be of order N/p, if N/p is an integer. Therefore, using Fermat primes, an integer α of order 2^m $m \le b$, is given by $3^{2^{b-m}} (\text{mod } F_t)$. The integer 2 is of order $2b = 2^{t+1}$. If α is taken as 2 or a power of 2, all the powers of α would be some powers of 2, and, for these cases, as discussed in the beginning of this section, the FNT can be computed very efficiently and is called the RT.

There are no other known Fermat primes. For digital filtering applications, $F_5 (b = 32)$ and $F_6 (b = 64)$ seem to be most practical. Lucas [13] has proven that every prime factor of composite F_t is of the form $K \cdot 2^{t+2} + 1$. Therefore, 2^{t+2} divides $0(F_t)$, for $t > 4$. In particular it can be verified that for F_5 and F_6, $0(F_t) = 2^{t+2}$. Therefore, for these choices of Fermat numbers, the maximum possible transform length is $2^{t+2} = 4b$. Also, we assert that α_p given by (23) is of order 2^{t+2} mod F_t, $t \ge 2$.

$$\sqrt{2} \triangleq \alpha_p = 2^{2^{t-2}} (2^{2^{t-1}} - 1). \tag{23}$$

We denote this α_p as $\sqrt{2}$ because

$$\alpha_p^2 = 2 \text{ mod } F_t.$$

The proof that α_p given by (23) is of order 2^{t+2} with respect to any factor of F_t is given in Appendix C. Any odd power of $\sqrt{2}$ will also be of order 2^{t+2}. By raising $\sqrt{2}$ to 2^{t+2-m}th power, we obtain an integer α of order 2^m, $m \le t + 2$.

Table I gives values of N for the two most important values of α and also gives the maximum possible N for the most practical values of b. The Fermat number transform has also been defined by Rader [10], using $\alpha = 2$ but his development does not indicate the possibility of using $\sqrt{2}$. Using the results of Section III, we have found the maximum possible length N for which an FNT exists, for the most practical values of b, and have found the corresponding integer α. Using $\alpha = \sqrt{2}$, the maximum possible lengths for these transforms have increased by a factor of 2, which makes them more useful. In the next section we discuss a two-dimensional convolution scheme whereby very long one-dimensional sequences can also be convolved using a two-dimensional FNT.

TABLE I
PARAMETERS FOR SEVERAL POSSIBLE IMPLICATIONS FOR THE FNT'S

t	b	F_t	N[a] $\alpha = 2$	N $\alpha = \sqrt{2}$	N_{max}	α for N_{max}
3	8	$2^8 + 1$	16	32	256	3
4	16	$2^{16} + 1$	32	64	65536	3
5	32	$2^{32} + 1$	64	128	128	$\sqrt{2}$
6	64	$2^{64} + 1$	128	256	256	$\sqrt{2}$

[a] This case corresponds to the RT.

The basis functions for the FNT are integer exponentials mod F_t, in comparison to the complex exponentials for the DFT. These integer exponentials fold over after they cross $F_t/2$. Because of the nature of modulo arithmetic, the FNT coefficients do not seem to have any physical meaning. Although the signal for which the FNT is being taken may be very small, its FNT coefficients may lie anywhere between 0 and $F_t - 1$. As a matter of fact, during various stages of the computation each accumulation of the signal "overflows" many times. But still the end result of the convolution would be exact if the input signals are properly bounded. FNT's follow all the properties mentioned in Appendix A.

Example

To make the ideas of this section more clear, we now present an example. This example will illustrate several points, treatment of negative values in the data, the structure of the transform and the inverse transform matrix, negative powers of α, frequent "overflow" during computation, meaningless of the transform values, and exactness of the final answer. This example will not demonstrate the efficient implementation of the RT using the binary arithmatic. That will be illustrated in Section VI on implementation of the RT.

Consider two sequences $x = (2,-2,1,0)$ and $h = (1,2,0,0)$, whose convolution is desired. From the overflow consideration, it is sufficient if we work modulo $F_2 = 17$. We want $N = 4$, for F_2 the integer 2 is of order 8, therefore $2^2 = 4$ is an α of order 4. The transformation matrix T is given by

$$T = \begin{bmatrix} 1 & 1 & 1 & 1 \\ 1 & 4 & 4^2 & 4^3 \\ 1 & 4^2 & 4^4 & 4^6 \\ 1 & 4^3 & 4^6 & 4^9 \end{bmatrix} \equiv \begin{bmatrix} 1 & 1 & 1 & 1 \\ 1 & 4 & -1 & -4 \\ 1 & -1 & 1 & -1 \\ 1 & -4 & -1 & 4 \end{bmatrix}$$

$$= \begin{bmatrix} 1 & 1 & 1 & 1 \\ 1 & 4 & 16 & 13 \\ 1 & 16 & 1 & 16 \\ 1 & 13 & 16 & 4 \end{bmatrix} (\text{mod } 17).$$

Since $4^{-1} = -4 \pmod{17}$, the inverse transformation matrix T^{-1} is given by

$$T^{-1} = 4^{-1}\begin{bmatrix} 1 & 1 & 1 & 1 \\ 1 & 4^{-1} & 4^{-2} & 4^{-3} \\ 1 & 4^{-2} & 4^{-4} & 4^{-6} \\ 1 & 4^{-3} & 4^{-6} & 4^{-9} \end{bmatrix} = -4\begin{bmatrix} 1 & 1 & 1 & 1 \\ 1 & -4 & -1 & 4 \\ 1 & -1 & 1 & -1 \\ 1 & 4 & -1 & -4 \end{bmatrix}$$

$$= 13\begin{bmatrix} 1 & 1 & 1 & 1 \\ 1 & 13 & 16 & 4 \\ 1 & 16 & 1 & 16 \\ 1 & 4 & 16 & 13 \end{bmatrix} \pmod{17}.$$

The transforms of x and h are given by

$$\boldsymbol{X} = T\boldsymbol{x} = \begin{bmatrix} 1 & 1 & 1 & 1 \\ 1 & 4 & 16 & 13 \\ 1 & 16 & 1 & 16 \\ 1 & 13 & 16 & 4 \end{bmatrix}\begin{bmatrix} 2 \\ 15 \\ 1 \\ 0 \end{bmatrix}$$

$$= \begin{bmatrix} 18 \\ 78 \\ 243 \\ 213 \end{bmatrix} = \begin{bmatrix} 1 \\ 10 \\ 5 \\ 9 \end{bmatrix} \pmod{17}.$$

Note that in $\boldsymbol{x}$, -2 was represented by $-2 + 17 = 15$. Similarly,

$$H = (3,9,16,10) \quad \text{and} \quad Y = X \cdot H = (3,90,80,90) = (3,5,12,5) \pmod{17}.$$

Taking the inverse transform of Y,

$$y = (2,2,14,2) \pmod{17}.$$

According to our assumption, integers are supposed to lie between -8 and 8. Therefore, 14 must be represented as $14 - 17 = -3$. This gives $y = (2,2,-3,2)$, which is the correct answer. Also, note that y is a symmetric sequence, therefore Y is also a symmetric sequence. Other than this, the transform values have no significance.

V. TWO-DIMENSIONAL CONVOLUTION FOR CONVOLVING LONG SEQUENCES

Arithmetic modulo F_t can be implemented using a $b = 2^t$ bit representation of integers, as will be discussed later in Section VI, with some provision for representing 2^b. The maximum length of sequences which can be cyclically convolved using the FNT is $4b$, using $\alpha = \sqrt{2}$, as discussed above. Therefore, the length of sequences which can be convolved is proportional to the word length in bits. Thus, for long sequences, word length requirement may be excessive. Rader [10] suggested using a two-dimensional convolution scheme to convolve long one-dimensional sequences. Agarwal and Burrus [11], [12] presented such a two-dimensional convolution scheme. Using this scheme, cyclic convolution of length $N = LM$ is implemented as a two-dimensional cyclic convolution of length $(2L - 1)$ by M (or $2L$ by M). This two-dimensional cyclic convolution can be implemented using a two-dimensional FNT [11], [12] defined similar to the one-dimensional transform. Using this two-dimensional scheme, the word length required is proportional to the square root of the length of the sequences to be convolved which would give for a maximum sequence length, $8b^2$, rather than $4b$. If M is taken as the maximum possible length $4b$, and $2L - 1$ is a small integer, then either direct convolution or another high-speed algorithm could be employed [12], [15] to compute convolution along the short dimension and the one-dimensional FNT could be used along the long dimension. Computationally this combination can be very efficient as will be shown in the implementation in Section VIII. Table II lists the maximum lengths for two-dimensional transforms.

TABLE II
MAXIMUM ONE-DIMENSIONAL CYCLIC CONVOLUTION LENGTHS USING TWO-DIMENSIONAL FNT OR RT

Word length b	N for $\alpha = 2$	N for $\alpha = \sqrt{2}$
16	512	2048
32	2048	8192
64	8192	32768

VI. A SUGGESTED HARDWARE IMPLEMENTATION

Since the structure of the FNT is similar to that of the DFT, and N is a power of 2, a fast implementation of the FNT's similar to the FFT with radix 2 exists. As a matter of fact, all we have to do is replace $W = e^{-j2\pi/N}$ by α in any FFT algorithm [1].

In computing the FNT, arithmetic is done modulo $2^b + 1$. In this arithmetic the only allowed integers are $0,1,\cdots,2^b$ and all integers whose absolute values do not exceed 2^{b-1} can be represented unambiguously. Negative integers are represented by adding $2^b + 1$ to them; this is similar to twos complement and ones complement representation of negative integers. Using a b-bit register, all integers from 0 to $2^b - 1$ can be represented. The problem remains to represent 2^b. If the data are uncorrelated, the probability that this number will appear after an arithmetical operation is approximately 2^{-b}. For digital filtering applications, b would typically be 32 or 64; in these cases the probability of occurrence of 2^b is extremely small. If occasional error in convolution is permissible, for these cases, probably there is no need of any extra hardware to represent 2^b. If the need exists, an extra bit could be used to represent 2^b at the expense of a more complicated hardware. The following discussion is based on the b-bit representation of the integers.

Now, we discuss how various basic arithmetical operations can be performed modulo F_t.

A. Negation

Let

$$A = \sum_{i=0}^{b-1} a_i 2^i, a_i = 0 \text{ or } 1.$$

Then

$$\begin{aligned} -A &= -\sum_{i=0}^{b-1} a_i 2^i \\ &= \sum_{i=0}^{b-1} \bar{a}_i 2^i - (2^b - 1) \\ &= \sum_{i=0}^{b-1} \bar{a}_i 2^i - (2^b - 1) + 2^b + 1 \qquad \text{mod } F_t \\ &= \sum_{i=0}^{b-1} \bar{a}_i 2^i + 2 \qquad \text{mod } F_t. \end{aligned}$$

Thus to negate a number, we have to complement each bit and add 2 to the result. For example,

$$(\text{mod } 17): 4 = 0100; \; -4 = \left\{\begin{array}{r} 1011 \\ +\ \ 10 \\ \hline 1101 \end{array}\right\} = 1101 = 13;$$

$$13 + 4 = 17.$$

B. Addition

When we add two b-bit integers, we obtain a b-bit integer and possibly a carry bit. The carry bit represents $2^b = -1 \text{ mod } F_t$. To implement arithmetic modulo $2^b + 1$, we simply subtract the carry bit. Thus the hardware should be of the carry subtract type. For example,

$$(\text{mod } 17): 10 + 9 = 17 = 2(\text{mod } 17) \quad \begin{array}{r} 1010 \\ +1001 \\ \hline 10011 \\ \searrow 1 \\ \hline 0010 = 2 \end{array}$$

C. Subtraction

Subtraction is implemented as an addition by first negating the subtrahend and then adding them. Addition must be done according to B.

D. General Multiplication

When we multiply two b-bit integers, we get a $2b$-bit product. Let C_L be the b-bit low-order part of it and C_H be the b-bit high-order part of it, then

$$A \times B = C_L + C_H 2^b = C_L - C_H (\text{mod } F_t).$$

Thus, all we have to do is subtract the higher order register from the lower order register. The subtraction needs to be done according to C. For example,

$$(\text{mod } 17): 13 \times 9 = 117 = 15(\text{mod } 17); \; 117 = \frac{0111}{C_H}\,\frac{0101}{C_L}$$

$$\begin{array}{rl} C_L = & 0101 \\ (-C_H) = & 1010 \\ \hline & 1111 = 15. \end{array}$$

E. Multiplication by a Power of 2

If α is taken as 2 or a power of 2, RT's are obtained and the only multiplications involved in computing them are those by some powers of 2. These multiplications are particularly simple to implement in arithmetic modulo F_t. Suppose we need to multiply the contents of a register by 2^k, $0 < k < b$, all we need to do is left-shift the contents of the register by k bits and subtract the k overflow bits. A convenient way to do this is to append the data register on the left with a register of zeros, left-shift the double register by k positions, dropping the leading zeros, and then subtract the higher order register from the lower order register, as in D. If k is outside the range $0 < k < b$, we make use of the fact that $2^b = -1 \text{ mod } F_t$. Computation of the inverse transform required multiplications by negative powers of 2 which can be converted to positive powers by the following relationship, $2^{-k} = -2^{b-k} \text{ mod } F_t$. An alternate method to multiply by 2^{-k} is to load the data in the higher order register of a double register, filling the lower order register with zeros, then right-shift the double register by k positions and subtract the low-order register from the high-order register mod F_t. For fast implementation of the bit shift operation, shift amount k should be expressed in a binary form and at a clock pulse shifting should be done by a power of 2. For example,

$$(\text{mod } 17): 11 \times 2^3 = 88 = 3 \text{ mod } 17$$

$$11 = 0000\ 1011.$$

$$\text{Shift left 3 positions} \quad \frac{0101}{C_H}\,\frac{1000}{C_L}$$

$$\begin{array}{rr} C_L = & 1000 \\ (-C_H) = & +1100 \\ \hline & 1\ 0100 \\ & \searrow 1 \\ \hline & 0011 = 3 \end{array}$$

$$11 \times 2^{-3} = 11 \times (-2^{4-3}) = -22 = 12 \text{ mod } 17$$

$$11 = 1011\ 0000.$$

$$\text{Shift right 3 positions } \frac{0001}{C_H}\ \frac{0110}{C_L}$$

$$\begin{aligned} C_H &= \quad 0001 \\ (-C_L) &= +1011 \\ \hline &\quad\ 1100 = 12. \end{aligned}$$

For implementation of the fast RT, unlike the FFT, we do not need to store the powers of α. For serial arithmetic, we could have a register which stores the shift amount k, and as we go along the fast RT flow chart, we continually update the shift amount. All this is particularly simple to do in binary arithmetic. If α is taken as $\sqrt{2}$, only the odd powers of α require a two-bit representation. In the fast FNT algorithm, only one stage requires multiplications by odd powers of α. Multiplication by an odd power of $\sqrt{2}$ would probably be best done by first multiplying by the just lower even power of $\sqrt{2}$, which is a power of 2, then multiplying by $\sqrt{2}$, which is a two-bit number. The resulting increase in computation effort is very small.

VII. COMPARISON WITH THE FFT

As noted in the previous section, computing the RT is a very simple operation on a binary machine. Now let us compare the complexity of various basic operations involved in computing the RT vis-a-vis the FFT. If the two sequences $x(n)$ and $h(n)$ have b_1 and b_2 bit representations, respectively, and are of length N, then the output $y(n)$ would need no more than a $(b_1 + b_2 + \log_2 N)$ bit representation. To obtain the correct result $b \geq b_1 + b_2 + \log_2 N$. In Section III we have given a better bound on the output. Roughly speaking, we need twice the number of bits to carry out the convolution using the RT as compared to the fixed-point FFT implementation of the convolution. But in the DFT, every data point is treated as a complex number and therefore requires two words, one for the real part and one for the imaginary part. Thus, in effect, the hardware requirement for two transforms is about the same. Although for real data it is possible to make use of the symmetry properties of the DFT's they require extra computation and for the purpose of comparison it will be ignored, even though we have taken this into account for our IBM 370/155 implementation to be discussed in the next section. Therefore we shall assume, in the FFT implementation, each data point is represented by a $b/2$ bit real part and a $b/2$ bit imaginary part.

One $b/2$ bit complex addition is equivalent to two $b/2$ bit real additions, which are comparable to a b-bit addition modulo F_t. Thus, the complexity of addition/subtraction is the same in both the transforms. Similarly, it can be shown that a $b/2$ bit complex multiplication is comparable to a b-bit multiplication modulo F_t. Computation of the RT requires multiplications by powers of 2, which implemented as bit shifts and subtractions become much simpler operations compared to complex multiplications required in the FFT implementation.

To compute a length N fast RT, $N \log_2 N$ additions/subtractions, and $(N/2) \log_2 N/2$ multiplications by some powers of 2 are required which are implemented as bit shifts and subtractions. To compute the convolution using the FFT, most of the time is taken in computing the complex multiplications required to compute the transforms. A comparison with the RT reveals that these complex multiplications are replaced by bit shifts and subtractions which are much faster operations. This results in considerable computational savings in the implementation of convolution. This fact has been verified on a general purpose machine (IBM 370/155). The computation required to multiply the two transforms is about the same for both the implementations. To convolve long sequences using the two-dimensional FNT or RT, the computational effort increases by, at the most, a factor of 2. Still, the RT implementation of convolution is much faster as compared to the FFT implementation.

RT's have some additional advantages over the FFT. First, the FFT implementation requires storing all the powers of W requiring a significant amount of storage which may be an important factor for a small minicomputer or a special purpose hardware implementation. Second, fixed-point FFT implementation introduces a significant amount of the roundoff noise at the output, 6–8 bits depending on the data [2]. This degrades the signal-to-noise ratio during the filtering operations. The FNT or RT implementation is error free, the only source of error is input A/D quantization.

VIII. IMPLEMENTATION ON THE IBM 370/155

The word length of the IBM 360–370 series is 32 bits and, therefore, is well suited for the implementation of convolution using the FNT or RT modulo $F_5 = 2^{32} + 1$. It has two's complement representation of negative integers, i.e., $-x$ is represented as $2^{32} - x$. We want negative integers to be represented as the complement modulo $2^{32} + 1$, i.e., $-x$ is to be represented by $2^{32} + 1 - x$, and to accomplish this 1 is added whenever a negative integer is encountered in the data. As noted before, on this machine, we cannot represent -1. If a -1 is encountered in the data, it is rounded either to 0 or to -2. This is equivalent to introducing some initial quantization noise. If the data are uncorrelated, the probability that -1 will appear after an arithmetical operation during computation of the transform is roughly $2^{-32} \simeq 10^{-10}$ per operation. This error introduced during computation is rather serious and probably the corresponding output block will be meaningless. Assuming $N = 64$, the probability that a particular block is erroneous is roughly $10^{-6} \sim 10^{-7}$. For most filtering operations, this may be permissible.

Logical add (ALR) and subtract instructions (SLR) are used to add and subtract two integers modulo F_5. If a carry is detected after add, 1 is subtracted from the result, and if no carry is detected after subtract, 1 is added to the result. Multiplication by 2^k is done using the logical left shift operation (SLDL) and multiplication by 2^{-k} is

TABLE III
CYCLIC CONVOLUTION TIMINGS FOR LENGTH N REAL SEQUENCES

N	FFT ms	FNT or RT ms
32	16	3.3
64	31	7.4
128	60	16.6[a]
256	123	40.0[b]
256	123	80.0[c]
512	245	166.0[c]
1024	530	340.0[c]
2048	1260	720.0[c]

[a] Using $\alpha = \sqrt{2}$.
[b] Using 2 by 128 convolution.
[c] Using two-dimensional RT.

done using logical right shift operation (SRDL), $0 < k < 32$, as explained in Section VI. Similarly, multiplication of two integers modulo F_5 can also be done.

Two assembler subprograms were written to compute the fast RT and the inverse for any sequence length from 2^1 to 2^6, taking α as a power of 2. Two more subprograms were written to compute the fast FNT and its inverse for length 128 sequences, using $\alpha = \sqrt{2}$, given by (23). Out of the 7 stages of the fast algorithm, only one stage requires multiplication by odd powers of $\sqrt{2}$, which are implemented as suggested in Section VI.

To cyclically convolve sequences longer than 128, two-dimensional convolution schemes were used [12]. One such scheme was 2 by 128 convolution for length 256 sequences discussed in Section X of [12]. Another program was written to convolve long one-dimensional sequences using two-dimensional RT's, as discussed in Section III of [12]. Timings for various implementations of convolution and their comparison with the FFT implementations are shown in Table III. To compute these timings it is assumed that transforms of the h sequences have been precalculated. Thus, timings are for computing the x transforms, multiplications of the transforms, and their inverse transforms. For the FFT implementation, a very efficient algorithm [17] was used, which employed both radix 4 and radix 2 steps and took into account the fact that the data are real. For sequences up to length 256, improvements of the order of 3 to 5 were obtained over the FFT implementations. Since the hardware of IBM 360/370 series is not designed to do arithmetic modulo Fermat numbers, extra branch instructions were necessary after add or subtract operations to take into account the carry bit. If the hardware is designed to do arithmetic modulo Fermat numbers, these instructions could be avoided, resulting in a significant reduction in the computation time. Copies of the assembler subprograms are available from the authors.

IX. GENERALIZATIONS AND OTHER APPLICATIONS

In this paper we have discussed convolution in the rings of integers modulo of Fermat numbers. These numbers seem to be the best choice for implementation on digital computers. Nevertheless, any integer F, for which $0(F) > 1$ can be used, was discussed in Section IV. Rader [10] proposed the use of Mersenne numbers M_p ($2^p - 1$, p a prime). MNT's have $\alpha = -2$ and $N = 2p$, and therefore do not have an FFT-type fast computational algorithm.

In many situations the quantization and overflow considerations may require arithmetic to be done using 10 to 20-bit arithmetic. For this situation, working in the integer ring modulo $F_4(2^{16}+1)$ would be insufficient and at the same time, computing the FNT modulo $F_5(2^{32}+1)$ would require excessive number of bits. In this situation, we may perform convolution modulo $2^{24}+1$. Also, many computers have 24-bit word length. Transforms similar to FNT's exist for this situation. We could take $F = 2^{24}+1$ and then $\alpha = 8$ gives $N = 16$, and $\alpha = 8^{1/2} = 2^7(2^{12}-1)$ gives $N = 32$ as the maximum length of the one-dimensional transform.

In general one could take $F = 2^{s2^t}+1$, s is an odd integer. In that case, $\alpha = 2^s$ would give $N = 2^{t+1}$, and $\alpha = (2^s)^{1/2} = 2^{[(s-1)/2+s2^t-2]}(2^{s2^{t-1}}-1)$ would give $N = 2^{t+2}$. In many situations it may be possible to have transforms of greater length than 2^{t+2}. For example, taking $F = 2^{40}+1$, the maximum possible transform length is 256, and, taking $F = 2^{80}+1$, the maximum possible transform length is 1024, but the corresponding α's may not be simple.

As discussed earlier, the FNT implementation requires roughly twice the number of bits for the arithmetic. Most minicomputers have short word lengths and in that situation it may not be possible to efficiently use FNT's, but there is a solution to this problem. We can split the data bits into two halves and then convolve them.

$$x(n) = x_2(n) + x_1(n)2^k, \qquad |x_2(n)| < 2^k$$

$$h(n) = h_2(n) + h_1(n)2^k, \qquad |h_2(n)| < 2^k \tag{24}$$

$$y = x*h = x_1*h_1\cdot 2^{2k} + (x_1*h_2 + x_2*h_1)2^k + x_2*h_2. \tag{25}$$

Now, since x_1, h_1, x_2, and h_2 have roughly half the number of bits, it should be possible to convolve them using approximately half the number of bits. If necessary, a more precise analysis of the above situation could be easily performed. In (25), the last term, in comparison to the first term, is very small and can be neglected. We need to take two transforms for x and two transforms for h, the summation shown within the parentheses can be performed in the transform domain. Finally, we need to take two inverse transforms, one for x_1*h_1 and the other for $(x_1*h_2 + x_2*h_1)$.

FNT's can also be used for convolutions required in block recursive realizations of recursive digital filters [16]. For these realizations, convolutions form the main computational task. Since convolution with the FNT's can be performed with both efficiency and with perfect accuracy, it should be very attractive for this application. For very high-order recursive filters this may reduce roundoff noise problems associated with these filters. Other applications of number theoretic transforms are discussed by Pollard [4] and Knuth [3], [7]. They include fast multiplication of large integers and multiplication of polynomials.

APPENDIX A

PROPERTIES OF THE TRANSFORMS HAVING THE CYCLIC CONVOLUTION PROPERTY

Below we summarize the definition and some important properties of the transforms having the cyclic convolution property. It is assumed that all arithmetical operations are performed in a ring. Most of these properties are similar to the DFT properties [1].

A. Definition

Transform:

$$F(k) = \sum_{n=0}^{N-1} f(n)\alpha^{nk} \qquad k = 0,1,\cdots,N-1. \quad \alpha^N = 1 \tag{A1}$$

Inverse Transform:

$$f(n) = N^{-1} \sum_{k=0}^{N-1} F(k)\alpha^{-nk} \qquad n = 0,1,\cdots,N-1. \tag{A2}$$

B. Basis Functions

$$\varphi_n(k) = \alpha^{-nk} \qquad n,k = 0,1,\cdots,N-1. \tag{A3}$$

C. Orthogonality

The inner product of two basis functions are given by [see (12) and (13)]

$$\langle\varphi_n,\varphi_m\rangle = \sum_{k=0}^{N-1} \varphi_n(k)\varphi_m^{-1}(k) = \sum_{k=0}^{N-1} \alpha^{(m-n)k}$$

$$= \begin{cases} N & \text{if } m = n \bmod N \\ 0 & \text{otherwise} \end{cases} \tag{A4}$$

This implies the orthogonality of the basis functions.

D. Periodicity

$F(k)$ can also be periodically extended, similar to the extension of $f(n)$,

$$f(n+N) = f(n)$$
$$F(K+N) = F(K). \tag{A5}$$

E. Symmetry Property

If the signal is symmetric, i.e.,

$$f(n) = f(-n) = f(N-n)$$

the transformed sequence is also symmetric, i.e.,

$$F(K) = F(-K) = F(N-K). \tag{A6}$$

If the signal is antisymmetric, i.e.,

$$f(n) = -f(-n) = -f(N-n),$$

the transformed sequence is also antisymmetric, i.e.,

$$F(K) = -F(-K) = -F(N-K). \tag{A7}$$

F. Symmetry of the Transformed Pair

$$T\{T\{f(n)\}\} = Nf(-n). \tag{A8}$$

G. Shift Theorem

If

$$T\{f(n)\} = F(K)$$
$$T\{f(n+m)\} = F(K)\alpha^{-mK}. \tag{A9}$$

H. Fast Computational Algorithm

If N can be factored as

$$N = r_1 \cdot r_2 \cdots r_m$$

then a fast computational algorithm similar to the FFT can be used to compute the transform which requires on the order of $N(r_1 + r_2 + \cdots + r_m)$ arithmetical operations.

I. Convolution Property

Transform of cyclic convolution of two sequences in the product of their transforms, i.e., if

$$y = x * h$$

then

$$T\{y\} = T\{x\} \cdot T\{h\}. \tag{A10}$$

J. Parseval's Theorem

Let

$$X(k) = T\{x(n)\}$$
$$H(k) = T\{h(n)\}.$$

Then

$$N \sum_{n=0}^{N-1} x(n)h(n) = \sum_{k=0}^{N-1} X(k)H(-k) \tag{A11}$$

and

$$N \sum_{n=0}^{N-1} x(n)h(-n) = \sum_{k=0}^{N-1} X(k)H(k). \tag{A12}$$

For the particular case of $x(n) = h(n)$

$$N \sum_{n=0}^{N-1} x^2(n) = \sum_{k=0}^{N-1} X(k)X(-k) \tag{A13}$$

and

$$N \sum_{n=0}^{N-1} x(n)x(-n) = \sum_{k=0}^{N-1} X(k)^2. \tag{A14}$$

Note that in this case, the conventional Parseval's theorem

$$N \sum_{n=0}^{N-1} |x(n)|^2 = \sum_{k=0}^{N-1} |X(k)|^2$$

does not hold, because in a ring magnitude is undefined.

K. Stretch Property

The stretch property exists if the transform for the longer length sequence exists in that ring.

L. Sampling Property

The sampling property exists.

APPENDIX B
PROOF OF THEOREM I

As shown before, the existence of such a transform depends on the existence of an α of order N with respect to each prime power factor moduli, $p_i^{r_i}$. According to Euler's Theorem [14], N, order of α, must divide $\varphi(p_i^{r_i})$, where $\varphi(p_i^{r_i})$ is the Euler's φ function given by

$$\varphi(p_i^{r_i}) = p_i^{r_i-1}(p_i - 1). \tag{B1}$$

Thus, $N \mid \varphi(p_i^{r_i})$, and since p_i is not a factor of N, $N \mid (p_i - 1)$, $i = 1,2,\cdots,l$ or

$$N \mid gcd\{p_1 - 1, p_2 - 1, \cdots, p_l - 1\}$$

or

$$N \mid 0(F). \tag{B2}$$

This proves the necessary part of the theorem. To prove sufficiency we have to show the existence of an α of order N if $N \mid 0(F)$. Note that (B2) rules out F being an even integer because in that case $0(F) = 1$. If $N \mid 0(F)$, $N \mid (p_i - 1)$ or $N \mid \varphi(p_i)$, therefore, one can find integers β_i of order N modulo $p_i^{r_i}$ [14], $i = 1,2,\cdots,l$. Then, by the Chinese Remainder Theorem, one can find an α of order N modulo $F = p_1^{r_1}\cdots p_l^{r_l}$, such that

$$\alpha = \beta_i (\text{mod } p_i^{r_i}), \qquad i = 1,2,\cdots,l. \tag{B3}$$

This is our desired α of order N. Since N and p_i are relatively prime, N^{-1} exists in this ring. This completes the proof of the theorem.

APPENDIX C

To prove that α_p of (23) is of order $2^{t+2} \bmod F_t$, we have to prove that it is of order 2^{t+2} with respect to each prime power factor of F_t (20). Let $p_i^{r_i}$ be a prime power factor of F_t, then

$$\alpha_p^{2^{t+2}} = 2^{2^{t+1}} = 1 \bmod F_t$$
$$= 1 \bmod p_i^{r_i}. \tag{C1}$$

Let N_i be the order of α_p with respect to $p_i^{r_i}$, then by Euler's theorem,

$$N_i \mid 2^{t+2} \tag{C2}$$

therefore N_i must be a power of 2. Also,

$$\alpha_p^{2^{t+1}} = 2^{2^t} = -1 \bmod F_t$$
$$= -1 \bmod p_i^{r_i}. \tag{C3}$$

From (C1) and (C3), it is obvious that

$$N_i = 2^{t+2}$$

completing the proof.

ACKNOWLEDGMENT

The authors wish to thank C. M. Rader of Lincoln Laboratories, Cambridge, Mass., for his comments and discussions.

REFERENCES

[1] B. Gold and C. M. Rader, *Digital Processing of Signals*. New York: McGraw-Hill, 1969.

[2] A. V. Oppenheim and C. Weinstein, "Effects of finite register length in digital filtering and the fast Fourier transform," *Proc. IEEE*, vol. 60, pp. 957–976, Aug. 1972.

[3] D. E. Knuth, *The Art of Computer Programming*, vol. 2, *Seminumerical Algorithms*. Reading, Mass.: Addison-Wesley, 1969, p. 555 and pp. 259–262.

[4] J. M. Pollard, "The fast Fourier transform in a finite field," *Math. Comput.*, vol. 25, pp. 365–374, Apr. 1971.

[5] I. J. Good, "The relation between two fast Fourier transforms," *IEEE Trans. Comput.*, vol. C-20, pp. 310–317, Mar. 1971.

[6] A. Schonhage and V. Strassen, "Fast multiplication of large numbers," (in German) *Comput.*, vol. 7, pp. 281–292, 1971.

[7] D. E. Knuth, "The art of computing programming—errata et addenda," Comput. Sci. Dep., Stanford University, Stanford, Calif., Rep. STAN-CS-71-194, pp. 21–26, Jan. 1971.

[8] P. J. Nicholson, "Algebraic theory of finite Fourier transforms," *J. Comput. Syst. Sci.*, vol. 5, pp. 524–547, 1971.

[9] C. M. Rader, "The number theoretic DFT and exact discrete convolution," IEEE Arden House Workshop on Digital Signal Processing, Harriman, N. Y., Jan. 11, 1972.

[10] ——, "Discrete convolution via Mersenne transforms," *IEEE Trans. Comput.*, vol. C-21, pp. 1269–1273, Dec. 1972.

[11] R. C. Agarwal and C. S. Burrus, "Fast digital convolution using Fermat transforms," in *Southwest IEEE Conf. Rec.*, Houston, Tex., Apr. 1973, pp. 538–543.

[12] ——, "Fast one-dimensional digital convolution by multidimensional techniques," *IEEE Trans. Acoust., Speech, and Signal Processing*, vol. ASSP-22, pp. 1–10, Feb. 1974.

[13] L. E. Dickson, *History of the Theory of Numbers*, vol. I. Washington, D. C.: Carnegie Institute, 1919, p. 376.

[14] O. Ore, *Number Theory and Its History*. New York: McGraw-Hill, 1948.

[15] D. A. Pitassi, "Fast convolution using Walsh functions," in *Proc. Conf. Applications of Walsh Functions*, Washington, D. C., April 1971, pp. 130–133.

[16] C. S. Burrus, "Block realization of digital filters," *IEEE Trans. Audio Electroacoust.*, vol. AU-20, pp. 230–235, Oct. 1972.

[17] R. C. Singleton, "An algorithm for computing the mixed radix fast Fourier transform," *IEEE Trans. Audio Electroacoust.*, vol. AU-17, pp. 93–103, June 1969.

Reprinted by permission from IEEE TRANSACTIONS ON ACOUSTICS, SPEECH, AND SIGNAL PROCESSING
Vol. ASSP-22, No. 2, April 1974, pp. 87-97

Fast One-Dimensional Digital Convolution by Multidimensional Techniques

RAMESH C. AGARWAL and CHARLES S. BURRUS, Member, IEEE

Abstract—This paper presents two formulations of multidimensional digital signals from one-dimensional digital signals so that multidimensional convolution will implement one-dimensional convolution of the original signals. This has reduced an important word length restriction when used with the Fermat number transform. The formulation is very general and includes block processing and sectioning as special cases and, when used with various fast algorithms for short length convolutions, results in improved multiplication efficiency.

I. Introduction

There are several advantages to formulating a one-dimensional digital convolution as a two- or higher dimensional problem. The first involves the Mersenne and Fermat number transforms which have recently been defined [1], [2] and which seem to have some advantages over the discrete Fourier transform (DFT) for implementing convolution on a digital computer. They can be computationally faster than the fast Fourier transform (FFT) implementation of the DFT and result in no roundoff error. These transforms have one limitation: the number of bits required for each word in an implementation is proportional to the length of the sequences to be convolved [1], [2]. It is the purpose of this paper to present a scheme whereby long sequences can be convolved by a two-dimensional convolution as mentioned by Rader [1]. This two-dimensional convolution can be implemented by a two-dimensional transform to allow a high-speed error-free convolution with the word length proportional to the square root of the length of the sequences.

This formulation also allows use of special short convolution algorithms similar to those proposed by Pitassi [7], Rayner [8], Davis [9], and Allwright [10]. These can be extended and combined with others to provide a very general and versatile format for doing fast digital convolution. The case of sequences of different lengths and the case where only part of the output sequence is desired are covered as special cases of the general problem.

Manuscript received March 15, 1973; revised July 10, 1973 and September 20, 1973. This work was supported by the National Science Foundation under Grant GK-23697.

The authors are with the Department of Electrical Engineering, Rice University, Houston, Tex. 77001.

II. Two-Dimensional Convolution Based on Overlap-Save

Consider the cyclic convolution of two sequences, $x(n)$ and $h(n)$, giving an output sequence $y(n)$, all of length N.

$$h(n) * x(n) = y(n). \tag{1}$$

This is defined by

$$y(n) = \sum_{q=0}^{N-1} h(n-q)\, x(q)$$

$$n = 0, 1, \cdots N-1 \tag{2}$$

where $h(n)$ and $x(n)$ are periodically extended outside their original domains of definition (or their indices evaluated modulo N).

In order to convert this one-dimensional problem to two dimensions a change of variables is made. First, it must be possible to factor N into integer factors

$$N = LM. \tag{3}$$

If a change of variables is made such that

$$\left.\begin{aligned} n &= l + mL \\ q &= k + pL \end{aligned}\right\} \quad \left\{\begin{aligned} k, l &= 0, 1, \cdots L-1 \\ p, m &= 0, 1, \cdots M-1 \end{aligned}\right.$$

then (2) becomes

$$y(l+mL) = \sum_{k=0}^{L-1} \sum_{p=0}^{M-1} h(l+mL-k-pL)\, x(k+pL). \tag{4}$$

Define now a two-dimensional $L \times M$ array $\hat{X}$ from the original length $N = LM$ signal, $x(n)$, by

$$\hat{X}(l, m) = x(l+mL) \tag{5}$$

where columns of $\hat{X}$ are the sections or blocks of x and the rows are samples of x taken every L values of n. In a similar way $\hat{H}$ and $\hat{Y}$ are defined by

$$\hat{H}(l, m) = h(l+mL)$$

$$\hat{Y}(l, m) = y(l+mL). \tag{6}$$

In terms of the two-dimensional signals, (4) becomes

$$\hat{Y}(l, m) = \sum_{p=0}^{M-1} \sum_{k=0}^{L-1} \hat{H}(l-k, m-p)\, \hat{X}(k, p) \tag{7}$$

which is a two-dimensional convolution. Note that values of $\hat{H}$ outside the $L \times M$ array are required. These values of $\hat{H}$ that are needed can be seen from (7) where $1-L \leqslant l-k \leqslant L-1$ and $1-M \leqslant m-p \leqslant M-1$. Values of $\hat{H}(l, m)$ outside the $L \times M$ array are defined by (6). We therefore define suitably extended arrays $\hat{\hat{H}}$ and $\hat{\hat{X}}$ so that two-dimensional convolution will give the desired answer. The extension of $\hat{H}$ is analogous to the overlap extensions used in the overlap-save algorithm for doing sectioning or block processing [3], [4]. Note that along the m dimension the values of $\hat{H}$ are periodic with period M, i.e., the desired convolution in (7) is cyclic in the m dimension (this is because the original desired convolution in (2) was defined as cyclic). Considering the values of $\hat{H}$ along l shows the convolution in that dimension is not cyclic.

An important factor when considering multidimensional convolution is the number of multiplications required in an implementation. If in (7) m and p are held constant then (7) becomes a scalar convolution along the l dimension of a length-L sequence of x with a length-$(2L - 1)$ sequence of h giving a length-L sequence of the output. There will be one of these length-L scalar convolutions for each value of m and p in (7) so that along the m dimension each "operation" will be a length-L convolution rather than a simple scalar multiplication. This is a sort of convolution of convolutions [3]. To count the total number of multiplications to compute (7) with l and k constant are found. But, with l and k not constant, this will be the number of length-L convolutions necessary and the total number of multiplications will be the number of length-L convolutions times the number of multiplications for a length-L convolution. In this case along the m dimension the number of multiplications is M^2 and along the l dimension it is L^2, so the total is $M^2 L^2$ or N^2, which is the same as a direct calculation of (2) would require. We will later use transforms and other schemes to reduce this number. Note the convolution can be carried out in either order.

If (7) is to be carried out by transforms and therefore cyclic convolution, $\hat{X}$ must be augmented with zeros so that aliasing of the noncyclic convolution along l does not occur and so that all arrays are the same size.

Consider the $2L - 1 \times M$ array $\hat{\hat{X}}$ formed by appending $(L - 1)$ rows of zeros to the bottom of $\hat{X}$.

$$\hat{\hat{X}} = \begin{bmatrix} x(0) & x(L) \cdots & x(N-L) \\ x(1) & x(L+1) & \\ \vdots & \vdots & \\ x(L-1) & & x(N-1) \\ 0 & 0 & 0 \\ 0 & 0 & 0 \\ \vdots & \vdots & \vdots \end{bmatrix} \tag{8}$$

$\hat{\hat{H}}$ is formed so that the columns of $\hat{\hat{H}}$ contain the periodic extension of the original $h(n)$ with period N.

$$\hat{\hat{H}} = \begin{bmatrix} h(N-L+1) & & h(N-2L+1) \\ \vdots & \vdots & \\ h(N-1) & h(L-1) & \\ h(0) & h(L) \cdots & h(N-L) \\ h(1) & h(L+1) & \\ \vdots & \vdots & \\ h(L-1) & & h(N-1) \end{bmatrix} \tag{9}$$

If two-dimensional cyclic convolution is carried out we have

$$\hat{\hat{Y}} = \hat{\hat{H}} * \hat{\hat{X}} \tag{10}$$

where the lower $L \times M$ partition of $\hat{\hat{Y}}$ is $\hat{Y}$ and the columns of $\hat{Y}$ are the desired blocks of $y(n)$ in (2). Because of ease in implementation with transforms, the arrays would usually be extended one additional row to be $2L \times M$ rather than the minimum $(2L - 1) \times M$.

III. Two-Dimensional Transform Implementation

The two-dimensional transform of $\hat{\hat{X}}$ is defined as

$$T\{\hat{\hat{X}}\} = F(j,k) = \sum_{m=0}^{M-1} \sum_{l=0}^{2L-1} \hat{\hat{X}}(l,m)\, \alpha_{2L}^{lj}\, \alpha_M^{mk} \tag{11}$$

and the inverse transform

$$T^{-1}\{F\} = \hat{\hat{X}}(l,m) = (2N)^{-1} \sum_{k=0}^{M-1} \sum_{j=0}^{2L-1} F(j,k)\, \alpha_{2L}^{-lj}\, \alpha_M^{-mk} \tag{12}$$

where α_M is of an order M (i.e., M is the least positive integer such that $(\alpha_M)^M = 1$ [2]). Applying the transform to (10) it can be shown that

$$T\{\hat{\hat{Y}}\} = T\{\hat{\hat{H}}\}\, T\{\hat{\hat{X}}\} \tag{13}$$

so that, similar to the one-dimensional case, (10) can be carried out by

$$\hat{\hat{Y}} = T^{-1}\left[T\{\hat{\hat{H}}\}\, T\{\hat{\hat{X}}\}\right]. \tag{14}$$

If the transform is the DFT, then

$$\alpha_M = e^{-j2\pi/M}$$

To compare multiplication efficiencies we assume the DFT of $\hat{\hat{H}}$ is already known and the number of multiplications for one $2L \times M$ transform, one $2L \times M$ complex multiplication, and one $2L \times M$ inverse transform are calculated. The number of complex multiplications is approximately $(2N \log 2N + 2N)$ as compared to $(N \log N + N)$ for a one-dimensional implementation and therefore one would not use the two-dimensional approach with the DFT for improved multiplication efficiency.

The computational advantage appears when used with the Fermat number transform [2] where word length requirements are a possible restriction. The Fermat number transform is defined in [2] and, although not named, is defined in a restricted form in the latter part of [1, eq. (38)]. It is a transform defined in a finite ring of integers with arithmetic performed modulo Fermat numbers ($2^b + 1$, $b = 2^t$), with $\alpha_{2b} = 2$ and $\alpha_{4b} = \sqrt{2}$, and having the property that multiplication of Fermat number transforms corresponds to conventional cyclic convolution (2) modulo $2^b + 1$. To perform convolution with the transform requires N real multiplications and a number of additions and word shifts proportional to $N \log N$. Unfortunately the transform requires word lengths proportional to the length of the sequences to be convolved [1], [2]. Since the lengths of the two dimensions are $2L$ and M rather than $N = LM$ for the one-dimensional signal, the word-length requirement using the two-dimensional transform is proportional to the square root of N rather than to N as for the one-dimensional problem. It is this reduction in the necessary word-length that makes two-dimensional formulation attractive with the Fermat number transform.

The consequences of this reduction in word length is of considerable practical importance. For example, using a word length of 16 b and the Fermat number transform [2] with $\alpha = 2$ to compute the complete noncyclic convolution of two sequences of equal length, the one-dimensional implementation restricts the sequence length to a maximum of 16 where the two-dimensional implementation increases the maximum to 256. A summary of the restrictions on sequence lengths is shown in Table I for the most practical word lengths and for two values of α. Note that one-dimensional length restrictions would be too severe for many applications but the two-dimensional restrictions would include most practical filters. If cyclic convolution is desired, then all the length restrictions are doubled since the addition of zeros to prevent aliasing is unnecessary.

The two-dimensional transform—or inverse transform—can be taken in either order. There is, however, a computational advantage in taking the transform first along the m direction (length M) and then along the l direction (length $2L$); half the $\hat{\hat{X}}$ sequences along the m direction are zero and half the $\hat{\hat{H}}$ sequences along the m direction are cyclically shifted by one position of the other half sequences. Also, while taking the inverse transform, there is an advantage to first taking the inverse transform along l, then along m, because we need only half the $\hat{\hat{Y}}$ sequences and therefore only half the sequences need be inverted along m.

TABLE I
Sequence Length Restrictions for One- and Two-Dimensional Implementation of Noncyclic Convolution by the Fermat Number Transform

Word Length (Bits)	Transform Basis α	Maximum Sequence Length—$N/2$ 1-D	2-D
16	2	16	256
16	$\sqrt{2}$	32	1024
32	2	32	1024
32	$\sqrt{2}$	64	4096
64	2	64	4096
64	$\sqrt{2}$	128	16 384

IV. Generalizations and the Inverse Problem

A generalization of the approach in this paper to higher orders is fairly obvious. For example, N could be factored into three integer factors $N = LMP$, as was done with two factors in (3). The signals $x(n)$ and $h(n)$ would then be redefined as three-dimensional $L \times M \times P$ arrays and (2) converted to a three-dimensional convolution by a change of variables as was done in two dimensions in (4)–(7). For x this would be

$$\tilde{X}(l, m, p) = x(l + mL + pML) \tag{15}$$

and with similar definitions for $\tilde{H}$ and $\tilde{Y}$ and after augmentation to prevent aliasing, (2) would become a three-dimensional convolution.

$$\tilde{\tilde{\tilde{Y}}} = \tilde{\tilde{\tilde{H}}} * \tilde{\tilde{\tilde{X}}}. \tag{16}$$

We would then have one-dimensional cyclic convolution of N-length sequences being carried out by three-dimensional cyclic convolution with dimensions of lengths $2L$, $2M$, and P. Use of orders higher than two does not seem needed with the Fermat number transform at this time, but will be exploited with other schemes in the next sections.

Still another variation would apply to the case where the filter is periodically time-varying. This, when converted to a two-dimensional problem with L equal the period of the filter, becomes time-invariant along one dimension, m, and time-varying along the other, l [5]. The DFT or Fermat transform could be applied to the two-dimensional signal along the time-invariant dimension and either direct calculation or another type transform applied to the other dimension.

The inverse of the above problem can be considered where one wishes to implement two-dimensional convolution by one-dimensional methods. This involves reversing the process presented in Section II by constructing one-dimensional sequences from the given arrays. MacAdams [6] has presented a scheme for doing this that can be seen to be the inverse of the problem addressed in this paper and can be applied to cyclic or noncyclic two-dimensional convolution.

V. *M*-Dimensional Convolution

If the length of the signals to be cyclically convolved in (2) can be written $N = 2^M$, then the methods used in (5) and (15) can be extended to define an M-dimensional signal with each dimension of length two. As before, this is done by a change of variables.

$$\widetilde{X}(l, m, p, \cdots) = x(l + 2m + 4p + \cdots), \qquad l, m, p, \cdots = 0, 1. \quad (17)$$

After a similar definition for $\widetilde{H}$ and $\widetilde{Y}$ cyclic convolution in (2) becomes

$$\widetilde{Y}(l, m, p, \cdots) = \sum_{k=0}^{1} \sum_{j=0}^{1} \sum_{q=0}^{1} \cdots \cdot \widetilde{H}(l-k, m-j, p-q, \cdots)\widetilde{X}(k, j, q, \cdots) \quad (18)$$

for $l, m, p, \cdots = 0, 1$.

Both $\widetilde{X}$ and $\widetilde{Y}$ are M-dimensional with each dimension of length two and $\widetilde{H}$ is also M-dimensional but must be defined with dimensions of length three in order to carry out (18). This is the same as was required in two dimensions where $\hat{\hat{H}}$ was defined in (9). Since the original convolution in (2) was defined as cyclic, the convolution along the last dimension in (18) is also cyclic so that $\widetilde{H}$ need not be extended along that dimension but can be evaluated for this index modulo 2.

This is an extremely general and versatile formulation for the original problem that can be used to improve computational efficiency. The convolution along each dimension of length two can be written in terms of scalar variables as

$$\begin{bmatrix} h_0 & h_{-1} \\ h_1 & h_0 \end{bmatrix} \begin{bmatrix} x_0 \\ x_1 \end{bmatrix} = \begin{bmatrix} y_0 \\ y_1 \end{bmatrix}. \quad (19)$$

The convolution of (18) can be viewed as M nested length 2 convolutions, each separately of the form shown in (19) requiring four multiplications. Using the same reasoning that was explained for the two-dimensional formulation, the total number of multiplications is $F = 4^M$. Since $N = 2^M$, this becomes $F = N^2$, which is again the same as would be required by directly calculating (2).

VI. A High Speed Algorithm

With this formulation of scalar convolution in terms of multidimensional short convolutions, various tricks for efficient short convolution can be used. Consider one of several possible algorithms suggested by Pitassi [7] where, rather than directly calculating (19), three intermediate numbers are found:

$$\begin{aligned} g_0 &= (h_0 + h_{-1})x_1 \\ g_1 &= h_0(x_0 - x_1) \\ g_2 &= (h_1 + h_0)x_0. \end{aligned} \quad (20)$$

The desired outputs are then calculated by

$$\begin{aligned} y_0 &= g_0 + g_1 \\ y_1 &= g_2 - g_1. \end{aligned} \quad (21)$$

This approach uses three multiplications in comparison with four multiplications for a direct calculation.

Using this result, the total number of multiplications to calculate (18) becomes

$$F = 3^M. \quad (22)$$

If the fact pointed out earlier, that the convolution along the last dimension is cyclic, is used, then a further reduction is possible.

Length-two cyclic convolution is given by

$$\begin{bmatrix} h_0 & h_1 \\ h_1 & h_0 \end{bmatrix} \begin{bmatrix} x_0 \\ x_1 \end{bmatrix} = \begin{bmatrix} y_0 \\ y_1 \end{bmatrix}. \quad (23)$$

This can be calculated from two intermediate values by

$$\begin{aligned} f_0 &= (h_0/2 + h_1/2)(x_0 + x_1) \\ f_1 &= (h_0/2 - h_1/2)(x_0 - x_1) \end{aligned}$$

to give

$$\begin{aligned} y_0 &= f_0 + f_1 \\ y_1 &= f_0 - f_1 \end{aligned} \quad (24)$$

requiring two multiplications (assuming the factors of $\frac{1}{2}$ are either precalculated or obtained by shifting). Using this result on the last dimension reduces (22) to

$$F = 2 \cdot 3^{M-1} \quad (25)$$

which is the same number obtained by Pitassi [7] and Rayner [8].

Pitassi developed his algorithm by relating the cyclic convolution of two sequences to the convolution of subsequences in the same manner that the FFT can be developed based on decimation in time. This was extended by Davis [9] to an approach similar to decimation in frequency where the subsequences are halves of the original sequences. Both of these are special cases of the multidimensional formulation since the values along any dimension are samples or blocks of the original sequence.

Another form of convolution that is sometimes desired is of the same form as (2) but with $h(n)$ not being periodically extended, rather having independent values for negative indices. The transmission matrix formulation for $N = 3$ is

$$\begin{bmatrix} h_0 & h_{-1} & h_{-2} \\ h_1 & h_0 & h_{-1} \\ h_2 & h_1 & h_0 \end{bmatrix} \begin{bmatrix} x_0 \\ x_1 \\ x_2 \end{bmatrix} = \begin{bmatrix} y_0 \\ y_1 \\ y_2 \end{bmatrix} \tag{26}$$

and because of its structure the operator is called a constant diagonal convolution matrix. This requires $2N - 1$ values of $h(n)$ from $n = -N + 1$ to $n = N - 1$ and N values of $x(n)$ to give N value of $y(n)$.

The same approach that was used for cyclic convolution is applicable here except the reduction described in (23) and (24) does not work since the M-dimensional formulation of (18) is no longer cyclic along any dimension. Therefore the number of multiplications necessary to implement a length-N constant diagonal convolution is

$$F = 3^M \tag{27}$$

which is the same as obtained by Allwright [10], using a matrix factorization approach.

Note that cyclic convolution in (2) can be viewed as a special case of constant diagonal convolution. Causal noncyclic convolution where $h(n)$ is zero for $n < 0$ is also a special case.

$$\begin{bmatrix} h_0 & 0 & 0 \\ h_1 & h_0 & 0 \\ h_2 & h_1 & h_0 \end{bmatrix} \begin{bmatrix} x_0 \\ x_1 \\ x_2 \end{bmatrix} = \begin{bmatrix} y_0 \\ y_1 \\ y_2 \end{bmatrix}. \tag{28}$$

The most common form of convolution desired is causal noncyclic convolution where all of the output sequence is obtained rather than only the first N points as in (28). For this case two length-N sequences are convolved to give a length-$(2N - 1)$ output. The transmission matrix formulation for $N = 3$ is

$$\begin{bmatrix} h_0 & 0 & 0 \\ h_1 & h_0 & 0 \\ h_2 & h_1 & h_0 \\ 0 & h_2 & h_1 \\ 0 & 0 & h_2 \end{bmatrix} \begin{bmatrix} x_0 \\ x_1 \\ x_2 \end{bmatrix} = \begin{bmatrix} y_0 \\ y_1 \\ y_2 \\ y_3 \\ y_4 \end{bmatrix}. \tag{29}$$

To apply the results from cyclic convolution, all sequences are extended with zeros to length $2N$. Cyclic convolution then gives the desired output of (29) and uses

$$F = 2 \cdot 3^M \tag{30}$$

multiplications.

A further reduction is possible by recognizing that along the last dimension only one multiplication is necessary rather than two for the cyclic case that gave (25) or three for the constant diagonal noncyclic case that gave (27). Consider the $(M + 1)$-dimensional convolution of (18) from the extended sequences with all indices except the last one held constant. The resulting length-two scalar convolution is

$$\hat{y}(p) = \sum_{j=0}^{1} \hat{h}(p - j)\hat{x}(j), \qquad p = 0, 1 \tag{31}$$

which becomes

$$\begin{aligned} \hat{y}_0 &= \hat{h}_0 \hat{x}_0 + \hat{h}_{-1} \hat{x}_1 \\ \hat{y}_1 &= \hat{h}_1 \hat{x}_0 + \hat{h}_0 \hat{x}_1. \end{aligned} \tag{32}$$

The $\hat{x}_1$ terms are always zero since the last half of the length-$2N$ sequence $x(n)$ has been added as zeros. Close examination shows that because $h(n)$ has also been extended with zeros, either $\hat{h}_0$ or $\hat{h}_1$ will always be zero, depending on values of the constant indices. Therefore the length-two convolution of (31) will require only one multiplication and the resulting total number of multiplications necessary to noncyclically convolve two length-N sequences giving a length-$2N$ output is

$$F = 3^M \tag{33}$$

which is the same as for the length-N constant diagonal convolution with a length-N output.

VII. Multidimensional Convolution Based on Overlap-Add

Another two-dimensional formulation can be developed that is a generalization of the overlap-add algorithm [3], [4]. In (7) the $\hat{H}$ function was augmented so that the desired output $y(n)$ is in blocks that are the columns of $\hat{Y}$. For this formulation $\hat{H}$ will not be augmented but $\hat{Y}$ will be and the desired $y(n)$ will be obtained by adding the overlapped columns of $\hat{Y}$. This formulation is based on noncyclic convolution given by

$$y(n) = \sum_{q=0}^{N-1} h(n - q)x(q), \qquad n = 0, 1, \cdots 2N - 1 \tag{34}$$

where $h(n)$ and $x(n)$ are sequences of length N and $y(n)$ of length $2N - 1$ and all are defined to be zero outside these lengths. The transmission matrix formulation for $N = 3$ is given by (29).

Using a similar factoring of N and change of variables as was done in (3) and (5) we have (34) becoming

$$\begin{aligned} \hat{Y}(l, m) &= \sum_{p=0}^{M-1} \sum_{k=0}^{L-1} \hat{H}(l - k, m - p)\hat{X}(k,) \\ l &= 0, 1, \cdots 2L - 2 \\ m &= 0, 1, \cdots 2M - 2. \end{aligned} \tag{35}$$

In this case both $\hat{X}$ and $\hat{H}$ are $L \times M$ arrays and $\hat{Y}$ is $(2L - 1) \times (2M - 1)$ with $\hat{X}$ and $\hat{H}$ having their columns the blocks of $x(n)$ and $h(n)$ but the columns of $\hat{Y}$ and the blocks of $y(n)$ have a more complicated relation than in (6). Both $\hat{X}$ and $\hat{H}$ are defined to be zero outside the domain of definition. Here we can show

$$y(l + mL) = \hat{Y}(l, m) + \hat{Y}(l + L, m - 1) \tag{36}$$

for

$$l = 0, 1, \cdots L - 1$$
$$m = 0, 1, \cdots 2M - 1$$

and $\hat{Y}(l, m) = 0$ outside its domain of definition of

$$l = 0, 1, \cdots 2L - 2$$
$$m = 0, 1, \cdots 2M - 2. \tag{37}$$

This formulation gives the implementation of one-dimensional convolution by the sum of overlapped columns of an array obtained by noncyclic two-dimensional convolution. This can be viewed as a generalization of the overlap-add algorithm [3], [4] used for sectioning or block processing.

If $N = 2^M$, the extension to M-dimensional convolution is similar to that done for the overlap-save technique in (17) and (18). Both $\tilde{X}$ and $\tilde{H}$ are M-dimensional with each dimension of length two and $\tilde{Y}$ is M-dimensional but with each of length three. For

$$\tilde{X}(l, m, p, \cdots) = x(l + 2m + 4p + \cdots)$$
$$\tilde{H}(l, m, p, \cdots) = h(l + 2m + 4p + \cdots) \tag{38}$$

(34) becomes

$$\tilde{Y}(l, m, p, \cdots) = \sum_{k=0}^{l} \sum_{j=0}^{m} \sum_{q=0}^{p} \cdots$$
$$\cdot \tilde{H}(l - k, m - j, p - q, \cdots) \tilde{X}(k, j, q, \cdots) \tag{39}$$

for $l, m, p, \cdots = 0, 1, 2$.

The calculation of $y(n)$ from $\tilde{Y}$ is a bit complicated but is a generalization of (36) and involves only additions. For example if $N = 2^3 = 8$ the $\tilde{Y}$ function would have three dimensions with a total of 27 elements. Along the l dimension there would be nine length-three blocks that, when overlapped and added along the m dimension, would give three length-seven blocks. These, when overlapped and added, would give the single length-fifteen sequence that would be $y(n)$.

An example for $N = 4$ would have a two-dimensional three × three $\tilde{Y}$

$$\tilde{Y} = \begin{bmatrix} \tilde{y}_{00} & \tilde{y}_{01} & \tilde{y}_{02} \\ \tilde{y}_{10} & \tilde{y}_{11} & \tilde{y}_{12} \\ \tilde{y}_{20} & \tilde{y}_{21} & \tilde{y}_{22} \end{bmatrix} \tag{40}$$

and $y(n)$ would be found by

$$\begin{aligned} y_0 &= \tilde{y}_{00} \\ y_1 &= \tilde{y}_{10} \\ y_2 &= \tilde{y}_{20} + \tilde{y}_{01} \\ y_3 &= \tilde{y}_{11} \\ y_4 &= \tilde{y}_{21} + \tilde{y}_{02} \\ y_5 &= \tilde{y}_{12} \\ y_6 &= \tilde{y}_{22}. \end{aligned} \tag{41}$$

Along each dimension of (39) the matrix formulation of the length-two convolution is

$$\begin{bmatrix} h_0 & 0 \\ h_1 & h_0 \\ 0 & h_1 \end{bmatrix} \begin{bmatrix} x_0 \\ x_1 \end{bmatrix} = \begin{bmatrix} y_0 \\ y_1 \\ y_2 \end{bmatrix} \tag{42}$$

which, if done directly, requires four multiplications and one addition. If an intermediate step is added for y_1, then

$$\begin{aligned} y_0 &= h_0 x_0 \\ y_1 &= (h_0 + h_1)(x_0 + x_1) - y_0 - y_1 \\ y_2 &= h_1 x_1 \end{aligned}$$

gives the three outputs with three multiplications and four additions/subtractions. Using this algorithm on each dimension of (39) and counting the number of required multiplications by the method used to find (33) we find that (39) can be computed with

$$F = 3^M \tag{44}$$

multiplications which are the same as that obtained by the overlap-save method in (33).

VII. Sequences of Different Lengths and Partial Outputs

Two modifications of the usual formulation of convolution are often desired. The first occurs when the two sequences are significantly different in lengths and the second when one desires only a portion of the output rather than all of it. Both cases can be formulated in terms of multidimensional convolutions and computational savings can be realized.

If the $h(n)$ is assumed to be the shorter sequence and if its length can be expressed as $L = 2^S$ the length of the longer sequence $x(n)$ will be expressed as $R = 2^S \cdot M$. It may be necessary to add zeros to both x and h. The multidimensional signals are formed to give S length-two dimensions and the last dimension of length M. If m is the index for the length-M dimension, then $\tilde{H}$ is formulated to be zero for all m other than $m = 0$. Therefore, if the fast algorithm of (20) or (43) is used along the S length-two dimensions, there will be only M multiplications required along the M dimension. The total number of multiplications is then

$$F = 3^S \cdot M. \tag{45}$$

This could also be seen by considering the problem as requiring M length-2^S noncyclic convolutions with the outputs overlapped and added.

If only a portion of the total output from the convolution is desired, then a similar saving can be obtained. Assume that both $h(n)$ and $x(n)$ are of length $N = LM$, with a noncyclic output $y(n)$ of length $2N - 1$. Out of this only a block of $y(n)$ of length L is desired. First, N extra zeros are appended to both the sequences and cyclic convolution of length $2N$ is formulated. This one-dimensional cyclic convolution is reformulated as a two-dimensional convolution as in (5)-(7). The first dimension is of length L and the convolution in this dimension is noncyclic and can be carried out as cyclic convolution of length $2L - 1$. In the second dimension the convolution is cyclic and of length $2M$. The two-dimensional arrays are formulated so that the desired length L block of $y(n)$ appears as a column of $\hat{Y}$ in (6). Thus, $\hat{Y}$ along the second dimension has to be computed only for one index. This replaces cyclic convolution of length $2M$ with a summation of $2M$ terms along the second dimension. The desired output block can be written as a summation of $2M$ convolutions of length L, where each convolution represents convolution of two sequences of lengths $2L - 1$ and L, respectively, giving a sequence of length L. Because zeros were appended to both $x(n)$ and $h(n)$, out of these $2M$ convolutions, between one and M convolutions would be nonzero, depending on the second index of $\hat{Y}$ for which the output is desired.

The convolutions along the first dimension of length L can be carried out either by transform techniques or multidimensional techniques discussed in this paper. If the transform techniques are used, the summation of convolutions can be carried out in the transform domain, thus requiring only one inverse transform to obtain the desired output. Rader [11] has discussed a similar technique for the particular case of estimation of autocorrelation function for the first few lag values. If $L = 2^S$ and the multidimensional methods are used, the number of multiplications are at most the same as given in (45).

The formulation just discussed can be extended for the situation where sampled output is desired, sampled at every Lth value. In this case for the two-dimensional formulation, the output appears as a row of $\hat{Y}$, and as before, the two-dimensional convolution again reduces to a summation of convolutions. If the output $y(n)$ is a narrow-band signal as compared to the sampling frequency, to reconstruct the analog signal, the samples of $y(n)$ at a lower sampling rate are sufficient. In this situation the formulation discussed here can result in computational savings.

If a multidimensional formulation is considered, we can obtain partial outputs as combinations of blocks and samples.

IX. Relations to Logical Convolution and Walsh Transforms

Consider logical convolution [12] of two sequences $x(n)$ and $h(n)$ of length $N = 2^M$, each giving an output sequence $y(n)$ also of length N. Logical convolution is defined similar to the cyclic convolution of (2), but the addition and subtraction of indices is done differently. All indices are represented in the binary form as an M-bit index. When indices have to be added or subtracted, they are added or subtracted bit by bit, modulo 2. Note that in logical convolution, addition and subtraction of indices are equivalent. We can convert this one-dimensional logical convolution problem to an M-dimensional convolution as in (17) and (18). Along any dimension, convolution appears as in (19), but since logical convolution is desired, $h(-1) = h(1)$. Therefore, if (18) is implemented as length-two cyclic convolutions along all the dimensions, $y(n)$ thus obtained is logical convolution of $x(n)$ and $h(n)$. Alternatively, if (18) is implemented as a noncyclic convolution along all the dimensions, we obtain noncyclic convolution of $x(n)$ and $h(n)$ as in (26).

Length-two cyclic convolution can be implemented using just two multiplications as in (24), which is a length-two DFT implementation. Thus (18) can be implemented as an M-dimensional cyclic convolution using M-dimensional DFT's of $\tilde{X}$ and $\tilde{H}$, where each dimension is of length two. Alternatively length-N logical convolution can also be implemented using length-N Walsh transforms of $x(n)$ and $h(n)$ [12]. The preceding development shows that length-($N = 2^M$) Walsh transform is equivalent to the M-dimensional DFT. Thus the M-dimensional approach establishes the logical convolution theorem for the Walsh transforms and it also establishes the fast Walsh transform algorithm as an M-dimensional DFT. These facts have been noted before in the literature.

For a particular formulation of (20), Pittasi [7] and Davis [9] have shown that some of the intermediate products correspond to multiplication of the Walsh transforms of the two sequences.

X. Generalizations and Applications

There are several modifications and generalizations that are possible with the formulation used here. The first will illustrate that fast algorithms exist for sequences of length other than two.

Consider the noncyclic convolution of two length-three sequences.

$$\begin{bmatrix} h_0 & 0 & 0 \\ h_1 & h_0 & 0 \\ h_2 & h_1 & h_0 \\ 0 & h_2 & h_1 \\ 0 & 0 & h_2 \end{bmatrix} \begin{bmatrix} x_0 \\ x_1 \\ x_2 \end{bmatrix} = \begin{bmatrix} y_0 \\ y_1 \\ y_2 \\ y_3 \\ y_4 \end{bmatrix}. \tag{46}$$

This would normally require nine multiplications and four additions. If six intermediate variables are calculated by

$$\begin{aligned} g_0 &= h_0 x_0 & g_3 &= (h_0 + h_1)(x_0 + x_1) \\ g_1 &= h_1 x_1 & g_4 &= (h_1 + h_2)(x_1 + x_2) \\ g_2 &= h_2 x_2 & g_5 &= (h_2 + h_0)(x_2 + x_0) \end{aligned} \quad (47)$$

then the desired output can be obtained by

$$\begin{aligned} y_0 &= g_0 \\ y_1 &= g_3 - g_0 - g_1 \\ y_2 &= g_5 + g_1 - g_0 - g_2 \\ y_3 &= g_4 - g_1 - g_2 \\ y_4 &= g_2 \end{aligned} \quad (48)$$

requiring six multiplications.

If two sequences of length $N = 3^M$ were to be convolved, then using an implementation with M dimensions of length three would require for total multiplications

$$F = 6^M. \quad (49)$$

For short sequences of this length, this approach is faster than adding zeros and using the next larger power of two-length with (39).

Similar results are possible with other lengths and a very general scheme can be developed for sequences of length

$$N = 2^M \cdot 3^S \cdots \quad (50)$$

using different algorithms along different dimensions in a manner similar to using a multiple radix FFT.

Because of the generality of the multidimensional formulation there are mixtures of high speed techniques that can be used.

In some situations, it may be advantageous to combine transform techniques with the fast algorithms for short convolutions discussed in this paper. One such situation uses the Fermat number transforms [2]. As discussed in Section III, using a b-bit implementation of Fermat number transforms, the maximum cyclic convolution length is $4b$. To cyclically convolve sequences longer than this, we need to formulate a two-dimensional convolution as in Section II. Assume $x(n)$, $h(n)$, and $y(n)$ are sequences of length-$(N = 8b)$ each. Consider the formulation of (7), with $M = 4b$ and $L = 2$. Equation (7) can be rewritten as

$$\hat{Y}(0, m) = \sum_{p=0}^{M-1} \hat{H}(0, m-p)\hat{X}(0, p) + \sum_{j=0}^{M-1} \hat{H}(-1, m-j)\hat{X}(1, j)$$

$$\hat{Y}(1, m) = \sum_{p=0}^{M-1} \hat{H}(1, m-p)\hat{X}(0, p) + \sum_{j=0}^{M-1} \cdot \hat{H}(0, m-j)\hat{X}(1, j), \quad m = 0, 1, \cdots, M-1. \quad (51)$$

In these equations, each summation represents cyclic convolution of length $4b$, which can be carried out by Fermat transforms. Taking length $4b$ Fermat transforms of all the sequences in (51), we obtain

$$\begin{aligned} T\{\hat{Y}(0, l)\} &= T\{\hat{H}(0, l)\} T\{\hat{X}(0, l)\} + T\{\hat{H}(-1, l)\} T\{\hat{X}(1, l)\} \\ T\{\hat{Y}(1, l)\} &= T\{\hat{H}(1, l)\} T\{\hat{X}(0, l)\} + T\{\hat{H}(0, l)\} T\{\hat{X}(1, l)\}. \end{aligned} \quad (52)$$

These equations are similar to (19) and we could employ the tricks of (20) and (21) along the first dimension, giving

$$\begin{aligned} T\{\hat{Y}(0, l)\} &= [T\{\hat{H}(0, l)\} + T\{\hat{H}(-1, l)\}] T\{\hat{X}(1, l)\} + [T\{\hat{X}(0, l)\} - T\{\hat{X}(1, l)\}] T\{\hat{H}(0, l)\} \\ T\{\hat{Y}(1, l)\} &= [T\{\hat{H}(0, l)\} + T\{\hat{H}(1, l)\}] T\{\hat{X}(0, l)\} - [T\{\hat{X}(0, l)\} - T\{\hat{X}(1, l)\}] T\{\hat{H}(0, l)\}. \end{aligned} \quad (53)$$

Note that $\hat{H}(-1, m) = \hat{H}(1, m-1)$, thus $\hat{H}(-1, l)$ is the cyclic shift by one position of $\hat{H}(1, l)$. We could employ the cyclic shift theorem to compute $T\{\hat{H}(-1, l)\}$ from $T\{\hat{H}(1, l)\}$. Assuming the $\hat{H}$ transforms are precalculated, this method requires computation of 2 length-$(4b)$ $\hat{X}$ transforms, $12b$ multiplications and $12b$ additions/subtractions to compute the $\hat{Y}$ transforms using (53), and two length-$(4b)$ $\hat{Y}$ inverse transforms. This is efficient because the only extra computation is $4b$ extra multiplications and is better than the two-dimensional Fermat number transform approach, which requires roughly twice the amount of computation. This use of $2 \times M$ convolution has a small computational advantage even when used with the FFT where, in effect, the last stage of the FFT algorithm is replaced by one of the fast length-two algorithms.

There are many other possibilities of combinations of Fermat and Fourier transforms of various lengths and dimensions with short convolution algorithms or with the use of special hardware. The arbitrary order of the various operations can also be used to advantage. As pointed out by Rader [11], and illustrated in our discussion of partial outputs and in (53), it is often more efficient to take transforms along one dimension than convolve along another before taking the inverse transform.

To illustrate the efficiencies of some of the techniques of this paper, a comparison of a particular case will be made with direct and FFT implementations. Consider the problem of noncyclic convolution of two length-$(N = 2^M)$ sequences to give an output of length $2N - 1$ as described in (34). The number of multiplications per output point will be calculated for three implementations. First, consider a direct implementation which requires N^2 multiplication for

$2N - 1$ output points (we will approximate this by $2N$ for simplification). This gives for multiplication per output point

$$F_0 = \tfrac{1}{2}N.$$

If the FFT is used in an efficient way, taking advantage of the fact that the data are real and using a mixed radix algorithm (2, 4, and 8), the number of multiplications necessary per output point can be calculated as described by Singleton [13] and is denoted F_1.

Using the M-dimensional formulation with the fast algorithm of (33) or (44) gives for the multiplications per output

$$F_2 = \frac{1}{2N} 3^M.$$

Table II compares these functions for various lengths up to 1024. Note the multidimensional implementation is more efficient than a direct implementation for all lengths and requires fewer multiplications than the FFT for lengths up to 128. If a less efficient FFT implementation requiring $4 \log N + 4$ multiplications per output point is used, the crossover length is above 2048.

The multidimension approach can also be used in the same way the FFT is used to implement on going processing or sectioning [4]. In contrast, to use with the FFT, this approach is most efficient when the length of the section or block is the same as the length of the convolution operator.

Initial observations indicate that block implementation of recursive filters [3] becomes more attractive when used with the techniques described in this paper. To illustrate this, we will again consider the multiplication efficiencies for three realizations. First consider a recursive filter with an equal order numerator and denominator of $N = 2^M$. The multiplications per output points for a direct implementation

$$F_0 = 2N.$$

Using efficient FFT algorithms, the results given by the methods in [3] are denoted F_1. Using three convolutions by (33) or (44) to implement the block recursive filter with the block length equal the order, gives as the multiplications per output point

$$F_2 = \frac{1}{N} \cdot 3^{M+1}.$$

Table III compares these multiplication efficiencies for orders up to 128. Note the multidimensional approach is more efficient than the direct for orders above three and more efficient than the FFT for orders up to about 256. This is yet to be explored in detail.

TABLE II
A Comparison of Multiplication Efficiencies for Three Implementations of Length-N Convolution

N	F_0	F_1	F_2
2	1	1.5	0.75
4	2	2.25	1.12
8	4	4.12	1.70
16	8	5.06	2.53
32	16	6.03	3.80
64	32	8.01	5.70
128	64	9.00	8.54
256	128	10.00	12.81
512	256	12.00	19.22
1024	512	13.00	28.83

TABLE III
A Comparison of Multiplication Efficiencies for Three Implementations of Block Recursive Filters of Order N

N	F_0	F_1	F_2
2	4	14	4.5
4	8	21	6.7
8	16	31	10.1
16	32	38	15.2
32	64	45	22.8
64	128	53	34.2
128	256	58	51.2

XI. Conclusions

This paper has presented two formulations of convolution in terms of multidimensional convolution—one based on a generalization of the overlap-save algorithm for sectioning and the other on the overlap-add algorithm. The first proved to be well suited for cyclic and constant-diagonal convolution and the second for noncyclic convolution. Fast algorithms were developed based on length-two and -three convolution that lead to an improvement in multiplication efficiency. The reduction in required word lengths proved to be a valuable feature when used with the Fermat number transform. The formulation proved to be well suited for the special cases where unequal-length sequences were convolved or where only a portion of the output was desired. It was further shown that various mixtures of algorithms could be used along the different dimensions to achieve certain advantages or to fit particular requirements. Finally, examples were presented to compare the multiplication efficiencies of a few implementations.

The formulation is so general that a complete and systematic investigation of all possible applications is difficult. The main ideas and relations to other works that we know of have been presented here. The investigation of word length and storage requirements and a more complete consideration of recursive implementations is still to be made.

Acknowledgment

The authors would like to thank R. A. Meyer for valuable discussions.

References

[1] C. M. Rader, "Discrete convolutions via Mersenne transforms," *IEEE Trans. Comput.*, vol. C-21, pp. 1269–1273, Dec. 1972.

[2] R. C. Agarwal and C. S. Burrus, "Fast digital convolutions using Fermat transforms," in *Southwestern IEEE Conf. Rec.*, Houston, Tex., Apr. 1973, pp. 538–543.

[3] C. S. Burrus, "Block realization of digital filters," *IEEE Trans. Audio Electroacoust.*, vol. AU-20, pp. 230–235, Oct. 1972.

[4] B. Gold and C. M. Rader, *Digital Processing of Signals.* New York: McGraw-Hill, 1969, pp. 208–211.

[5] R. A. Meyer and C. S. Burrus, "Certain properties of periodically time-varying digital filters," in *Southwestern IEEE Conf. Rec.*, Houston, Tex., Apr. 1973, pp. 529–535.

[6] D. P. MacAdam, "Image restoration by constrained deconvolution," *J. Opt. Soc. Amer.*, vol. 60, pp. 1617–1627, Dec. 1970.

[7] D. A. Pitassi, "Fast convolution using the Walsh transforms," in *Proc. Conf. Applications of Walsh Functions*, Washington, D.C., Apr. 1971, pp. 130–133.

[8] P. J. W. Rayner, "A fast cyclic convolution algorithm," presented at Symp. Digital Filtering, Imperial College, London, England, Aug. 1971.

[9] W. F. Davis, "A class of efficient convolution algorithms," in *Proc. Symp. Applications of Walsh Functions*, Washington, D.C., Mar. 1972, pp. 318–329.

[10] J. C. Allwright, "Real factorization of noncyclic convolution operators with applications to fast convolution," *Electron. Lett.*, vol. 7, pp. 718–719, Dec. 1971.

[11] C. M. Rader, "An improved algorithm for high speed autocorrelation with applications to spectral estimation," *IEEE Trans. Audio Electroacoust.*, vol. AU-18, pp. 439–441, Dec. 1970.

[12] G. S. Robinson, "Logical convolution and discrete Walsh and Fourier power spectra," *IEEE Trans. Audio Electroacoust.*, vol. AU-20, pp. 271–280, Oct. 1972.

[13] R. C. Singleton, "An algorithm for computing the mixed radix fast Fourier transform," *IEEE Trans. Audio Electroacoust.*, vol. AU-17, pp. 93–103, June 1969.

Hardware Realization of a Fermat Number Transform

JAMES H. McCLELLAN, MEMBER, IEEE

Abstract—The hardware design and implementation of a Fermat number transform (FNT) is described. The arithmetic logic design is treated in detail and a new data representation for integers modulo a Fermat number is derived. In addition, the FNT is compared with the fast Fourier transform (FFT) on the basis of hardware required for a pipeline convolver.

I. Introduction

NUMBER THEORY TRANSFORMS are attractive for the implementation of digital convolution because it is possible to obtain a transform that involves no multiplications. Since multipliers are a major hardware expense in fast Fourier transform (FFT) pipeline convolvers, or in direct-form convolvers, the potential for building cheaper and/or faster high-speed convolvers should lie with number theory transforms. Many other factors cloud the picture and the savings in multiplier hardware can be offset by increased memory hardware and transform size in some cases. It is the purpose of this paper to examine some of these hardware issues for the Fermat number transform. (FNT).

In the realm of radar signal processing, the potential for higher throughput is worth exploring. For this reason a small prototype number theory transform was designed and constructed (Fig. 1). This machine is the implementation of a 64-point, 16-bit FNT. An important element in the design of the FNT is a new coding scheme for the data. This coding scheme simplifies the arithmetic operations modulo the Fermat number and results in an FNT butterfly that can be clocked reliably at 38 MHz.

The experience derived from designing and building this hardware serves as the basis for estimates of the size of large pipeline FNT convolvers for use in radar matched filtering. The result of the hardware comparison of the FNT versus a pipeline FFT indicates that the anticipated savings of the FNT can be realized for small systems (e.g., length 64 convolution) when the signal to be filtered is real. However, in larger systems where one must use two-dimensional convolution to implement one-dimensional convolution, the savings in multiplier hardware are offset by increased transform size and the corresponding increase in memory size and reference spectrum multiplier hardware. In this case, when the signal to be filtered is real, the FNT still offers a potential savings in the amount of hardware versus the FFT. When the signal is complex valued the amount of FNT hardware approximately doubles and is

Manuscript received June 10, 1975; revised December 10, 1975. This work was sponsored by the Ballistic Missile Defense Program of the Department of the Army.

The author was with the Lincoln Laboratory, Massachusetts Institute of Technology, Lexington, MA. He is now with the Department of Electrical Engineering and Computer Science, Massachusetts Institute of Technology, Cambridge, MA 02139.

Fig. 1. Photograph of the FNT hardware: length 64, 16-bit system. Arithmetic logic is contained on the left-hand board, the memory and control logic on the right-hand board. Total IC count 283; 90 IC's for the FNT butterfly.

much greater than a pipeline FFT. Finally, the implementation of two-dimensional convolution is not penalized due to increased memory size and the savings in multiplier hardware with the FNT will translate into savings for the overall convolver system.

II. Theory

In 1971 several authors [1]-[3] generalized the idea of convolution via the discrete Fourier transform (DFT) [4] to a large class of transforms where the underlying algebraic structure is a finite field or ring. In the ring of integers modulo a Fermat number, these transforms offer two advantages: no multiplications in the transform and decomposition into a fast algorithm analogous to the FFT.

The FNT of $\{x(k)\}$ is defined as

$$X(n) = \sum_{k=0}^{N-1} x(k)\,\alpha^{-\langle nk\rangle} \bmod F_t \tag{1}$$

where $F_t = 2^{2^t} + 1$, the tth Fermat number, N is a power of 2, and α is an Nth root unit (i.e., $\alpha^N = 1 \bmod F_t$ and $\alpha^m \neq 1$, $1 \leqslant m < N$). The notation $\langle nk\rangle$ means nk modulo N.

The only FNT's considered here are those in which α has a simple binary representation, so that the multiplications implied in (1) are easy to implement. It is possible to show that for $N = 2^{t+1}$, $\alpha = 2$ is the Nth root of unity [6]. In this case the multiplications in (1) become bit shifts. Agarwal and Burrus [7] showed that $\alpha = 2^{3 \cdot 2^{t-2}} - 2^{2^{t-2}}$ is an Nth root of unity for $N = 2^{t+2}$. Since $\alpha^2 = 2 \bmod F_t$, this α is usually referred to as the square root of 2 and written $\alpha = \sqrt{2}$. For general F_t $(t > 4)$, $N = 2^{t+2}$ is the largest power of 2 for which a transform can be defined. Since $\sqrt{2}$ has a two-bit representation, multiplication by $\sqrt{2}$ can be implemented with one subtraction. In addition, the case of N equal to a power of 2 is important because it allows factorization of the transform into a structure similar to the FFT algorithm for the DFT. These properties of the FNT are explained in detail in [7].

A fundamental constraint imposed by the transform definition is that the wordlength of the arithmetic (determined by F_t) is linearly related to the transform length. For the case $\alpha = \sqrt{2}$, $N = 4 \times$ wordlength. It is possible to ease this constraint by using a two-dimensional implementation of the one-dimensional convolution [8]. With $\alpha = \sqrt{2}$ an $N \times N$ two-dimensional transform can be defined with $N = 2^{t+2}$. Using this two-dimensional transform a one-dimensional circular convolution of length $\frac{1}{2}N^2$ can be implemented. A 50 percent loss in convolutional efficiency is incurred by using a two-dimensional transform, because with a length N^2 one-dimensional FFT, a length N^2 one-dimensional circular convolution can be realized.

Returning to the definition in (1), note that the transform is defined in the algebraic system of integers modulo F_t. Thus, the implementation of circular convolution using (1) will result in circular convolution modulo F_t. In particular, if the circular convolution

$$Y_n = \sum_{k=0}^{N-1} x_k h_{\langle n-k \rangle} \tag{2}$$

is computed using FNT's the result will be

$$\hat{Y}_n = \left\{ \sum_{k=0}^{N-1} x_k h_{\langle n-k \rangle} \right\} \bmod F_t. \tag{3}$$

It is possible to determine Y_n from $\hat{Y}_n$ if and only if Y_n is known *a priori* to lie in the set $[P, P + F_t - 1]$. Using *a priori* knowledge of the ranges of $\{x_k\}$ and $\{h_k\}$ an estimate of the output range Y_n can be made.

A consequence of the fact that the wordlength of the system must be a power of 2 is that it is not possible to tradeoff wordlength versus performance as is commonly done in the realization of digital filters. However, the computation of the convolution using the FNT (or any number theoretic transform) is exact. That is, after the quantization of the input data and the filter coefficients, no additional quantization noise is introduced in the filtering process. Thus, the need for simulations of the filtering process is reduced to determining two wordlengths as opposed to present efforts that involve all the internal precisions of the calculation [10].

A more complete discussion of all these theoretical issues can be found in [1]-[10]. In the following sections our attention will focus on the hardware issues encountered in realizing the FNT.

III. REPERESENTATION OF NUMBERS MODULO $2^t + 1$[1]

In this section a new data coding scheme is described for performing arithmetic modulo $2^t + 1$. The main result is that modulo $2^t + 1$ arithmetic can be implemented in a manner that is similar to 1's complement arithmetic (i.e., modulo $2^t - 1$ arithmetic). A description of the arithmetic operations of the FNT when the data are encoded in 2's complement notation can be found in [7]. The arithmetic operations of interest are addition, subtraction and multiplication by a power of 2, because these are the basic operations in the butterfly of the FNT using an FFT-like structure.

Recall that in the ring of integers mod $2^t + 1$, there are $2^t + 1$ elements. Thus, $t + 1$ bits are needed to represent all possible numbers. If a binary coding scheme were used, then whenever the most significant bit (MSB) was one, all the other bits would be zero. This combination would represent the number 2^t. Thus, the tth bit is used only in this one case. A new coding scheme is proposed in which the collection of $t + 1$ bits $[b_t b_{t-1} \cdots b_0]$ represents the number B in the following way.

1) If $b_t = 1$, then $B = 0$.

2) If $b_t = 0$, then $B = \sigma_{t-1} 2^{t-1} + \sigma_{t-2} 2^{t-2} + \cdots + \sigma_0$ where

$$\sigma_j = \begin{cases} 1 & \text{if } b_j = 1 \\ -1 & \text{if } b_j = 0. \end{cases}$$

Example 1: Letting $t = 4$

1 0 0 0 0 represents zero
0 1 0 1 0 $2^3 - 2^2 + 2 - 1 = 5$
0 0 0 1 1 $-2^3 - 2^2 + 2 + 1 = -9 = 8 \bmod 17$

and

1 0 1 0 1 is an illegal combination.

Ordinarily, the representation proposed would yield only odd numbers. However, the use of modulo $2^t + 1$ arithmetic allows both even and odd numbers to be represented. To see this, note that

$$\begin{aligned} B &= (2b_{t-1} - 1) 2^{t-1} + (2b_{t-2} - 1) 2^{t-2} + \cdots + (2b_0 - 1) \\ &= (b_{t-2} 2^{t-1} + \cdots + 2b_0 - b_{t-1} + 2) \bmod (2^t + 1). \end{aligned} \tag{4}$$

The term in brackets takes on all values from +1 to 2^t and the special case of zero is handled by $b_t = 1$, so all numbers are represented.

Consider arithmetic operations using this number representation. First of all, multiplication by a power of 2 is trivial. If the number is zero (i.e., $b_t = 1$), you do nothing. If the number is nonzero, the low-order t bits are circularly shifted to the left a number of places equal to the power of 2, and a bit is replaced by its complement as it enters the least significant bit (LSB) position. This is a consequence of the fact that $2^t = -1 \bmod (2^t + 1)$.

Example 2: Letting $t = 4$ and $2^t + 1 = 17$, 8 is represented as 0 0 0 1 1. Applying the above rule, $8 \times 2 = 16 =$ 0 0 1 1 1; and 0 0 1 1 1 $= 8 - 7 = -1 = 16 \bmod 17$. Further, $8 \times 8 =$ 0 1 1 1 0 $= 14 - 1 = 13 = 64 \bmod 17$.

In a hardware implementation the MSB is used as a control bit. If it is one then the number is zero and the rotation is inhibited. This is charcteristic of all operations using this new coding system.

[1] The representation offered here is a special case of a more general class of representation as noted by L. Leibowitz of NRL.

Another easy operation is that of forming the negative of a number. Obviously, this is done by complementing the low-order t bits except in the case where $b_t = 1$. Again, the MSB is a control bit that inhibits the operation if it is one. Since we know how to form the negative of a number and multiply by 2, the only operation left to consider is addition.

If either or both of the operands for addition are zero (i.e., $b_t = 1$), then there is no addition to take place; so these special cases can be sensed and the addition inhibited. Now consider the addition of two numbers A and B where $A \neq 0$ and $B \neq 0$. Let

$$A = a_t a_{t-1} \cdots a_0 \qquad \text{with } a_t = 0$$

and

$$B = b_t b_{t-1} \cdots b_0 \qquad \text{with } b_t = 0.$$

Interpret the t LSB's of A and B as the binary representation of $\hat{A}$ and $\hat{B}$, and form the sum of $\hat{A}$ and $\hat{B}$ using unsigned binary addition to obtain $\hat{C}$. That is,

$$\begin{array}{l} \hat{A} = a_{t-1} 2^{t-1} + a_{t-2} 2^{t-2} + \cdots + a_0 \\ +\hat{B} = b_{t-1} 2^{t-1} + b_{t-2} 2^{t-2} + \cdots + b_0 \\ \hline \hat{C} = c_t 2^t + c_{t-1} 2^{t-1} + \cdots + c_0. \end{array} \tag{5}$$

It is possible to deduce from $\hat{C}$ the desired sum $C = (A + B) \cdot \bmod (2^t + 1)$. Since $A = 2\hat{A} + 2 \bmod (2^t + 1)$ and $B = 2\hat{B} + 2 \bmod (2^t + 1)$, $C = 2\hat{A} + 2\hat{B} + 4 \bmod (2^t + 1)$. If C can be expressed as $C = 2\overline{C} + 2 \bmod (2^t + 1)$ with $\overline{C}$ a t-bit number, then the t bits of $\overline{C}$ are the t-LSB's of C. There are two cases, depending on the value of c_t. If $c_t = 1$, then $\hat{C} = +2^t + C' = C' - 1 \bmod (2^t + 1)$.

Thus, $C = (2\hat{A} + 2\hat{B} + 4) \bmod (2^t + 1) = (2\hat{C} + 4) \bmod (2^t + 1) = (2C' + 2) \bmod (2^t + 1)$, and $\overline{C} = C'$.

If $c_t = 0$ then $C = 2C' + 4 \bmod (2^t + 1)$ and the answer is $\overline{C} = C' + 1$. However, this results in an extra level of add as in the case of 1's complement arithmetic. In 1's complement the output carry is added to the LSB. In this new mod $2^t + 1$ arithmetic, one takes the output carry, complements it and adds it to the LSB. Thus, the initial claim that this new arithmetic is only as complex as 1's complement is justified. There is a small amount of additional complexity due to the control bit, but this acts only as an inhibit signal.

Example 3: Let $t = 4$ and $2^t + 1 = 17$.

1)
```
  0 1 0 1 0  =  5
  0 0 0 1 1  =  8
  ---------
    1 1 0 1  =  0 1 1 1 0  =  14 - 1 = 13
```

2)
```
  0 1 0 1 0  =  5
  1 0 0 0 0  =  0      MSB = 1 inhibits the addition
  ---------
  0 1 0 1 0  =  5
```

3)
```
  0 1 0 1 0  =  5
  0 0 1 0 1  =  5 - 10 = -5 = 12 mod 17
  ---------
  1 1 1 1 1
        + 1
  ---------
  1 0 0 0 0  =  0 mod 17.
```

In this last example note that the second add automatically produced the control bit indicator for the special case of zero. Now let us examine how one converts from a binary coded representation of numbers to this new representation.

The code conversion between a binary representation and this new code falls into two cases. Let B be a number which is represented in both codes. Let $b_t b_{t-1} \cdots b_0$ be the binary representation of B and $a_t a_{t-1} \cdots a_0$ the new representation. Also, let $\hat{B}$ be the number represented by interpreting $a_{t-1} a_{t-2} \cdots a_0$ as a binary code. The conversion rules are as follows.

1) If $B = 0$, then $a_t = 1$, $\hat{B} = 0$, and $a_k = 0$ for $k = 0, 1, \cdots t$. This is a special case and is done separately.

2) If $B \neq 0$, then $a_t = 0$ and $B = (2\hat{B} + 2) \bmod (2^t + 1)$. Conversion from the new code to binary is implemented by forming $2\hat{B} + 2$ and comparing this sum to 2^t. If the sum is larger than 2^t; then $2^t + 1$ is subtracted to give the proper binary representation of B.

If the binary representation is given, the sum $B + 2^t - 1$ is formed. If the result is odd, $2^t + 1$ is subtracted; and finally, this result is right-shifted one place. The resulting t bits are the t LSB's $a_{t-1} \cdots a_0$.

Example 4: Let $t = 4$ and $2^t + 1 = 17$.

1) $B = 1\ 0\ 0\ 0\ 0 = 16$

$$\hat{B} = \left(\frac{16 + 15 - 17}{2}\right) = 7.$$

The new representation is $A = 0\ 0\ 1\ 1\ 1$.

2) $A = 0\ 0\ 1\ 0\ 0$
$\hat{B} = 4$
$B = 2 \cdot 4 + 2 = 10 = 0\ 1\ 0\ 1\ 0.$

3) $B = 0\ 0\ 0\ 0\ 0$ if and only if
$A = 1\ 0\ 0\ 0\ 0.$

In the next section the hardware system will be described and this new coding scheme will be applied to the logic design of the butterfly of the FNT algorithm.

IV. System Description

The FNT prototype hardware is a realization of a 64-point transform in the finite field of integers modulo $2^{16} + 1$. The fast FNT algorithm implemented is a radix-2 constant geometry decimation in frequency (DIF) decomposition of the FNT. Fig. 2 shows a flow diagram of this algorithm for a 16-point transform [11]. Although the constant geometry structure does not allow in-place calculation of the transform, it simplifies the memory addressing because the addressing does not vary from stage to stage. The price one pays for this simplification is a doubling of the memory size required for the transform. However, 128 words per chip is a convenient level of integration with ECL, thus making the constant geometry structure attractive.

Fig. 3 is a block diagram of the complete system showing the four major subsystems. The computational element (CE) is a radix-2 DIF butterfly for the fast FNT algorithm; the memory element contains 128 17-bit words for use as intermediate storage during the transform; the control element is a hard-wired implementation of the fast FNT algorithm; and the

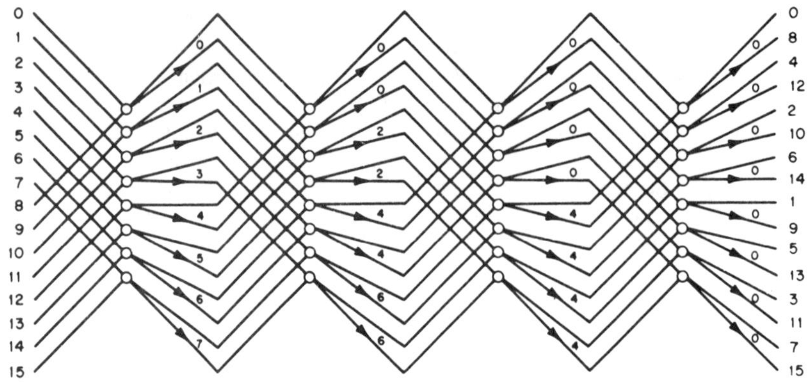

Fig. 2. Flow diagram of a radix-2, 16 point, constant geometry FFT algorithm using the decimation in frequency structure.

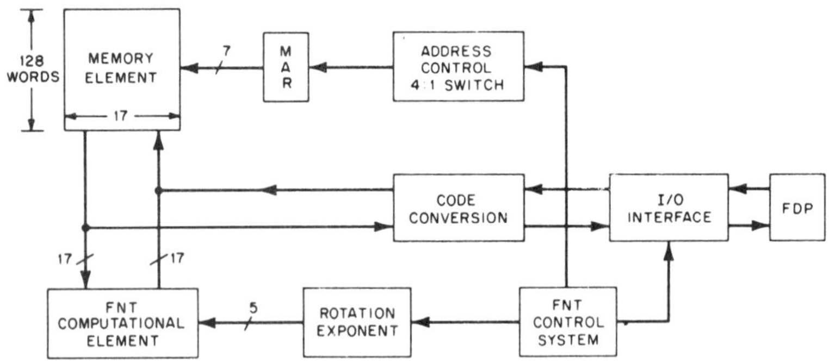

Fig. 3. Block diagram of the 64-point FNT hardware system illustrating the data flow between the major subsystems.

input/output (I/O) section provides the interface with the fast digital processor (FDP).

The goal in building this hardware was to construct a CE that would operate at a clock rate of 40 MHz. In order to achieve this speed, ECL 10 K circuits were used. The basic gate in this logic family as a propagation delay of 2 ns, and thus these circuits are well suited for very high-speed systems. Even with such high-speed logic circuits, two levels of reclock and fast carry addition were used in the CE to realize a working system that runs reliably at 38 MHz.

Fig. 4 shows a functional diagram of the butterfly which consists of an adder, a subtracter, a rotator, input buffer registers R_A and R_B, reclock registers R_W, R_X, and R_Y and an output register R_Z. Register transfers are made at each clock pulse, so that data are always flowing through the CE as would be the case in a pipelined fast FNT. As the timing diagram in Fig. 5 shows, the output of the butterfly is only written into memory from R_Z at t_4 and t_5. During the other clock epochs the contents of R_Z may be changing, but this does not affect the algorithm.

Adder (Subtracter) Logic: In Section III a nonstandard coding for data manipulation in the FNT was derived. Recall that the rule for addition of two nonzero numbers $A = [a_{16} a_{15} \cdots a_0]$ and $B = [b_{16} b_{15} \cdots b_0]$ is as follows.

Step 1: Add the 16 LSB's of A and B with the carry in equal to zero.

Step 2: Complement the carry out from Step 1 and add it to the sum from Step 1.

If either A or B is zero (i.e., b_{16} or $a_{16} = 1$), then the carry must be inhibited. Finally, if both A and B are zero the MSB of the sum is set to one. Fig. 6 shows a realization of the addition process. The structure of Fig. 6 is inefficient in two respects. First of all, two 16-bit adders are required, although the second one is simple because one input is zero. Secondly, the addition is very slow because the carry must propagate through both 16-bit adders. The use of carry lookahead (CLA) logic as in Fig. 7 will improve both situations. Now let us look at the details of the implementation using ECL building blocks.

The 16-bit adder can be realized using 4 MC10181 arithmetic logic units (ALU's). In addition to producing the sum outputs, the ALU's also produce carry propagation and carry generate information for use in a CLA block. Thus, the CLA logic of Fig. 7 is physically spread between the ALU's and a CLA block (MC10179). The addition process can be speeded up further if the carries into each ALU are formed in parallel using the output of the CLA block. Then the add time will be reduced by 3 X (propagation delay time from C_n to C_{n+4}). For the MC10181 this is approximately 10 ns which is significant for realizing a 25-ns clock epoch. Fig. 8 shows the final realization of the 25-ns adder. The logic expressions for the carries were derived from the fact that $C_{n+4} = P_n C_n + G_n$ for each ALU.

Worst-case add time was calculated to be 24 ns and was measured as 21 ns. The flip-flop setting time and setup time account for the remainder of the 25-ns epoch.

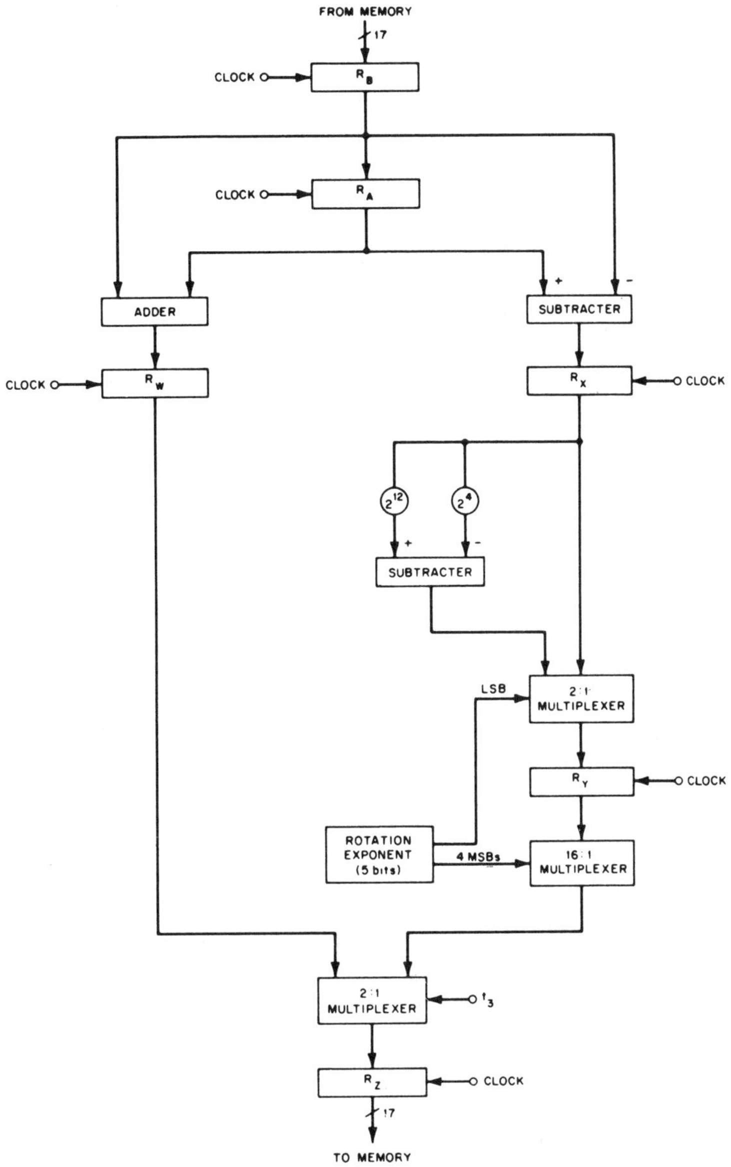

Fig. 4. Register transfer diagram of the CE of the FNT algorithm. Path on the left implements $A + B$ and the path on the right $(A - B)\sqrt{2}^K$.

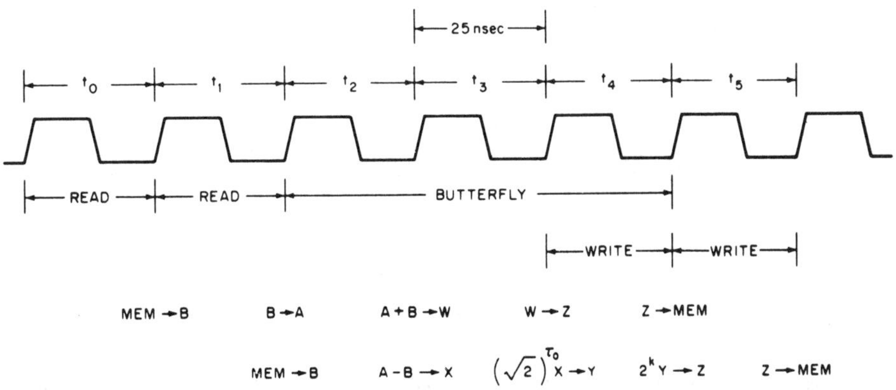

Fig. 5. Timing diagram for the internal clocking operation of the FNT butterfly. Register transfers at each clock pulse are also shown.

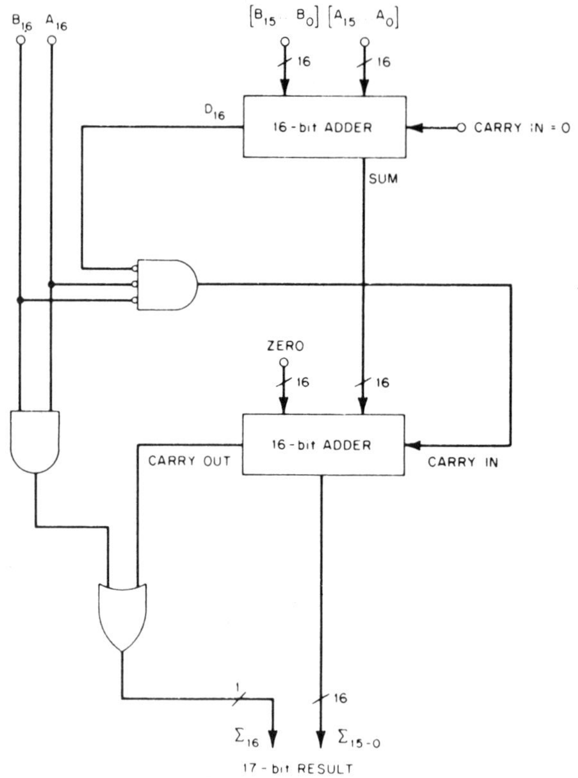

Fig. 6. Implementation of additon modulo the Fermat number $(2^{16} + 1)$ using two 16-bit adders and the new coding scheme.

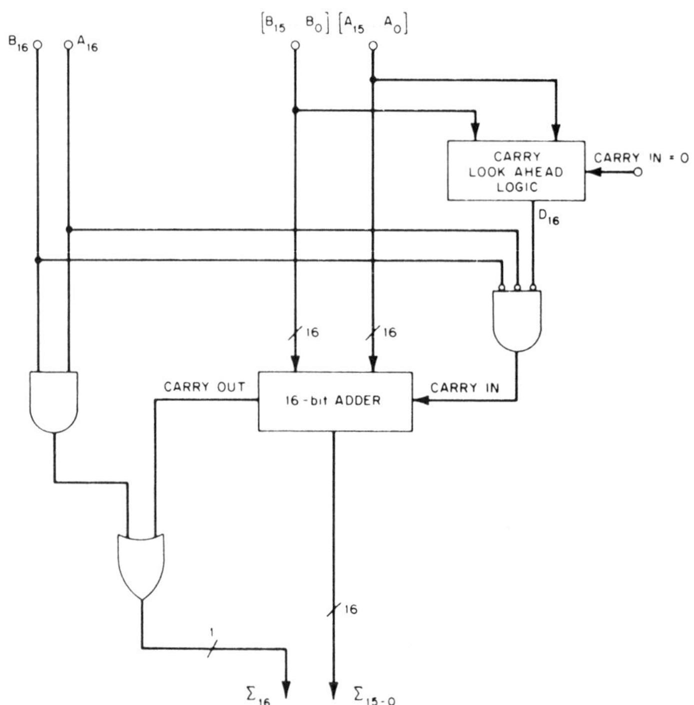

Fig. 7. Improvement of the Fermat number adder by the introduction of CLA logic to produce the end around carry required in Fig. 6. *Note:* D_{16} is the carry generated by adding the two numbers as in normal binary addition and must be complemented before being used as the end around carry.

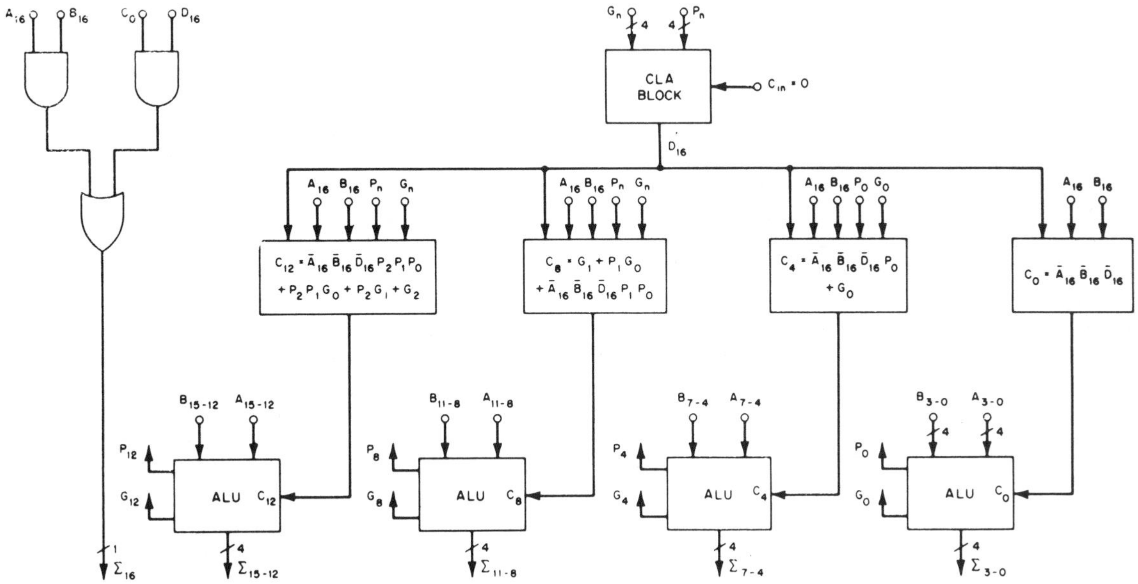

Fig. 8. Detailed logic diagram of the Fermat number adder using fast carry techniques. Addition time is approximately 21 ns.

As noted above, the subtracter, $A - B$ can be implemented by complementing B and adding it to A. The use of the MC10181 ALU allows the B input to be complemented internal to the 10181 via mode control. This results in a slight design change in the adder unless the complement is inhibited when B is zero.

The addition $A + B$ completes the calculation of one output of CE. The result is held in the register R_W and then is moved to R_Z to be written back into memory. For the other output of the CE, the quantity $A - B$ is stored in the reclock register R_X for subsequent rotation by a power of $\sqrt{2}$.

Rotation by $\sqrt{2^k}$: The rotation by $\sqrt{2^k}$ is split into two stages requiring a 25-ns epoch. In the first stage, the quantity $X = A - B$ is multiplied by $\sqrt{2}$ if k is odd. The $\sqrt{2}$ multiplier is merely a subtracter. However, since the output is zero whenever the input is zero, some simplification of the subtracter logic results, and the subtraction time is reduced. A 2:1 multiplexer at the output of the subtracter selects whether the input is to be multiplied by $\sqrt{2}$ or by 1, and is controlled by the LSB of k. The result of this calculation is stored in the reclock register R_Y. The second stage of the rotation is a multiplication by a power of 2, namely $[k/2]$. ([ ] denotes the greatest integer function.) This multiplication is implemented as a 16:1 multiplexer controlled by the upper four bits of the binary representation of the power of 2. Actually, the 16:1 multiplexer is realized as a cascade of two 4:1 multiplexers with inverters for the end around shifts. The shifting network is followed by a 2:1 multiplexer which selects which butterfly output ($A + B$ or $2^k \cdot Y$) is to be stored in R_Z and then written back into memory. This multiplexer is controlled by t_3 and its output is $(2^k \cdot Y) \cdot \bar{t}_3 + W \cdot t_3$.

The logic for the CE consists of 90 IC's, all of which are located on one board (18 × 16 in). Referring to Fig. 1, the total FNT system is packaged on two boards which have been strapped together. The CE logic resides on the left-hand board (with the 24 pin ALU's), and the memory and control are located on the other board (the memory chips are the white packages).

V. Hardware Comparison of the FNT Versus the FFT

Since one purpose in building a hardware prototype of the FNT was to gain a working knowledge of the amount of hardware needed to implement an FNT convolver, it is appropriate to compare this method with the standard FFT implementation of convolution. It is impossible to make a comparison that will apply in all cases, or even a majority of cases. Therefore, a specific case has been chosen for study; namely, digital filter implementation for radar signal processing. The problem areas to be described below are representative of the general problems associated with FNT convolver implementation.

The signal bandwidths encountered in radar signal processing (10-30 MHz) require a pipeline architecture for either the FNT or the FFT [12]. Typically, the overlap-save version of high-speed convolution is employed for real-time processing of the radar returns and a 50 percent overlap of the input data is common. Furthermore, the length of the convolution to be implemented is assumed to be large (e.g., 512 or greater). Two cases will be considered: a length 1024-linear convolution of real data and a length 1024 convolution of complex data. Four measures of hardware complexity are the basis of comparison: the number of butterflies per output point, the number of reference spectrum multiplies per output point, the total amount of interstage delay line memory in the forward and inverse transforms, and the total amount of reference spectrum memory. The FFT implementation will be considered first.

For either real or complex data, it is assumed that the FFT

implementation employs an 11-stage radix-2 pipeline FFT in both the forward and inverse transforms. (*Note:* For real data, it is possible to do a length N transform with one length $N/2$ transform and some overhead to combine the real and imaginary parts [13]. However, the overhead amounts to an additional butterfly so there is little, if anything to gain by using this fact in a pipeline FFT.) Much work has been done on the hardware complexity of pipeline FFT's and the four measures of complexity we are considering here were also used in [14]. For the case of a length 2048 pipeline FFT convolver the number of butterflies per output point is $2 \log_2 N = 22$, assuming 50 percent convolution overlap. Likewise, two reference spectrum multiplies must be done per output point. The amount of interstage delay line memory can be calculated from the formula

$$\text{IDM} = \frac{N}{2}(r+1) \tag{6}$$

where r is the radix of the transform. Thus, for two radix-2 pipelines, the total is $3N = 6144 = 6$ K words of memory. Finally, the reference spectrum requires 2 K words of memory. Now we turn our attention to the FNT.

A pipeline FNT structure is identical to the pipeline FFT except in the butterfly where rotation by $e^{j2\pi k/N}$ (in the FFT case) is replaced by multiplication by $\sqrt{2^k}$. Thus, many of the results quoted above are applicable to the FNT. Since the FNT naturally processes real input data, the cases of real and complex convolution require different realizations. In both cases, however, a two-dimensional implementation of the convolution is required [8]. The two arrays to be convolved are H and X, where

$$X = \underbrace{\begin{bmatrix} x(0) & x(L) & \cdots & x(N-L) \\ x(1) & & & \\ \vdots & & & \\ x(L-1) & x(2L-1) & \cdots & x(N-1) \\ 0 & 0 & \cdots & 0 \\ \vdots & \vdots & & \vdots \\ 0 & 0 & & 0 \end{bmatrix}}_{64} \; 64 \qquad L = 32 \tag{7a}$$

and

$$H = \begin{bmatrix} h(N-L+1) & h(1) & \cdots & h(N-2L+1) \\ \vdots & \vdots & & \\ h(N-1) & h(L-1) & & \vdots \\ h(0) & h(L) & & \vdots \\ h(1) & & & \\ \vdots & \vdots & & \\ h(L-1) & h(2L-1) & \cdots & h(N-1) \end{bmatrix} \tag{7b}$$

The length 1024 convolution of real signals can be implemented with a 64×64 transform. The input data is the array X of (7) and it is advantageous to exploit the fact that half of the X array is zero by doing the row transforms first. Hence, 96 length 64 FNT's must be computed for the entire two-dimensional transform. The total number of butterflies for the complete convolution is $2 \times 96 \times 32 \log_2 64 = 9 \times 2^{12}$, or 36 butterflies per output point. This is reflected in the structure of Fig. 9 where there are 36 butterflies working in parallel —six in each 64-point FNT. Since 2^{12} reference function multiplies must be done during each convolution, four reference multiplies per output point are required. The interstage delay memory requirement is calculated from the individual 64-point transforms. The first and last 64-point FNT's are computing 32 transforms simultaneously. This is accomplished by making all interstage delay lines 32 times as long as in a standard pipeline and by modifying the control to switch at 1/32 the speed of a normal pipeline. That is, the rotation exponents and commutator switches are only changed at every 32nd clock pulse. The other four FNT's employ a normal pipeline structure. Applying (6), the total interstage delay memory is $(2 \cdot 3 \cdot 32 \cdot 32) + (4 \cdot 3 \cdot 32) = 6.4$ K. Finally, it is easy to verify that the amount of reference function memory is 4 K.

If the signals to be convolved are complex; then one possible implementation is to handle the real and imaginary parts of the data separately. The number of butterflies per output point, the interstage delay memory, and the reference spectrum memory are all doubled. However, the number of real multipliers for the reference function multiply is quadrupled because the multiplication is now complex. Table I summarizes the three systems versus the four hardware measures. Notice that the FNT always requires more memory and more computational elements than the FFT. Hardware savings are possible because most of the hardware cost of the FFT is concentrated in the butterfly elements (up to 80 percent) and because the FNT butterfly requires from one-third to one-sixth the hardware of an FFT butterfly. These remarks apply to the FNT when the data to be convolved are real, but when the data are complex, the situation becomes much worse because all measures of hardware complexity are increased by a factor of 2 or 4. Therefore, we will concentrate on the case of real convolution in the following discussion.

In order to be more specific, let us divide the hardware cost of any convolver into two parts, the percentage of hardware for the butterfly elements and the percentage of hardware for the rest of the system. We obtain the following relations between the FFT and the FNT for length 1024 convolution:

$$B_{\text{FNT}} = 36/22\,\lambda B_{\text{FFT}} \tag{8}$$

$$B_{\text{FFT}} = \mu T_{\text{FFT}} \tag{9}$$

$$(T_{\text{FNT}} - B_{\text{FNT}}) = \nu(T_{\text{FFT}} - B_{\text{FFT}}) \tag{10}$$

where B denotes the butterfly cost and T the total convolver cost. λ is the ratio of FNT butterfly cost to FFT butterfly cost on a per unit basis, μ is the percentage of FFT hardware cost that is concentrated in the CE's, and ν is the ratio of the nonbutterfly cost in the FFT to the nonbutterfly hardware cost in the FNT.

Normalizing $T_{\text{FFT}} = 1$, the total hardware for the FNT is

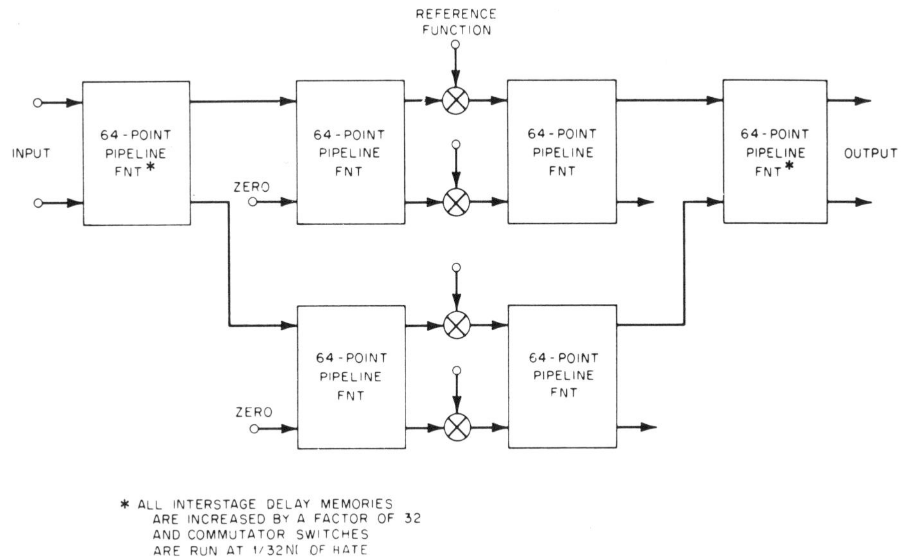

Fig. 9. Implementation of length 1024 aperiodic convolution of real signals via pipelined 64 × 64 two-dimensional FNT's. Zeros in the input array have been exploited to minimize the amount of hardware.

TABLE I
HARDWARE MEASURES FOR FFT AND FNT IMPLEMENTATIONS OF LENGTH 1024 APERIODIC CONVOLUTION OF REAL SIGNALS

	FFT Real or Complex Data	FNT Real Data	FNT Complex Data
Butterflies	22	36	72
Reference Multiplies	2	2	16
Interstage Delay Memory	6 K complex words[a]	6.4 K real words[b]	12.8 K real words
Reference Spectrum Memory	2 K complex words	4 K real words	8 K real words

[a]One complex word will contain approximately 27 bits for typical high-precision radar applications.
[b]One real word contains 33 bits for FNT.

$$T_{\text{FNT}} = \left(\frac{18}{11}\lambda - \nu\right)\mu + \nu \tag{11}$$

and the savings can be expressed as

$$S_{\text{FNT}} = 1 - T_{\text{FNT}} = (1 - \nu) + \mu\left(\frac{18}{11}\lambda - \nu\right). \tag{12}$$

Thus, the problem is now that of determining realistic values for the three ratios λ, μ, and ν. We will argue that a nominal value for ν is two, and then plot the savings S_{FNT} as a function of λ and μ.

If we assume that the accuracy requirements of the system are high, then the FFT implementation should use a hybrid floating point scheme as described in [10]. This study described a complex data format using 27 bits—11 bits each for the real and imaginary parts of the mantissa and 5 bits for a common exponent. In contrast, the FNT would employ modulo $2^{32} + 1$ arithmetic, implying a 33-bit data word. For the nonbutterfly hardware, three components must be considered. First, the numbers of reference multiplies being compared are two 11 × 11 complex multipliers and four 32 × 32 real multipliers. Since the hardware complexity of an $n \times n$ array multiplier (which would be required at radar speeds) is proportional to n^2, the four real multipliers amount to about $(4 \cdot 32^2)/(2 \cdot 4 \cdot 11^2) = 4\frac{1}{2}$ times the hardware of the two complex multipliers. Secondly, the interstage delay memory is 6 K words of 27 bits each for the FFT versus 6.4 K words of 33 bits each for the FNT, or a factor of 1.3 in favor of the FFT. Similarly, the reference function memory differs by a factor of 2.44 in favor of the FFT. Depending on the detailed logic realization, the value of ν will vary, but $\nu = 2$ conveys the fact that the two-dimensional convolution using the FNT requires twice as much nonbutterfly hardware as the FFT. This is a functional limitation of the FNT for long convolutions. At this point we can isolate the impacts of the butterfly hardware on the possible savings for the FNT. Fig. 10 shows a plot of hardware savings versus the percentage of butterfly hardware in the FFT (μ). Several curves are shown with λ as a parameter. λ is the ratio of FNT butterfly hardware to FFT butterfly hardware. The range of values for λ were chosen by comparing IC counts for the FNT butterfly design and a proposed FFT butterfly design used in a digital matched filter for radar signal processing [10]. The value of ν was assumed to be two and if it were larger, then the horizontal intercept of the lines would be moved to the right to a higher value of μ.

Independent of the exact values of λ, μ, and ν, Fig. 10 clearly shows that the FNT will provide hardware savings over the FFT only when the FFT hardware is dominated by the CE cost as in the case of a pipeline architecture.[2] Furthermore,

[2]*Note:* As transform size increases the memory cost of a pipeline increases faster than the CE cost and will become a significant fraction of the total cost for large transforms (e.g., length 16 K or 32 K).

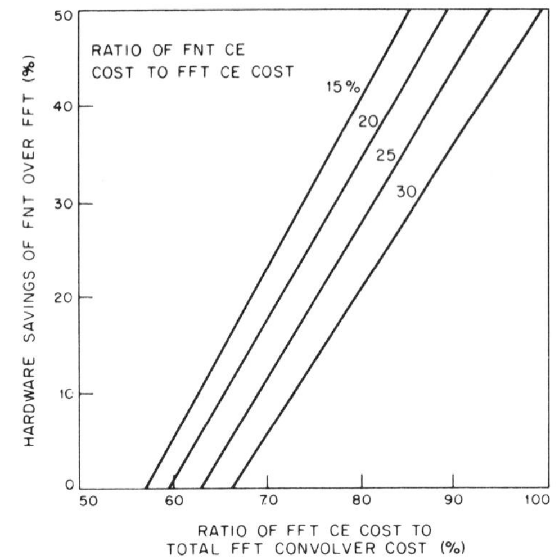

Fig. 10. Hardware savings tradeoff between the FNT and the FFT for a length 1024 aperiodic convolution of real signals. It is assumed that the nonbutterfly hardware required in the FNT is twice that of the FFT (i.e., $\nu = 2$).

the signals to be convolved must be real, because complex data essentially require two real FNT convolvers. Although short convolutions (e.g., length 64) have not been discussed here, it is worth mentioning that there is a good chance for significant hardware savings over the FFT because a one-dimensional FNT can be used. This means that the nonbutterfly hardware of the two systems will be approximately the same (i.e., $\nu \approx 1$) and the savings will begin at a value of μ near zero.

VI. Summary

A hardware implementation of the FNT has been built. In the course of designing this machine, a new representation for numbers modulo $2^t + 1$ was derived to facilitate the arithmetic operations in the FNT butterfly. The design goal of a 40 MHz clock rate through the butterfly was nearly achieved, with the final reliable clock rate being 38 MHz.

A comparison with the pipeline FFT for special purpose convolution hardware has been made based on the present FNT design. The conclusion of this study is that the FNT is a useful alternative to the FFT if the signal to be filtered is real and the computational elements are a major part of the overall convolver cost as in a pipeline architecture. For convolutions of modest length (e.g., length 64 convolution) the savings of the FNT are significant. However, for longer convolutions where one must use two-dimensional convolution for the implementation of one-dimensional convolution, the savings in multiplier hardware are offset by increased transform size and the corresponding increase in memory size and reference spectrum multiplier hardware. In this case, the FNT may still offer a small savings in the total amount of hardware versus the FFT, depending on the specific details of the hardware implementation. Finally in the implementation of two-dimensional convolution there is no penalty due to increased memory size and the savings in multiplier hardware will translate into savings for the overall convolver system.

Acknowlegment

Special thanks go to B. Gold and C. Rader who first introduced the author to this problem and to A. Huntoon and E. Aho whose hardware expertise contributed to the successful completion of the project.

References

[1] J. M. Pollard, "The fast Fourier transform in a finite field," *Math. Comp.*, vol. 25, pp. 365-374, Apr. 1971.

[2] A. Schonhage and V. Strassen, "Fast multiplication of large numbers" (in German), *Comput.*, vol. 7, pp. 281-292, 1971.

[3] P. J. Nicholson, "Algebraic theory of finite Fourier transforms," *J. Comput. Syst. Sci.*, vol. 5, pp. 524-547, 1971.

[4] T. G. Stockham, "High speed convolution and correlation," in *1966 Spring Joint Comput. Conf., AFIPS Conf. Proc.*, vol. 28. Washington, DC: Spartan, 1966, pp. 229-233.

[5] D. E. Knuth, *The Art of Computer Programming*, vol. 2, *Seminumerical Algorithms*. Reading, MA: Addison-Wesley, 1969, pp. 269-275 (2nd printing).

[6] C. M. Rader, "Discrete convolution via Mersenne transforms," *IEEE Trans. Comput.*, vol. C-21, pp. 1269-1273, Dec. 1972.

[7] R. C. Agarwal and C. S. Burrus, "Fast convolution using Fermat number transforms with applications to digital filtering," *IEEE Trans. Acoust., Speech, Signal Processing*, vol. ASSP-22, pp. 87-97, Apr. 1974.

[8] ——, "Fast one-dimensional digital convolution by multidimensional techniques," *IEEE Trans. Acoust., Speech, Signal Processing*, vol. ASSP-22, pp. 1-10, Feb. 1974.

[9] ——, "Number theoretic transforms to implement fast digital convolution," *Proc. IEEE*, vol. 63, pp. 550-560, Apr. 1975.

[10] R. J. Purdy *et al.*, "Digital signal processor designs for radar applications," Mass. Inst. Technol. Lincoln Laboratory, Lexington, MA, Tech. Note 1974-58, vols. 1 and 2, Dec. 1974.

[11] L. R. Rabiner and B. Gold, *Theory and Application of Digital Signal Processing*. Englewood Cliffs, NJ: Prentice-Hall, 1974.

[12] P. E. Blankenship and E. M. Hofstetter, "Digital pulse compression via fast convolution," *IEEE Trans. Acoust., Speech, Signal Processing*, vol. ASSP-23, pp. 189-201, Apr. 1975.

[13] J. W. Cooley, P. A. Lewis, and P. D. Welch, "The fast Fourier transform algorithm: programming considerations in the calculation of sine, cosine, and LaPlace transforms," *J. Sound Vib.*, vol. 12, pp. 315-337, July 1970.

[14] B. Gold and T. Bially, "Parallelism in fast Fourier transform hardware," *IEEE Trans. Audio Electroacoust.*, vol. AU-21, pp. 5-16, Feb. 1973.

[15] C. M. Rader, "On the application of the number theoretic transforms of high speed convolution to two-dimensional filtering," *IEEE Trans. Circuits and Systems*, vol. CAS-22, p. 575, June 1975.

A Simplified Binary Arithmetic for the Fermat Number Transform

LAWRENCE M. LEIBOWITZ, STUDENT MEMBER, IEEE

Abstract—A binary arithmetic that permits the exact computation of the Fermat number transform (FNT) is described. This technique involves arithmetic in a binary code corresponding to the simplest one of a set of code translations from the normal binary representation of each integer in the ring of integers modulo a Fermat number $F_t = 2^b + 1$, $b = 2^t$. The resulting FNT binary arithmetic operations are of the complexity of 1's complement arithmetic as in the case of a previously proposed technique which corresponds to another one of the set of code translations. The general multiplication of two integers modulo F_t required in the computation of FNT convolution is discussed.

Introduction

A SET of transforms described as number theoretic transforms (NTT's) with application to the efficient computation of discrete convolution has been described in [1]-[4]. In particular, the NTT with a modulus of the form of the tth Fermat number $F_t = 2^b + 1$, $b = 2^t$, referred to as the Fermat number transform (FNT), has been described in detail by Agarwal and Burrus [3]. The FNT of the time-domain sequence $x(n)$ is of the form

$$X(k) = \sum_{n=0}^{N-1} x(n)\,\alpha^{nk} \bmod F_t, \qquad k = 0, 1, \cdots, N-1$$

where N is the sequence length, α a root of unity of order N, and F_t the modulus describing the finite field or ring of integers [1]. The FNT leads to values of α, N, and F_t that result in transforms that can be efficiently evaluated on a digital computing device using a fast Fourier transform (FFT)-type algorithm [5] with each multiplication accomplished by a single binary word shift and subtraction.

In their software realization of the FNT, Agarwal and Burrus [3] define a binary arithmetic modulo F_t. The utilization of such a modulus requires $b + 1$ bits. The representation of the quantity $2^b = -1 \bmod F_t$ requires the $(b + 1)$th bit. In order to simplify modular arithmetic operations, Agarwal and Burrus limit their realization to b-bit arithmetic. This involves some input quantization error when -1 occurs as an input sample as well as the extremely small, but realistic, probability of a complete data block in error when a -1 occurs as the result of an FNT computation.

It is, of course, desirable to compute the FNT exactly. The difficulties in performing binary arithmetic modulo F_t become apparent when one considers, for example, multiplication or addition in a ring of integers modulo F_t involving the binary representation of -1, 2^b. For example, when $b = 4$ the product of 10000 (-1) with itself is 00001 (+1).

A technique for the exact computation of the FNT and its implementation are described by McClellan [6]. McClellan's approach involves the definition of a new binary code representation for the integers modulo F_t. Given a binary representation of $b + 1$ bits, $A = [a_b, a_{b-1}, \cdots, a_0]$, this new code is described as follows.

If $a_b = 1$, then $A = 0$.

If $a_b = 0$, then $A = \sigma_{b-1} 2^{b-1} + \sigma_{b-2} 2^{b-2} + \cdots + \sigma_0$

where

$$\sigma_j = \begin{cases} 1 & \text{if } a_j = 1 \\ -1 & \text{if } a_j = 0. \end{cases}$$

Thus McClellan's representation of numbers modulo F_t involves normal binary weighting with digits ± 1. This technique results in a representation of the integer 0 as 2^b in binary notation (for $b = 4$, 0 is represented as 10000) and thus the $(b + 1)$th bit is used only to represent 0. The presence of the extra bit in a multiplication or addition operation involving the number representation for 0 is detected and the normal steps in these operations are properly inhibited in order to produce a correct result. Thus binary arithmetic operations with a $b + 1$ bit word are effectively eliminated. McClellan shows that this number representation provides a binary arithmetic modulo F_t for negation, addition, and multiplication by integer powers of 2. This new arithmetic is shown to be similar to and as complex as ordinary 1's complement arithmetic.

In this paper a binary arithmetic modulo F_t of less mathematical complexity than that of [6] will be presented. Although the translations between the quantities in the finite ring and the number representation in this binary arithmetic are trivial, the resulting binary arithmetic operations are identical to those of McClellan. Thus the FNT system design presented in [6] is applicable to this new arithmetic with the exception of the code translation which is mathematically simpler. In addition to binary arithmetic operations involved in the computation of the FNT, this paper also considers general multiplication of integers modulo F_t as required in the overall computation of FNT convolution.

Code Translation

There is a set of code translations that each result in the simplified binary arithmetic modulo F_t to be described here. These translations involve the one-to-one representation of a

Manuscript received November 6, 1975; revised February 12, 1976. This paper represents a part of the research performed under the Edison Memorial Fellowship in partial fulfillment of the requirements for the Doctor of Science degree at the School of Engineering and Applied Science, George Washington University, Washington, DC.

The author is with the Naval Research Laboratory, Washington, DC 20375.

number A in the ring of integers modulo F_t as the binary number corresponding to RA-1 modulo F_t where R is any integer in the ring with an inverse. It can be shown that the code representation of McClellan [6] corresponds to the case of $R = 2^{b-1} + 1$ which for the case of $b = 4$ is $R = 9$. The simplest code translation obviously occurs for $R = 1$ for any value of b. The discussion in this paper will concentrate on the resulting binary arithmetic for the code translation corresponding to $R = 1$.

Diminished-1 Number Representation

To represent all integers in the ring of integers modulo F_t requires $b + 1$ bits. The additional bit is required in order to represent the number $2^b = -1 \bmod F_t$. In order to overcome the problem of performing binary arithmetic with this additional bit, let us describe a modified binary number system. To avoid additions and multiplications involving the additional bit, allow the additional bit to be a 1 only when the number to be represented is 0. This can be easily achieved by subtracting 1 from the normal binary representation of every integer in the ring and corresponds to the above set of code translations with $R = 1$. The normal representation and this diminished-1 representation are indicated in Table I for $b = 4$.

In the diminished-1 number representation the b least significant bits (lsb's) indicate the value of the number. The numbers from 1 to 2^b are represented in order by the binary numbers from 0 to $2^b - 1$. Using this representation, the arithmetic operations necessary to perform convolution by FNT's will be considered.

Binary Arithmetic Operations Modulo F_t

Negation

It can be seen from Table I that each of the negative numbers ($>2^{b-1}$) is the b-lsb complement of its positive counterpart. Thus with this system the negative of a nonzero number in diminished-1 representation is the complement of its b lsb's. Denoting the b-lsb complement by an overbar, this can be shown as follows:

$$\begin{aligned}\overline{A - 1} &= 2^b - 1 - (A - 1)\\ &= 2^b + 1 - A - 1\\ &= (-A) - 1.\end{aligned}$$

If the most significant bit (msb) is 1, the negation is inhibited.

Example 1

13 = 4 mod 17 = $00\bar{0}\bar{1}\bar{1}$ = 01100.

Addition

To perform addition of two numbers represented as $A - 1$ and $B - 1$,

$$(A - 1) + (B - 1) = (A + B - 1) - 1$$

and thus

$$(A + B - 1) = [(A - 1) + (B - 1)] + 1.$$

Since the $(b + 1)$th bit of the addends is used only to inhibit addition if an addend is zero, addition of nonzero addends involves only the b-lsb's. The above equations indicate that a 1 must be added to the sum of two diminished-1 numbers to provide a correct result. When a carry is generated from the b-bit sum, a residue reduction modulo F_t requires the subtraction of a 1 since $2^b = -1 \bmod F_t$ and no corrective addition is necessary.

Thus to add two numbers in diminished-1 representation:

If the msb of either addend is 1, inhibit the addition and the remaining addend is the sum.

If the msb of both addends are 0, ignoring the msb add the b lsb's, complement the carry and add it to the b lsb's of the sum.

The $(b + 1)$th bit or msb of the resulting sum, is the carry out of the bth bit.

TABLE I
CORRESPONDENCE BETWEEN NORMAL AND DIMINISHED-1 REPRESENTATIONS
($b = 4$)

Normal Value	Binary Representation	Diminished-1 Value
0	00000	1
1	00001	2
2	00010	3
3	00011	4
4	00100	5
5	00101	6
6	00110	7
7	00111	8
8	01000	9 (−8)
9 (−8)	01001	10 (−7)
10 (−7)	01010	11 (−6)
11 (−6)	01011	12 (−5)
12 (−5)	01100	13 (−4)
13 (−4)	01101	14 (−3)
14 (−3)	01110	15 (−2)
15 (−2)	01111	16 (−1)
16 (−1)	10000	0

Example 2

```
add   10000   0
      00100   5
      -----   -
      00100   5
```

Example 3

```
add   00111    8
      01101   14
      -----   --
      10100   22 = 5 mod 17
       \
        0
      -----
      00100
```

Example 4

```
add   00011    4
      00101    6
      -----   --
      01000   10
       \
        1
      -----
      01001
```

Example 5

```
add   00100    5
      01011   12      5 mod 17
      -----   --
      01111   17      0 mod 17
       \
        1
      -----
      10000
```

Code Translation

Using the above description of diminished-1 addition, we can present rules for the translation between a binary number B and its diminished-1 representation D.

From binary to diminished-1 representation, perform a diminished-1 addition of B and the binary representation of $2^b - 1$.

Example 6

```
B   01100   12
    01111
    -----
    11011
     \
      0
    -----
D   01011
```

Example 7

```
B   00000   0
    01111
    -----
    01111
     \
      1
    -----
D   10000
```

From diminished-1 to binary representation, complement the msb of D and add it to the b lsb's.

Example 8

```
D = 01011 = 12
      ↘1
B = 01100
```

Example 9

```
D = 10000 = 0
      ↘0
B = 00000
```

Subtraction

Having defined negation and addition in the diminished-1 number system, we can perform subtraction by negating the subtrahend and adding it to the minuend.

Example 10

```
subtract   7   00110
          -5   01011
           2   10001
                  ↘0
               00001 = 2
```

Multiplication by Powers of 2

In performing a multiplication if the multiplier or multiplicand are 0, as detected by the presence of a 1 in the $(b + 1)$th bit, the multiplication is inhibited and the product is zero. To perform multiplication of diminished-1 numbers by powers of 2, let us consider the following:

$$(A - 1) \cdot 2 = (2A - 1) - 1$$

and thus

$$(2A - 1) = (A - 1) \cdot 2 + 1.$$

Therefore, each multiplication by 2 involves a left shift, ignoring the msb, and a corrective addition of a 1. If the bit shifted out from the bth position is a zero, it is complemented and shifted into the lsb in order to accomplish the addition of a 1. If this bit is a 1, a subtraction of 1 is also required to accomplish a residue reduction. With the corrective addition of +1, these cancel out and a 0 is shifted into the lsb. Thus for each factor of 2, a left-circular shift of the b-lsb's is required and the bit circulated into the lsb is complemented.

Example 11

```
         8 × 11    2³ × 11 = 2 × 2 × 2 × 11    88    3 mod 17

     11  01010
 2 × 11  00100
 4 × 11  01001
 8 × 11  00010 = 3 mod 17
```

In order to realize the full efficiency of the FNT, it is necessary to accomplish multiplication by powers of 2 by a single multibit shift. The above procedure can be extended in a straightforward manner to provide such multiplication.

General Multiplication

The last operation required to carry out convolution with the FNT is a general multiplication by any two integers modulo F_t. To perform a multiplication of the numbers A and B, represented as $A - 1$ and $B - 1$ in the diminished-1 number system,

$$(A - 1)(B - 1) = A \cdot B - (A + B) + 1$$
$$= (A \cdot B - 1) - (A + B - 1) + 1$$

and the desired result is

$$(A \cdot B - 1) = (A - 1)(B - 1) + (A + B - 1) - 1.$$

Thus to carry out such an operation, ignoring the msb perform a binary multiplication of the diminished-1 representations of A and B, add this result to the b-lsb's of the diminished-1 addition of A and B and then perform a residue reduction by a diminished-1 subtraction of the b-msb's from the b-lsb's.[1] As discussed previously, if the msb of either multiplier or multiplicand is 1, the multiplication is inhibited and the result is set to zero.

Example 12

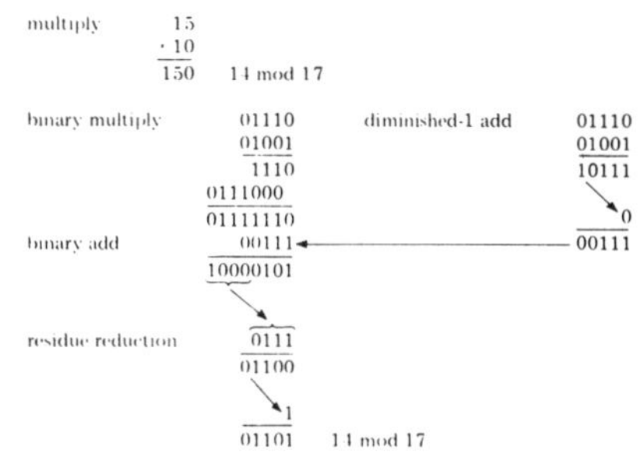

An alternate multiplication technique can also be considered. Assuming that the numbers are determined to be nonzero and a multiplication is required, a translation to normal binary coding is completed. Following a binary multiplication, a residue reduction by diminished-1 subtraction of the b msb's of the product from the b lsb's is performed. The result is the desired product. This is a realization of the equation $(A \cdot B - 1) = A \cdot B - 1$.

Example 13

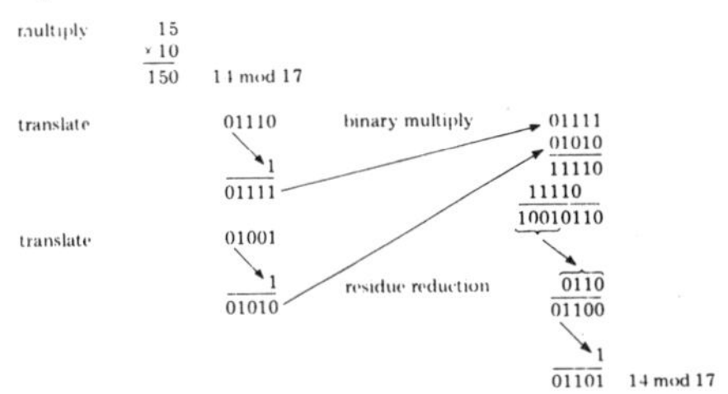

Another general multiplication technique involves a translation of either the multiplier or the multiplicand to normal binary form, the other remaining in diminished-1 representation. The operations involved in comparison to the previous techniques include one diminished-1 to binary translation and a general diminished-1 addition.

In comparing these general multiplication schemes, several factors should be considered. The basic tradeoff involves the general diminished-1 and binary additions as compared to the translation of numbers in diminished-1 representation to normal binary representation. The remaining operations of binary multiplication, residue reduction in diminished-1 notation, and multiplication by zero are identical in the various techniques. Since in most cases the translation from diminished-1 to binary will be simpler and faster than a general binary or diminished-1 addition, the second technique is the most preferable.

[1] This particular general multiplication scheme is not applicable with code translations other than that corresponding to $R = \pm 1$.

CONCLUSION

A binary arithmetic applicable to the exact computation of the FNT has been presented. This arithmetic involves operations on a binary representation of the integers modulo F_t diminished by 1. This diminished-1 representation is mathematically simpler than that of McClellan [6], both being among a general set of such binary code translations. This arithmetic provides simple realizations of all the operations required to compute the FNT. These operations are identical to those described and implemented by McClellan [6] and are of the complexity of 1's complement arithmetic. The general multiplication of two integers modulo F_t in diminished-1 representation can be accomplished in a straightforward manner.

ACKNOWLEDGMENT

The author wishes to express his appreciation to Prof. J. H. McClellan of the Department of Electrical Engineering and Computer Science, Massachusetts Institute of Technology, Cambridge, for his comments and suggestions.

REFERENCES

[1] J. M. Pollard, "The fast Fourier transform in a finite field," *Math. Comput.*, vol. 25, pp. 365-374, Apr. 1971.

[2] C. M. Rader, "Discrete convolutions via Mersenne transforms," *IEEE Trans. Comput.*, vol. C-21, pp. 1269-1273, Dec. 1972.

[3] R. C. Agarwal and C. S. Burrus, "Fast convolution using Fermat number transforms with applications to digital filtering," *IEEE Trans. Acoust., Speech, Signal Processing*, vol. ASSP-22, pp. 87-97, Apr. 1974.

[4] C. M. Rader, "Convolution and correlation using number theoretic transforms," in *Theory and Application of Digital Signal Processing*, L. R. Rabiner and B. Gold. Englewood Cliffs, NJ: Prentice-Hall, 1975, sect. 6.19.

[5] J. W. Cooley and J. W. Tukey, "An algorithm for the machine calculation of complex Fourier series," *Math. Comput.*, vol. 19, pp. 297-301, Apr. 1965.

[6] J. H. McClellan, "Hardware for the Fermat number transform," Lincoln Laboratory, Mass. Inst. Technol., Lexington, Tech. Note ESD-TR-75-149, Apr. 1975.

The Use of Finite Fields to Compute Convolutions

IRVING S. REED, FELLOW, IEEE, AND T. K. TRUONG

Abstract—A transform is defined in the Galois field of q^2 elements $GF(q^2)$, a finite field analogous to the field of complex numbers, when q is a prime such that (—1) is not a quadratic residue. It is shown that the action of this transform over $GF(q^2)$ is equivalent to the discrete Fourier transform of a sequence of complex integers of finite dynamic range. If q is a Mersenne prime, one can utilize the fast Fourier transform (FFT) algorithm to yield a fast convolution without the usual round-off problem of complex numbers.

I. Introduction

Recently Rader [1] defined transforms of the form

$$A_k \equiv \sum_{n=0}^{d-1} a_n 2^{nk} \bmod b$$

where b is either a Mersenne prime,

$$b = 2^p - 1, \qquad \text{for } p = 2,3,5,7,13,17,19,31,61$$

or a Fermat number,

$$b = 1 + 2^{(2^m)}, \qquad \text{for } m = 0,1,2,3,4.$$

He showed that the convolution of two finite sequences can be obtained as the inverse transform of the product of their transforms. The primary advantage of the Rader transform is that multiplications by powers of two are performed by simple bit rotations. Of course, this advantage must be weighed against the difficulties of computing the answer modulo b and of the numeric constraints relating word length, length of sequence d, and the compositeness of d, imposed by the two choices for b [1].

It is well known that Fourier-like transforms can be defined in a Galois field to yield circular convolutions [2]. Addition and multiplication in $GF(q^2)$, where the polynomial $x^2 + 1$ is irreducible, are the same as the operations for complex numbers. We restrict attention to this field. In order to define finite Fourier transforms in $GF(q^2)$, an algorithm for finding primitive elements in the multiplicative subgroups of $GF(q^2)$ is developed in this paper.

Transforms over the finite field $GF(q^2)$ can be used to compute the convolution of two d-point sequences of complex numbers. As will be seen, if q is a Mersenne prime, the arithmetic operations performing the transform require only addition modulo q, the complement operation, and bit rotations of the word. The advantage of this transform on $GF(q^2)$ over the usual discrete Fourier transform of complex numbers is that, for q sufficiently large, one can use this transform over $GF(q^2)$ to transform a sequence of complex integers a_n into the sequence A_k in $GF(q^2)$ for which the inverse transform of A_k on $GF(q^2)$ is precisely the original sequence of complex numbers a_n. Consequently, filtering operations or convolutions without round-off error can be obtained using this transform on a sequence of complex integers. If q is a Mersenne prime, the fast Fourier transform (FFT) algorithm also can be utilized to realize the needed transforms.

Manuscript received May 31, 1974; revised September 24, 1974. This work was supported in part by the U.S. Air Force Office of Scientific Research under Grant AFOSR 75-2798.

The authors are with the Department of Electrical Engineering, University of Southern California, Los Angeles, Calif. 90007.

II. Fourier Transforms on a Finite Field

Let $GF(q^n)$ be a Galois field, let q be a prime, and let the integer d divide $q^n - 1$. Also let the element $r \in GF(q^n)$ generate the cyclic subgroup of d elements, $G_d = (r,r^2,\cdots,r^{d-1},1)$ in the multiplicative group of $GF(q^n)$. Then a transform over this subgroup G_d [2] can be defined by the following

$$A_k = \sum_{n=0}^{d-1} a_n r^{kn}, \qquad \text{for } 0 \le k \le d-1 \tag{1}$$

where d divides $q^n - 1$ and $a_n \in GF(q^n)$, for $0 \le n \le d-1$.

To invert (1), observe first that all elements of G_d satisfy the equation

$$x^d - 1 = 0.$$

However, since $x^d - 1$ factors as

$$x^d - 1 = (x-1)\sum_{k=0}^{d-1} x^k,$$

one has

$$\sum_{k=0}^{d-1} x^k = 0, \qquad \text{for } x \ne 1 \text{ and } x \in G_d \subset GF(q^n)$$

$$\sum_{k=0}^{d-1} x^k = (d), \qquad \text{for } x = 1 \tag{2}$$

where (d) denotes the residue of d modulo the prime q. This formula is given by Pollard [2, eq. 8] and earlier by Reed and Solomon [3, eq. 3].

From (2) we now derive the discrete "delta" function needed to invert (1). Consider the sum of x^n over all the elements of the multiplicative subgroup G_d. This is

$$\sum_{x \in G_d} x^n = \sum_{k=0}^{d-1} (r^k)^n = \sum_{k=0}^{d-1} (r^n)^k.$$

However, this is in the form of (2) and r^n is an element of G_d, thus

$$\sum_{x \in G_d} x^n = \sum_{k=0}^{d-1} (r^n)^k = 0, \qquad \text{for } n \not\equiv 0 \bmod d$$
$$= (d), \qquad \text{for } n \equiv 0 \bmod d$$
$$= (d)\delta_d(n) \qquad (3)$$

where $\delta_d(n)$ is the delta function

$$\delta_d(n) = 0, \qquad \text{for } n \not\equiv 0 \bmod d$$
$$= 1, \qquad \text{for } n \equiv 0 \bmod d.$$

Since (d) is a nonzero element of the field $GF(q^n)$, the inverse $(d)^{-1}$ exists in $GF(q^n)$. Now, multiply A_k by $(d)^{-1}r^{-km}$ and sum on k, for $k = 0, 1, 2, \cdots, d-1$. This yields by (1) and (3),

$$(d)^{-1}\sum_{k=0}^{d-1} A_k r^{-km} = (d)^{-1}\sum_{k=0}^{d-1}\sum_{n=0}^{d-1} a_n r^{kn} r^{-km}$$
$$= (d)^{-1}\sum_{n=0}^{d-1} a_n \left(\sum_{k=0}^{d-1} r^{k(n-m)}\right)$$
$$= (d)^{-1}(d)\sum_{n=0}^{d-1} a_n \delta_d(n-m)$$
$$= a_m.$$

Thus,

$$A_k = \sum_{n=0}^{d-1} a_n r^{kn}$$
$$a_m = (d)^{-1}\sum_{k=0}^{d-1} A_k r^{-km} \qquad (4)$$

where d divides $q^n - 1$, a_m and A_k are elements of $GF(q^n)$, and r is a generator of the d element subgroup G_d, a multiplicative subgroup of the multiplicative group of $GF(q^n)$.

To show the circular convolution property of (4) let

$$A_k = \sum_{n=0}^{d-1} a_n r^{kn} \qquad B_k = \sum_{n=0}^{d-1} b_n r^{kn} \qquad C_k = A_k \cdot B_k.$$

Then by (4) the inverse transform of C_k, for $k = 0, 1, \cdots, d-1$, is

$$c_p = (d)^{-1}\sum_{k=0}^{d-1} C_k r^{-kp}$$
$$= (d)^{-1}\sum_{k=0}^{d-1}\sum_{n=0}^{d-1}\sum_{m=0}^{d-1} a_n b_m r^{k(m+n-p)}$$
$$= (d)^{-1}\sum_{n=0}^{d-1}\sum_{m=0}^{d-1} a_n b_m \sum_{k=0}^{d-1} r^{k(m+n-p)}$$
$$= \sum_{n=0}^{d-1}\sum_{m=0}^{d-1} a_n b_m \delta_d(m+n-p) = \sum_{n=0}^{d-1} a_n b_{(p-n)} \qquad (5)$$

where $(p-n)$ denotes the residue of $p-n$ modulo d.

If a_n and b_n are restricted to $GF(q)$, the integers modulo the prime q, then the convolution (5), the inverse transform of $C_k = A_k \cdot B_k$, will also be in $GF(q)$. Hence, if $r \in GF(q^n)$ can be found so that multiplications by powers of r are simple in either hardware or software, then the aforementioned convolution in $GF(q^n)$ would have practical value in finding convolutions of integers. In this paper attention is restricted to the case, $n = 2$.

Rader [1] showed that binary arithmetic is easy to perform if b is a Mersenne or Fermat prime. Here we restrict b to be a Mersenne prime q.

If

$$x^2 \equiv -1 \bmod q \qquad (6)$$

is not solvable, then the nonsolvability of (6) is equivalent to the statement: -1 is a quadratic nonresidue [4, p. 82]. By Euler's theorem [4, p. 80] this is further equivalent to

$$\left(\frac{-1}{q}\right) = (-1)^{(q-1)/2} = (-1)^{(2^p-2)/2}$$
$$= (-1)^{(2^{p-1}-1)} = -1$$

where $\left(\frac{a}{q}\right)$ is the Legendre symbol defined by

$$\left(\frac{a}{q}\right) = +1, \qquad \text{if } a \text{ is a quadratic residue mod } q$$
$$= -1, \qquad \text{if } a \text{ is a quadratic nonresidue mod } q.$$

Thus -1 is a quadratic nonresidue modulo q and (6) is not solvable in $GF(q)$; the polynomial

$$p(x) = x^2 + 1$$

is then irreducible in $GF(q)$. A root, say, $\hat{\imath}$, of

$$p(x) = x^2 + 1 = 0 \qquad (7)$$

exists and can be found in the extension field $GF(q^2)$. $GF(q^2)$ is composed of the set

$$GF(q^2) = \{a + \hat{\imath}b \mid a, b \in GF(q)\} \qquad (8)$$

where $\hat{\imath}$ is a root of (7), satisfying

$$\hat{\imath}^2 = -1 \qquad (9)$$

where $-1 \equiv (q-1) \bmod q$.

If $x^2 + 1 = 0$ is not solvable in $GF(q)$, $\hat{\imath} \in GF(q^2)$, plays a similar role over the finite field $GF(q)$ that $\sqrt{-1} = i$ plays over the field of rational numbers. For example, suppose $a + \hat{\imath}b$ and $c + \hat{\imath}d$ are elements of $GF(q^2)$, then by (9)

$$(a + \hat{\imath}b) \pm (c + \hat{\imath}d) = (a \pm c) + \hat{\imath}(b \pm d)$$
$$(a + \hat{\imath}b)(c + \hat{\imath}d) = ac + \hat{\imath}^2 bd + \hat{\imath}bc + \hat{\imath}ad$$
$$= ac - bd + \hat{\imath}(bc + ad).$$

These are the exact analogues of what one might expect if $a + \hat{\imath}b$ and $c + \hat{\imath}d$ were complex numbers. In applications to radar and communications systems, one generally wants to take convolutions of complex numbers. It can be shown similarly that this analogue does not hold in $GF(q^2)$, where q is a Fermat prime.

If -1 is quadratic nonresidue mod q, then convolution (5) of the complex integers can be performed with transforms of type (4) on a Galois field $GF(q^2)$ where a_n and b_n are restricted to $GF(q^2)$. In other words, if $a_n, b_n \in$

$GF(q^2)$, for $n = 0, 1, 2,\cdots, d-1$, the transforms are

$$A_k = \sum_{n=0}^{d-1} a_n r^{kn}$$

$$B_k = \sum_{n=0}^{d-1} b_n r^{kn}, \qquad \text{for } k = 0,1,\cdots,d-1$$

where r is a generator of a d-element subgroup G_d of the multiplicative group of $GF(q^2)$. Then taking the inverse transform on $C_k = A_k \cdot B_k$, for $k = 0, 1,\cdots, d-1$, one obtains

$$c_n = (d)^{-1} \sum_{k=0}^{d-1} C_k r^{-kn}$$

where $c_n \in GF(q^2)$ and d divides $q^2 - 1$.

If q is a Mersenne prime, the order t of the multiplicative group with generator α of $GF(q^2)$ factors as follows [5]:

$$t = (2^p - 1)^2 - 1 = 2^{p+1}(2^{p-1} - 1).$$

Since t has the factor $d = 2^{p+1}$, the usual FFT algorithm can be used to calculate transforms of as many as $d = 2^{p+1}$ points. If $d = 2^k$, $1 \le k \le p+1$, and α is a primitive element of $GF(q^2)$, then the generator of G_d, a multiplicative subgroup of order d, is evidently $r = \alpha^{2^{p+1-k}(2^{p-1}-1)}$.

It is of interest to demonstrate an important property of the Galois field $GF(q^2)$ which the field of complex rational numbers does not have. If $x = a + \hat{\imath}b \in GF(q^2)$, $x \ne 0$, then

$$x^{q^2-1} = (a + \hat{\imath}b)^{q^2-1} = 1$$

which is a special case of the result, $x^{q^n-1} = 1$, for all nonzero $x \in GF(q^n)$.

III. Algorithm for Finding Primitive Elements in $GF(q^2)$ Where q is a Mersenne Prime

In (1) if one wants to take transforms over $GF(q^2)$ of of 2^{p+1} point sequences of complex integers, $a_n \in GF(q^2)$, then one needs to find a primitive element in $G_{2^{p+1}}$ of $GF(q^2)$. To achieve this, the following theorems are useful.

Theorem 1: If q is a Mersenne prime and $d = 2^k$, where $1 \le k \le p+1$, then $r = a + \hat{\imath}b$ is a primitive element in G_d of $GF(q^2)$, if and only if $r^{d/2} \equiv -1 \bmod q$.

Proof: If $r^d \equiv 1 \bmod q$, then $r^{d/2} \equiv \pm 1 \bmod q$. By hypothesis the order of r is d so that $r^{d/2} \not\equiv 1 \bmod q$. Therefore,

$$r^{d/2} = (a + \hat{\imath}b)^{d/2} \equiv -1 \bmod q. \tag{10}$$

Assume d is not the smallest positive number for which this is true. Then there exists $l = 2^n < d$, $n < k$, such that $r^l \equiv 1 \bmod q$. Let $t = (d,l)$, the greatest common divisor in d and l. Then $dx + ly = t$, for some x, y, and

$$r^t = r^{dx+ly} = (r^d)^x (r^l)^y = (1)^x (1)^y = 1.$$

Since $t < d$, t divides $d/2$. Therefore, $r^{(d/2)} \equiv 1 \bmod q$. This contradicts (10).

Theorem 2: For Mersenne primes, $q > 3$, the first quadratic nonresidue modulo q in the sequence $1, 2, 3,\cdots, q-1$, is 3.

Proof: Consider the sequence $1, 2,\cdots, q-1$. We consecutively check the numbers in sequence until the first quadratic nonresidue is found. One is a quadratic residue. By [4, p. 81], $\left(\frac{2}{2^p-1}\right) = (-1)^{((2^p-1)^2-1)/8} = (-1)^{(2^{p+1}(2^{p-1}-1))/8} = 1$, and 2 is a quadratic residue. By the quadratic reciprocity law,

$$\left(\frac{3}{2^p-1}\right) = \left(\frac{(2^p-1) \bmod 3}{3}\right)(-1)^{(3-1)/2} = (\tfrac{1}{3})(-1).$$

However $\left(\frac{1}{3}\right) = 1$. Thus $\left(\frac{3}{2^p-1}\right) = -1$.

To find a primitive element in $G_{2^{p+1}}$ of $GF(q^2)$, one can use the following procedure. Assume an element $r = a + \hat{\imath}b$ is of order 2^{p+1} in $GF(q^2)$. Then $(a + \hat{\imath}b)^{2^p}$ factors as

$$(a + \hat{\imath}b)^{2^p} = (a + \hat{\imath}b)(a + \hat{\imath}b)^{2^p-1}. \tag{11}$$

Employing the binomial theorem,

$$(a + \hat{\imath}b)^{2^p-1} = a^{2^p-1} + \binom{2^p-1}{1} a^{2^p-1-1}(\hat{\imath}b) + \cdots + \binom{2^p-1}{r} a^{2^p-1-r}(\hat{\imath}b)^r + \cdots + (\hat{\imath}b)^{2^p-1}$$

but the binomial coefficients

$$\binom{2^p-1}{k} = \frac{(2^p-1)(2^p-1-1)(2^p-1-2)\cdots(2^p-1-k+1)}{1\cdot 2\cdot 3\cdots k}, \qquad \text{for } 1 \le k \le 2^p - 2$$

are all divisible by $2^p - 1$ since $2^p - 1$ is a prime. Hence

$$(a + \hat{\imath}b)^{2^p} \equiv (a + \hat{\imath}b)(a + \hat{\imath}^{2^p-1}b) \bmod q.$$

However

$$\hat{\imath}^{2^p-1} = \hat{\imath}(-1)^{(2^p-2)/2} = -\hat{\imath}$$

so that (11) becomes

$$(a + \hat{\imath}b)^{2^p} \equiv (a^2 + b^2) \bmod q.$$

By Theorem 1, it follows that

$$(a + \hat{\imath}b)^{2^p} \equiv a^2 + b^2 \equiv -1 \bmod q. \tag{12}$$

Let $X \equiv a^2 \bmod 2^p - 1$ and let $Y \equiv -b^2 \bmod 2^p - 1$. Then (12) becomes

$$X + 1 \equiv Y \bmod q. \tag{13}$$

By definition in (13), X is a quadratic residue. For Y,

$$\left(\frac{Y}{q}\right) = \left(\frac{-b^2}{q}\right) = \left(\frac{-1}{q}\right)\left(\frac{b^2}{q}\right) = \left(\frac{-1}{q}\right).$$

However, $\left(\frac{-1}{q}\right) = (-1)^{((2^p-1-1)/2)} = (-1)^{2^{p-1}-1} = -1$.

Thus, $Y = X + 1$ is a quadratic nonresidue. Hence, one way to choose the numbers X and Y is to let X and Y be two consecutive numbers from the set of integers, $1, 2,\cdots, 2^p - 2$, such that the first number X is a square and the second number is a nonsquare. By Theorem 2, for $q > 3$,

$Y = 3$ is nonsquare and the preceding element $X = 2$ is a square. Thus it is sufficient to let

$$a^2 \equiv X \equiv 2 \bmod 2^p - 1 \tag{14a}$$

$$b^2 \equiv -X - 1 \equiv -Y \equiv -3 \bmod 2^p - 1. \tag{14b}$$

To find the solutions of congruence (14a), we use the procedure described in [4, p. 187, solution to problem 2, a]. This is as follows.

Since $\left(\dfrac{2}{2^p - 1}\right) = 1$, by Euler's theorem

$$2^{((2^p-1)-1)/2} \equiv 1 \bmod (2^p - 1).$$

Multiplying both sides of the congruence by 2, then

$$2^{((2^p-2)/2)+1} = 2^{2^{p-1}} \equiv 2 \bmod 2^p - 1.$$

Hence $(2^{2^{p-2}})^2 \equiv 2 \bmod 2^p - 1$. Thus the solution of (14a) is

$$a \equiv \pm 2^{2^{p-2}} \bmod 2^p - 1. \tag{15a}$$

Using the same procedure for (14b), b is given by

$$b \equiv \pm(-3)^{2^{p-2}} \bmod 2^p - 1. \tag{15b}$$

In G_d of $GF(q^2)$, there always exists a primitive element, $r = a + \hat{\imath}b$. By Theorem 1,

$$(a + \hat{\imath}b)^{d/2} \equiv -1 \bmod q. \tag{16}$$

Raising both sides of (16) to the jth power, (16) becomes

$$((a + \hat{\imath}b)^{d/2})^j \equiv (-1)^j \bmod q.$$

By Theorem 1, $((a + \hat{\imath}b)^{d/2})^j$ is a primitive element only when j is an odd number. The elements are distinct and include all primitive elements of G_d. In other words, in the cyclic subgroup of G_d of $GF(q^2)$, if $a + \hat{\imath}b$ is a primitive element, then $(a + \hat{\imath}b)^j$ is also a primitive element for $j = 1, 3, 5, \cdots, d - 1$.

Assume α is of order 2^{p+1} in $GF(q^2)$. If d divides 2^{p+1}, then $r = \alpha^{2^{p+1}/d}$ is of order d in $GF(q^2)$. For the transform in (1), with d a factor of 2^{p+1}, one can take $r = \alpha^{2^{p+1}/d}$ as the primitive element and d as the transform length.

In order to perform multiplications by powers of r in hardware, it might be desirable to represent the p-bit words, a and b, where $r = a + \hat{\imath}b$, as a minimal sum of powers of two. Then, for example, $r^{n+1} \bmod q$ can be obtained by multiplying r^n by r modulo q recursively using a minimal number of bit rotations, p-bit word complements, and additions [1].

Example: Find all primitive elements expressed as a sum of powers of two in the subgroup G_d of the multiplicative group of $GF(q^2)$, where $p = 5$, $q = 2^p - 1 = 31$, and $d = 2^3$.

To do this, first find an element with order $2^{p+1} = 2^{5+1} = 64$ in $GF(31^2)$. According to Theorem 1, if $r = (a + \hat{\imath}b)$ is a primitive element in G_{2^6} of $GF(31^2)$, then

$$(a + \hat{\imath}b)^{2^5} \equiv -1 \bmod 31. \tag{17}$$

By (12), (17) becomes

$$a^2 + b^2 \equiv -1 \bmod 31. \tag{18}$$

TABLE I
THE PRIMITIVE ELEMENTS IN G_8 OF $GF(31^2)$

$(27 + \hat{\imath}4) = (2^5 - 2^2 - 2^0) + \hat{\imath}2^2$
$(27 + \hat{\imath}4)^3 \bmod 31 = 4 + \hat{\imath}4 = 2^2 + \hat{\imath}2^2$
$(27 + \hat{\imath}4)^5 \bmod 31 = 4 + \hat{\imath}27 = 2^2 + \hat{\imath}(2^5 - 2^2 - 2^0)$
$(27 + \hat{\imath}4)^7 \bmod 31 = 27 + \hat{\imath}27 = (2^5 - 2^2 - 2^0) + \hat{\imath}(2^5 - 2^2 - 2^0)$

By Theorem 2, 3 is nonsquare and 2 is a square. By (15a)

$$a \equiv 2^{2^{5-2}} \equiv 2^8 \equiv 8 \bmod 31.$$

By (15b),

$$b \equiv (-3)^{2^{5-2}} \equiv 20 \bmod 31.$$

Thus 64 is the smallest positive integer such that

$$(8 + \hat{\imath}20)^{64} \equiv 1 \bmod 31.$$

An element with order $2^{p+1}/d = 64/8 = 8$ is

$$(8 + \hat{\imath}20)^8 \equiv (27 + \hat{\imath}4) \bmod 31$$

and the four primitive elements in subgroup G_8 of $GF(31^2)$ are shown in Table I.

In Table I, $r = 4 + \hat{\imath}4$ has the shortest number of terms of power of 2. This primitive element is such that multiplications by powers of r in G_8 of $GF(31^2)$ may in some cases yield the least hardware. Such an r will be called the simplest primitive element. Using the same procedure as Example 1 and an assembler program, the simplest primitive elements for different Mersenne primes were found. These are shown in Table II. These elements were chosen from $d/2$ primitive elements in G_d for different d and q.

IV. FAST FOURIER TRANSFORM IN $GF(q^2)$

The usual discrete Fourier transform is

$$F_k = \sum_{n=0}^{d-1} a_n \omega^{nK} \tag{19}$$

where ω is a dth root of unity.

The dth root of unity $\omega = e^{2\pi i/d}$ in (19) generally cannot be expressed exactly on a computer in a finite word length. Therefore, transform (19) and its inverse transform will be in error for most cases of interest. In a transform of type (4) on the Galois field $GF(q^2)$, where q is the Mersenne prime $2^p - 1$ and p is a prime, a primitive element, $r = a + \hat{\imath}b$, where $a, b \in GF(q)$, can be expressed exactly in two p-bit words [1]. If $a_n \in GF(q^2)$, for $n = 0, 1, 2, \cdots, d - 1$, the transform over $GF(q^2)$ of the a_n is

$$A_k = \sum_{n=0}^{d-1} a_n r^{kn} = (\alpha_k + \hat{\imath}\beta_k), \qquad \text{where } \alpha_k, \beta_k \in GF(q).$$

Then, by (4), the inverse transform over $GF(q^2)$ of the sequence A_k equals the original sequence of a_n exactly. If one performs this procedure repeatedly, the original numbers a_n would still be obtained without error.

It is well known [2] that the FFT algorithm over finite fields is similar to the FFT algorithm over the field of complex numbers [6, p. 179]. Assume $a(t)$ is the input data in time domain, $B(f)$ is output data in frequency domain, and $d = 2^N$. Let t and f denote binary integers and let r be a primitive element in G_d. The FFT over $GF(q^2)$ algorithm is composed then of the following N stages of computation:

Stage 1:

$$B_1(f_1 t_{N-1}\cdots t_1) = \sum_{t_N=0}^{1} (r^{2^{N-1}})^{t_N f_1} a(t_N t_{N-1}\cdots t_1) \tag{20}$$

$$\vdots$$

Stage j:

$$B_j(f_1\cdots f_j t_{N-j}\cdots t_1) = \sum_{t_{N-j+1}=0}^{1} (r^{2^{N-j}})^{t_{N-j+1}(f_j\cdots f_1)} \cdot B_{j-1}(f_1\cdots f_{j-1} t_{N-j+1}\cdots t_1)$$

$$\vdots$$

Stage N:

$$B_N(f_1\cdots f_N) = \sum_{t_1=0}^{1} r^{t_1(f_N\cdots f_1)} B_{N-1}(f_1\cdots f_{N-1} t_1)$$

where the bits $f_1\cdots f_j$ of f are in bit-revered order for $j = 1,2,\cdots,N$.

The inverse algorithm can be obtained by substituting r^{-1} for r and multiplying $B_N(f)$ by $(d)^{-1}$.

V. Complex Number Convolutions with $GF(q^2)$ Transforms

In order to carry out operations in $GF(q^2)$ which arrive at the correct arithmetic convolutions, one must usually restrict the operating ranges of the complex variables. For example, to compute the circular convolution (5), i.e.,

$$\begin{aligned} c_p &= \sum_{n=0}^{d-1} a_n b_{(p-n)} \\ &= \sum_{n=0}^{d-1} ((\alpha_n x_n - \beta_n y_n) + \hat{\imath}(\alpha_n y_n + \beta_n x_n)) \end{aligned} \tag{21}$$

of the complex numbers $a_n = \alpha_n + \hat{\imath}\beta_n$ and $b_n = x_n + \hat{\imath}y_n$, one requires the final convolution to lie in the same "dynamic range" as the complex integers a_n and b_n. That is,

$$\left|\sum_{n=0}^{d-1} (\alpha_n x_n - \beta_n y_n)\right| \le \sum_{n=0}^{d-1} (|\alpha_n|\cdot|x_n| + |\beta_n|\cdot|y_n|) \le (q-1)/2 \tag{22}$$

and

$$\left|\sum_{n=0}^{d-1} (\alpha_n y_n + \beta_n x_n)\right| \le \sum_{n=0}^{d-1} (|\alpha_n|\cdot|y_n| + |\beta_n|\cdot|x_n|) \le (q-1)/2.$$

If $\max|\alpha_n| = \max|x_n| = \max|\beta_n| = \max|y_n| = A$, A is called the dynamic range of α_n, y_n, β_n, and x_n. Then by (22) the largest value of A is given by

$$A = \left[\sqrt{\frac{q-1}{4d}}\right] \tag{23}$$

where $[x]$ denotes the greatest integer less than x. Assuming (23), which for many practical applications is pessimistic, one needs to choose q sufficiently large so that the integer components of a_n,b_n lie in the interval $-A \le \alpha_n$, β_n, x_n, $y_n \le A$ and the circular convolution satisfies inequality (22). For example, let $q = 2^{31} - 1$ and let $d = 2^8$. By (23),

$$A = \left[\sqrt{\frac{2^{31}-1}{4x2^8}}\right] \cong 2^{10}.$$

If a_n and b_n are constrained to the interval

$$-2^{10} \le \alpha_n, \beta_n, x_n, y_n \le 2^{10}$$

we are guaranteed to keep the components c_p in the interval

$$-(2^{31}-1)/2 \le x \le (2^{31}-1)/2.$$

In order to compare the present method for computing a convolution of complex integers with the usual FFT over the complex numbers, a convolution program for complex integers in $GF(q^2)$ was written in an assembler language for $q = 2^{31} - 1$. As a simple example, the step function defined by

$$\begin{aligned} a(t) &= 0 + \hat{\imath}0, && \text{for } 0 \le t \le 2^7 \\ &= 10 + \hat{\imath}0, && \text{for } 2^7 < t < 2^8 \end{aligned}$$

was convolved with itself. This convolution is shown in Fig. 1 and is without error. The convolution of $a(t)$ with itself was also performed with a standard IBM FFT subroutine package to an accuracy of 24 bits. Because of both truncation error and round-off error, the latter was in error with magnitude $E(t)$ as shown in Fig. 2. Though this example is only illustrative it suggests that digital filtering over $GF(q^2)$ where q is a Mersenne prime can be accomplished without round-off error.

VI. Summary

Transforms over $GF(q^2)$ can be found which do not introduce round-off error and which can be used to compute information-lossless convolutions of sequences of complex numbers by the FFT method. Methods for finding the primitive elements of G_d in $GF(q^2)$ were also developed. A disadvantage of this transform is that multiplication by powers of the primitive element is not as simple as that developed previously by Rader (see Table II). This disadvantage is offset by the fact that here, for the case of practical interest, the FFT algorithm applies on $GF(q^2)$ when q is a Mersenne prime.

VII. Acknowledgment

The authors wish to thank Prof. L. R. Welch for his mathematical suggestions and Dr. B. Gold of M.I.T. Lincoln Laboratory for his early support and encouragement of the research which led to this paper.

References

[1] C. M. Rader, "Discrete convolution via Mersenne transforms," *IEEE Trans. Comput.*, vol. C-21, Dec. 1972.

[2] J. M. Pollard, "The fast Fourier transform in a finite field," *Math. Comput.*, vol. 25, no. 114, Apr. 1971.

[3] I. S. Reed and G. Solomon, "A decoding procedure for polynomial codes," M.I.T. Lincoln Lab., Lexington, Mass., Group Rep. 47.24, ASTIA No. 24-1701, March 6, 1959.

[4] I. M. Vinogradov, *Elements of Number Theory.* New York: Dover, 1954.

[5] I. S. Reed, "The use of finite fields and rings to compute convolutions," M.I.T. Lincoln Lab., Lab. Tech. Memo. 24L-0012, Oct. 1973.

[6] B. Gold and C. M. Rader, *Digital Processing of Signals.* New York: McGraw-Hill. 1969.

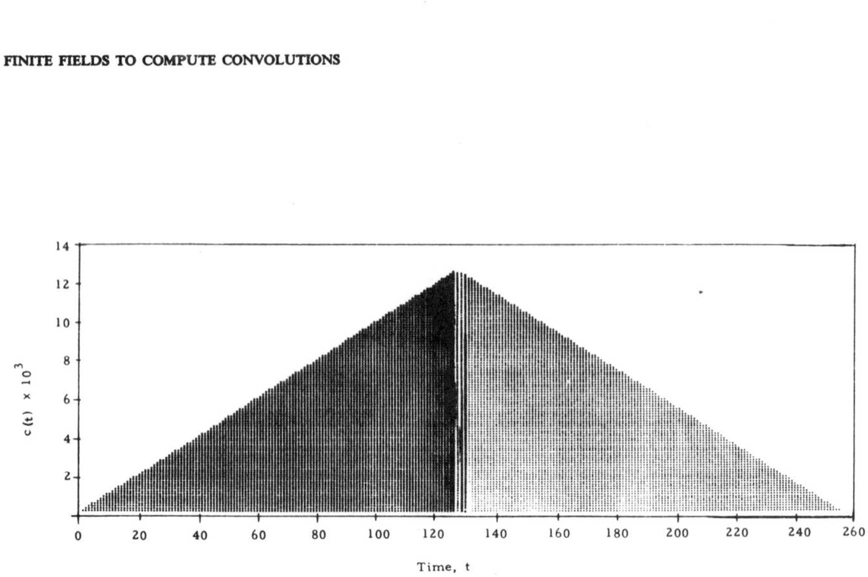

Fig. 1. Convolution of two rectangular pulses in $GF(q^2)$.

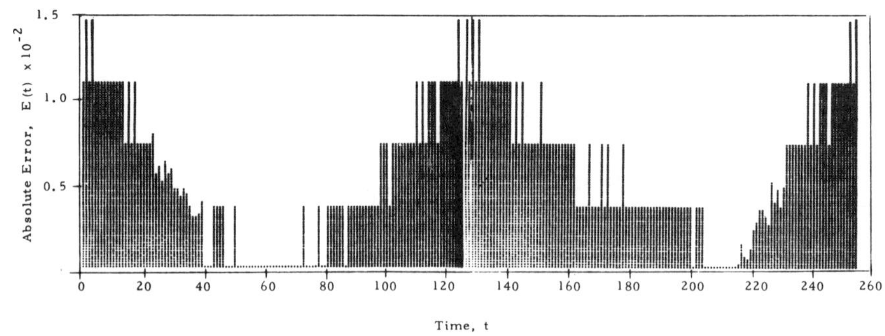

Fig. 2. Error produced by convolution of rectangular pulses in complex number field using Harm subroutine.

TABLE II
SIMPLEST PRIMITIVE ELEMENTS FOR DIFFERENT q AND d

	$d=2^8$	$d=2^9$	$d=2^{10}$	$d=2^{11}$	$d=2^{12}$
$q=2^{31}-1$	$(2^{27}+2^{25}+2^{21}+2^{19}+2^{16}-2^{13}-2^{10}+2^4-2^1)+\hat{i}(2^{27}+2^{23}-2^{19}-2^{17}+2^{12}+2^{10}+2^8+2^3)$	$(2^{30}-2^{28}+2^{25}+2^{19}+2^{18}-2^{16}+2^{13}+2^{10}+2^7)+\hat{i}(2^{30}-2^{28}+2^{22}+2^{20}-2^{14}+2^8+2^2)$	$2^{31}-2^{28}+2^{25}-2^{18}-2^{13}+2^{10}+2^5)+\hat{i}(2^{30}-2^{25}-2^{19}+2^{16}-2^{13}-2^7+2^4-2^2)$	$(2^{28}+2^{24}+2^{18}-2^{16}+2^{14}-2^6+2^4+2^0)+\hat{i}(2^{27}-2^{24}+2^{17}-2^{14}+2^{11}+2^{10}-2^0)$	$(2^{25}+2^{22}+2^{21}-2^{16}+2^{13}+2^5)+\hat{i}(2^{30}+2^{25}-2^{22}+2^{14}+2^{11}+2^6+2^2)$
$q=2^{19}-1$	$(2^{15}+2^{12}-2^7+2^0)+\hat{i}(2^{15}+2^{13}+2^{11}+2^7+2^4+2^3-2^0)$	$(2^{17}+2^{15}+2^{12}+2^9-2^4+2^0)+\hat{i}(2^{15}-2^{10}-2^3)$	$(2^{12}+2^{10}-2^1)+\hat{i}(2^{19}-2^{17}+2^{14}+2^{13}-2^7-2^2)$	$(2^{19}-2^{17}+2^{12})+\hat{i}(2^{16}+2^{11}-2^8+2^2+2^0)$	$(2^{13}+2^8-2^5+2^2)+\hat{i}(2^{12}+2^9+2^6+2^3)$
$q=2^{17}-1$	$(2^8+2^5)+\hat{i}(2^{16}+2^{12}+2^{11}-2^9+2^7-2^4+2^2)$	$(2^{12}+2^8)+\hat{i}(2^{17}-2^{11}+2^8+2^5+2^2)$	$(2^{11}+2^8-2^2)+\hat{i}(2^{15}+2^{12}+2^7-2^0)$	$(2^7+2^3)+\hat{i}(2^{12}+2^{11}-2^9+2^7+2^5-2^0)$	$(2^{10}-2^2+2^0)+\hat{i}(2^{14}-2^6)$
$q=2^{13}-1$	$(2^7-2^4+2^2)+\hat{i}(2^{11}+2^7-2^3+2^0)$	$(2^{11}-2^7-2^5+2^3)+\hat{i}(2^8+2^5)$	$(2^9-2^5+2^0)+\hat{i}(2^9-2)$	$(2^9+2^7+2^4+2)+\hat{i}(2^{10}+2^5)$	$(2^{11}+2^7+2^4+2^2)+\hat{i}(2^9+2)$

Reprinted with permission from IBM Journal of research and development

Vol. 20 | No. 5 | September 1976

H. J. Nussbaumer

Digital Filtering Using Complex Mersenne Transforms

H. J. Nussbaumer

Digital Filtering Using Complex Mersenne Transforms

Abstract: Complex Mersenne Transforms are defined in a ring of integers modulo a Mersenne or pseudo-Mersenne number and can be computed without multiplications. It is shown that under certain conditions, these transforms can be computed by means of fast transform algorithms and permit the evaluation of digital convolutions with better efficiency and accuracy than does the Fast Fourier Transform.

Introduction

With the rapid advances in large scale integration, a growing number of complex digital signal processing applications are becoming economically feasible. In most cases the bulk of the processing workload appears to consist of digital filter computation. Future progress in digital signal processing, either towards high speed, real time operation or increased sophistication, thus largely depends on increased efficiency in digital filter computation. This can be achieved not only by implementing improved filter circuits but also by using better computation algorithms, as will be discussed in this paper.

Rader [1] and Agarwal and Burrus [2, 3] have recently introduced Mersenne Transforms and Fermat Number Transforms. These two transforms have the circular convolution property and therefore can be used for evaluating digital filters in the same way as Discrete Fourier Transforms (DFT) [4, 5].

Mersenne Transforms and Fermat Number Transforms are very promising for digital filter computation because they can be calculated without multiplications. Their main drawback is a rigid relationship between transform length and word length, caused by the fact that all operations are performed in a finite ring with arithmetic carried out modulo an integer p. Another difficulty arises because it is not possible to achieve simultaneously optimum efficiency in reducing the number of operations and in implementing arithmetic operations. This is so because Fermat Number Transforms are amenable to a fast transform algorithm, and Mersenne Transforms are not, whereas arithmetic operations can be implemented more efficiently modulo a Mersenne number than modulo a Fermat number.

In this paper we consider complex transforms defined in the ring of integers modulo a Mersenne number. We show that these Complex Mersenne Transforms can be partly computed by a fast transform technique and have a maximum length up to four times that of conventional Mersenne Transforms. We discuss the use of such transforms for filtering complex signals and show that calculations in the first stages of the direct transforms can be performed on words of reduced length. We then extend these results to cover the case of pseudo-Mersenne numbers. We show that this leads to a definition of pseudo-Mersenne Transforms, which can be completely computed by means of a fast transform algorithm but in which the advantage of performing arithmetic operations modulo a Mersenne number is retained.

Complex Mersenne Transform

Let q be a prime and $p = 2^q - 1$. Mersenne numbers p are primes for $q = 2, 3, 5, 7, 13, 17, 19, 31, 61 \cdots$ [6].

The Mersenne Transform of an integer sequence $\{a_n\}$ having q terms is defined by

$$A_k = \left(\left(\sum_{n=0}^{q-1} a_n 2^{nk}\right)\right) \qquad k = 0, 1 \cdots q-1, \tag{1}$$

where any quantity enclosed by superfluous double parentheses is to be replaced by its value modulo p.

The Inverse Mersenne Transform is defined by:

$$a_m = \left(\left(R \sum_{k=0}^{q-1} A_k 2^{-mk}\right)\right) \qquad m = 0, 1 \cdots q-1, \tag{2}$$

where R is such that $((R \cdot q)) = 1$, all exponents and indices being taken modulo q and all operations being performed modulo p.

It can be demonstrated easily [1] that the Mersenne Transform satisfies the convolution theorem; that is to say, if $\{X_k\}$ is the Mersenne Transform of $\{x_n\}$, then with $Z_k = ((A_k \cdot X_k))$, the Inverse Mersenne Transform $\{z_m\}$ of $\{Z_k\}$ is given by

$$z_m = \left(\left(\sum_{n=0}^{q-1} a_n\, x_{m-n}\right)\right). \tag{3}$$

If $\{a_n\}$ and $\{x_n\}$ are properly bounded [1], z_m is equal to the output of the ordinary cyclic convolution with

$$z_m = \sum_{n=0}^{q-1} a_n\, x_{m-n}. \tag{4}$$

Under these conditions, digital filtering of real integer sequences can be performed by dividing the sequences into blocks, padding the blocks with zeros [4] to prevent folding, and aliasing and computing the cyclic convolutions by means of Mersenne Transforms. This provides a very efficient way of computing digital filters because calculations of Mersenne Transforms reduce to one's complement additions and circular shifts. Because Mersenne Transforms are evaluated without multiplications, computation of a time-invariant circular convolution having q points reduces to one multiplication per output sample, as opposed to q multiplications with direct calculation. This corresponds, in the case of a nonrecursive digital filter having N taps, to a number of multiplications per output sample that is small and essentially independent of the number of taps. Direct computation would require N multiplications in the general case and implementation by means of the Fast Fourier Transform (FFT) would require $K \log_2 N$ multiplications.

The main limitations of the Mersenne Transform approach are related to the fact that the number of transform terms q is a prime. This means that calculations of the transforms cannot be simplified by an FFT-type algorithm and that the number of transform terms is equal to the word size. These limitations can be slightly alleviated [1] by using a root -2 instead of 2 in (1) and (2). The maximum transform length then becomes $2q$. It is also possible to increase the maximum convolution size by resorting to multidimensional convolutions [1, 2, 7]. Unfortunately, this result is achieved at the expense of increased requirements for computation and storage.

We show now that by defining Complex Mersenne Transforms, it is possible to achieve higher computation efficiency while increasing both maximum transform length and convolution length.

In many instances of digital signal processing, digital filtering of complex signals is required. Modem equalizers for phase modulated signals are a good example of such applications [8]. In that case, a complex integer sequence $\{x_n = y_n + j\hat{y}_n\}$ is to be filtered by a complex sequence $\{a_n = b_n + j\hat{b}_n\}$, having N terms, to produce a complex output sequence $\{z_m = u_m + j\hat{u}_m\}$ with

$$z_m = \sum_{n=0}^{N-1} a_n x_{m-n} = \sum_{n=0}^{N-1} \left(b_n y_{m-n} - \hat{b}_n \hat{y}_{m-n}\right) + j \sum_{n=0}^{N-1} \left(\hat{b}_n y_{m-n} + b_n \hat{y}_{m-n}\right) \qquad j = \sqrt{-1}. \tag{5}$$

The complex convolution (5) can be calculated by means of real transforms provided that computation on real and imaginary numbers is carried out in separate transforms. It is even possible, with Fermat Number Transforms, to reduce the number of multiplications by taking advantage of the real number representation of $j = \sqrt{-1}$ in a Fermat number system [9].

It is more natural, however, to compute complex convolutions with complex transforms such as the Fourier transform or Complex Number Theoretic Transforms [3, 10, 11]. In particular, complex transforms can be defined in a Mersenne ring. Along these lines, we have proposed a digital filter based on Complex Mersenne Transforms [12]. The author thanks the referees for bringing to his attention the independent work of Vegh and Leibowitz on the same subject [13].

In the following we restrict our discussion to Complex Mersenne Transforms that have simple roots and can be computed without multiplications. In a Mersenne ring, with $p = 2^q - 1$, 2 and -2 are respectively roots of unity of orders q and $2q$, corresponding to transforms of lengths q and $2q$, respectively. Since q is a prime, 2^d and -2^d are also roots of orders q and $2q$, provided d is not a multiple of q. This implies that $2j$ is a root of order $4q$ and $1 + j$ or $1 - j$ are roots of order $8q$. Higher-order complex roots do not have a simple structure and therefore will generally not be of practical interest.

Under these conditions, a Complex Mersenne Transform having $4q$ terms can be defined by

$$A_k = \left(\left(\sum_{n=0}^{4q-1} a_n j^{nk} 2^{nk}\right)\right) \qquad j = \sqrt{-1},\ k = 0, 1 \cdots 4q - 1. \tag{6}$$

Because q has an inverse modulo p and the inverse of 4 is 2^{q-2}, $4q$ has an inverse R such that $((4qR)) = 1$ and the inverse transform of A_k is

$$a_m = \left(\left(R \sum_{k=0}^{4q-1} A_k j^{-mk} 2^{-mk}\right)\right) \qquad m = 0, 1 \cdots 4q - 1, \tag{7}$$

where all exponents and indices are taken modulo $4q$.

We demonstrate that this Complex Mersenne Transform satisfies the convolution theorem. Taking the complex transform $\{X_k\}$ of the complex sequence $\{x_l\}$, performing the term-by-term complex multiplications $X_k \cdot A_k$ and taking the inverse transform yields 499

$$z_m = \left(\left(\sum_{n=0}^{4q-1} \sum_{l=0}^{4q-1} a_n x_l R \left(\sum_{s=0}^{3} j^{rs} 2^{rs}\right) \sum_{k=0}^{q-1} 2^{4rk}\right)\right) \tag{8}$$

with $l + n - m = r$ and k replaced by $4k + s$.

As in the case of conventional Mersenne Transforms, the sum $\Sigma_{k=0}^{q-1} 2^{4rk}$ is equal to zero for $(r)_{\text{mod } q} \neq 0$ and equal to q for $(r)_{\text{mod } q} = 0$. For $(r)_{\text{mod } q} = 0$, $r = q\,t$ and $((\Sigma_{s=0}^{3} j^{rs} 2^{rs}))$ becomes $((\Sigma_{s=0}^{3} j^{qts}))$. Because q is an odd prime, this sum is different from zero only when t is a multiple of 4 and is equal to 4 when t is a multiple of 4. Under these conditions, the product of the two sums is different from zero only for $(r)_{\text{mod } 4q} = 0$ and z_m becomes

$$z_m = \left(\left(\sum_{n=0}^{4q-1} a_n\, x_{(m-n)_{\text{mod } 4q}}\right)\right), \tag{9}$$

which shows that two complex sequences of length $4q$ can be cyclically convolved by means of Complex Mersenne Transforms modulo $p = 2^q - 1$. In such an approach, all arithmetic operations are performed as in normal, complex arithmetic with $j^2 = -1$ and real and imaginary parts treated separately modulo p.

Using a root $j + 1$ leads to a definition of a Complex Mersenne Transform having $8q$ terms with

$$A_k = \left(\left(\sum_{n=0}^{8q-1} a_n (j+1)^{nk}\right)\right) \qquad k = 0, 1 \cdots 8q - 1, \tag{10}$$

and, with R such that $((8qR)) = 1$, an inverse transform

$$a_m = \left(\left(R \sum_{k=0}^{8q-1} A_k (j+1)^{-mk}\right)\right)$$

$$m = 0, 1 \cdots 8q - 1, \tag{11}$$

with all exponents and indices taken modulo $8q$. It can be demonstrated easily, by using the same development as given above, that this transform permits evaluation of convolutions of length $8q$ with word lengths of q bits.

Fast computation of Complex Mersenne Transforms

We have so far defined Complex Mersenne Transforms that can be computed without multiplications and have a length up to four times that of conventional Mersenne Transforms. The calculation of these transforms can be partly simplified by an FFT-type algorithm because the number of terms is no longer a prime.

Using either decimation in time or decimation in frequency [14] allows a decomposition into transforms having four q terms and transforms having eight q terms in the case of roots $2j$ and $j + 1$, respectively. These FFT decompositions reduce the number of real operations to $8q(q + 1)$ additions and $8q(q - 1)$ shifts for a complex transform having $4q$ points, and to $16\,q(q - 1) + 52q$ additions and $16(q - 1)^2 + 24(q - 1)$ shifts for a complex transform having $8q$ points. If we assume, for the sake of comparison, the existence of hypothetical real Mersenne Transforms having $4q$ points and $8q$ points, computing the transforms of the real and imaginary parts of a complex sequence would require $4q(4q - 1)$ additions and $(4q - 1)^2$ shifts for $4q$-point transforms and $8q(8q - 1)$ additions and $(8q - 1)^2$ shifts for $8q$-point transforms. This means that in the practical range of interest for q (where $q = 31$), substituting Complex Mersenne Transforms for conventional Mersenne Transforms results approximately in an eightfold reduction in the number of operations.

Computation of a complex convolution by means of Complex Mersenne Transforms is carried out as with DFT, with real and complex parts being evaluated modulo p separately. In order to avoid errors due to overflow, the amplitudes of real and imaginary outputs must be bounded to $(p - 1)/2$. This means that usually the word length of real and imaginary parts of input sequences $\{a_n\}$ and $\{x_n\}$ is less than half that of output sequences. In other words, all computations are carried out modulo p on q-bit words, yielding q-bit word outputs, and the input sequences are represented by words of less than $(q - 1)/2$ bits. It is possible to take advantage of the input word length limitations to further reduce the processing workload. This is done by selecting a fast transform decomposition such that the first two transform stages can be computed without multiplication by powers of two. In such an approach, a Complex Mersenne Transform having $4q$ points, with root $2j$, is decomposed into q four-point transforms with

$$A_k = \left(\left(\sum_{n=0}^{3} a_{qn} j^{qnk} + j^k 2^k \sum_{n=0}^{3} a_{qn+1} j^{qnk} + \cdots + j^{(q-1)k}\, 2^{(q-1)k} \sum_{n=0}^{3} a_{qn+q-1} j^{qnk}\right)\right). \tag{12}$$

The q four-point transforms corresponding to the first two stages of the decomposition require only multiplications by ± 1 and $\pm j$ and can therefore be computed in normal, complex arithmetic, on words of length approximately half that of the final result.

If the two sequences $\{y_n\}$ and $\{a_n\}$ to be convolved are real, the full benefit of using Complex Mersenne Transforms can be retained by processing simultaneously two successive blocks of the sequence $\{y_n\}$ by means of the same Complex Mersenne Transforms. This is done by computing the complex convolution $\{z_m\}$ of the sequence $\{a_n\}$ with the auxiliary complex sequence $\{x_n = y_n + jy_{n+8q}\}$. The real part $\{u_m\}$ and the imaginary part $\{\hat{u}_m\}$ of $\{z_m\}$ yield respectively the convolutions of $\{y_n\}$ and of the next block $\{y_{n+8q}\}$ with $\{a_n\}$.

Complex pseudo-Mersenne Transforms

We have seen above that a conventional q-point Mersenne Transform could be extended to $8q$ points by oper-

Table 1 Maximum odd length and corresponding power-of-2 roots for real transforms modulo $p = 2^q - 1$ with q odd and p composite.

q	*Prime factorization of* $p = 2^q - 1 \quad p_1, p_2 \cdots p_i$	*Prime factorization of* $p_i - 1$	*Max. odd length*	*Power of 2 roots*
15	$7 \cdot 31 \cdot 151$	$2 \cdot 3 \quad 2 \cdot 3 \cdot 5 \quad 2 \cdot 3 \cdot 5^2$	3	/
21	$7^2 \cdot 127 \cdot 337$	$2 \cdot 3 \quad 2 \cdot 3^2 \cdot 7 \quad 2^4 \cdot 3 \cdot 7$	3	/
23	$47 \cdot 178481$	$2 \cdot 23 \quad 2^4 \cdot 5 \cdot 23 \cdot 97$	23	2
25	$31 \cdot 601 \cdot 1801$	$2 \cdot 3 \cdot 5 \quad 2^3 \cdot 3 \cdot 5^2 \quad 2^3 \cdot 3^2 \cdot 5^2$	15	/
27	$7 \cdot 73 \cdot 262657$	$2 \cdot 3 \quad 2^3 \cdot 3^2 \quad 2^9 \cdot 3^3 \cdot 19$	3	/
29	$233 \cdot 1103 \cdot 2089$	$2^3 \cdot 29 \quad 2 \cdot 19 \cdot 29 \quad 2^3 \cdot 3^2 \cdot 29$	29	2
33	$7 \cdot 23 \cdot 89 \cdot 599479$	$2 \cdot 3 \quad 2 \cdot 11 \quad 2^3 \cdot 11 \quad 2 \cdot 3 \cdot 11 \cdot 31 \cdot 293$	/	/
35	$31 \cdot 71 \cdot 127 \cdot 122921$	$2 \cdot 3 \cdot 5 \quad 2 \cdot 5 \cdot 7 \quad 2 \cdot 3^2 \cdot 7 \quad 2^3 \cdot 5 \cdot 7 \cdot 439$	/	/
37	$223 \cdot 616318177$	$2 \cdot 3 \cdot 37 \quad 2^5 \cdot 3 \cdot 37 \cdot 167 \cdot 1039$	37 111	2 /
39	$7 \cdot 79 \cdot 8191 \cdot 121369$	$2 \cdot 3 \quad 2 \cdot 3 \cdot 13 \quad 2 \cdot 3^2 \cdot 5 \cdot 7 \cdot 13 \quad 2^3 \cdot 3 \cdot 13 \cdot 389$	3	/
41	$13367 \cdot 164511353$	$2 \cdot 41 \cdot 163 \quad 2^3 \cdot 41 \cdot 59 \cdot 8501$	41	2
43	$431 \cdot 9719 \cdot 2099863$	$2 \cdot 5 \cdot 43 \quad 2 \cdot 43 \cdot 113 \quad 2 \cdot 43 \cdot 3^2 \cdot 2713$	43	2
45	$7 \cdot 31 \cdot 73 \cdot 151 \cdot 631 \cdot 23311$	$2 \cdot 3 \quad 2 \cdot 3 \cdot 5 \quad 2^3 \cdot 3^2 \quad 2 \cdot 3 \cdot 5^2 \quad 2 \cdot 3^2 \cdot 5 \cdot 7 \quad 2 \cdot 3^2 \cdot 5 \cdot 7 \cdot 37$	3	/
47	$2351 \cdot 4513 \cdot 13264529$	$2 \cdot 5^2 \cdot 47 \quad 2^5 \cdot 3 \cdot 47 \quad 2^4 \cdot 31 \cdot 47 \cdot 569$	47	2
49	$127 \cdot 4432676798593$	$2 \cdot 3^2 \cdot 7 \quad 2^7 \cdot 3^2 \cdot 7^2 \cdot 43 \cdot 337 \cdot 5419$	63	/

ating in the complex number field and replacing the root 2 of order q, by a complex root $(1 + j)$ of order $8q$. Up to now, the discussion has been restricted to Mersenne numbers, that is, to numbers $p = 2^q - 1$ such that q is a prime.

If p is not a prime, its prime factorization is given by

$$p = p_1^{d_1} \cdot p_2^{d_2} \cdots p_i^{d_i}. \tag{13}$$

An M-point real transform having the circular convolution property can be defined in the ring of integers modulo p provided M-point transforms can be defined separately in the fields $p_1, p_2 \cdots p_i$. This follows directly from the Chinese remainder theorem [3, 15], and leads to the condition for the existence of an M-point transform in the ring p that M must simultaneously divide $p_1 - 1$, $p_2 - 1, \cdots p_i - 1$. When p is a prime, the maximum length of the transform is $M = p - 1$. Transforms in a ring p, with p nonprime, are therefore proportionally shorter than transforms defined modulo a prime number. If p and q are composites with $q = q_1 \cdot q_2$ and q_1 prime, $2^{q_1} - 1$ divides p and the maximum transform length is governed by that possible for $2^{q_1} - 1$. This has led us to consider that the only transforms of interest in a ring modulo $2^q - 1$ are Mersenne transforms.

The situation, however, changes noticeably if we consider Complex Mersenne Transforms. In that case, the transform length becomes $8M$ if 2 does not divide M. This means that even if M is too small to yield a real transform of practical interest, useful Complex Mersenne Transforms of length $8M$ can still be defined.

Let us define these points precisely. Assuming we restrict our discussion to transforms that can be computed without multiplications, and therefore have roots 2^W of order M, we have by definition $2^{WM} \equiv 1$, with 2^{nW} taking M distinct values for $n = 0, 1 \cdots M - 1$. The exponents Wn are taken modulo MW with $(MW)_{\text{mod } q} = 0$. If we consider now a root $(j + 1)^W$, this root will be of order $8M$ in the ring p if $(j + 1)^{8MW} \equiv 1$ and if $(j + 1)^{Wn}$ takes $8M$ distinct values for n varying from 0 up to $8M - 1$. We first notice that as $(j + 1)^8 = 2^4$, $(j + 1)^{8MW} = 2^{4MW} \equiv 1$. If W is odd, $(j + 1)^{nW}$ takes eight distinct complex values for n varying from $n = 0$ to $n = 7$ and if $q = WM$ is odd, $(((j + 1)^{8nW})) = ((2^{4nW}))$ takes M distinct values for n varying from 0 to $M - 1$.

Under these conditions, $(j + 1)^W$ is a root of order $8M$ in the ring modulo $p = 2^q - 1$ if q, M, and W are odd. Moreover, the existence of an M-point real transform in the ring modulo p implies that M has an inverse, R modulo p. As the inverse of 8 modulo p is 2^{q-3}, $8M$ always has an inverse, $2^{q-3}R$. This means that, given an M-point real transform of root 2^w with p composite and q, w, M odd integers, we can define $8M$-point complex transforms in the ring modulo p with

$$A_k = \left(\left(\sum_{n=0}^{8M-1} a_n (j + 1)^{wnk}\right)\right)$$

$$j = \sqrt{-1}$$

$$k = 0, 1 \cdots 8M - 1, \tag{14}$$

$$a_m = \left(\left(2^{q-3} R \sum_{k=0}^{8M-1} A_k (j + 1)^{-wnk}\right)\right)$$

$$m = 0, 1 \cdots 8M - 1. \tag{15}$$

Table 2 Length and roots for real and complex transforms in the ring $(2^q - 1)/p_i^{d_i}$.

q	*Transform ring*	*Real transform Length*	*Real transform Root*	*Complex transform Length*	*Complex transform Root*	*Approximate word length Nb of bits*
15	$\frac{2^{15}-1}{7}$	5	2^3	40 $(2^3 \cdot 5)$	$2(j-1)$	12
21	$\frac{2^{21}-1}{7^2}$	7	2^3	56 $(2^3 \cdot 7)$	$2(j-1)$	15
25	$\frac{2^{25}-1}{31}$	25	2	200 $(2^3 \cdot 5^2)$	$j+1$	20
27	$\frac{2^{27}-1}{7.73}$	27	2	216 $(2^3 \cdot 3^3)$	$j+1$	18
35	$\frac{2^{35}-1}{31.127}$	35	2	280 $(2^3 \cdot 5 \cdot 7)$	$j+1$	23
35	$\frac{2^{35}-1}{127}$	5	2^7	40 $(2^3 \cdot 5)$	$2^3(1-j)$	28
35	$\frac{2^{35}-1}{31}$	7	2^5	56 $(2^3 \cdot 7)$	$-2^2(1+j)$	30
45	$\frac{2^{45}-1}{7.73}$	5	2^9	40 $(2^3 \cdot 5)$	$2^4(1+j)$	36
49	$\frac{2^{49}-1}{127}$	7	2^7	56 $(2^3 \cdot 7)$	$2^3(1-j)$	42
49	$\frac{2^{49}-1}{127}$	49	2	392 $(2^3 \cdot 7^2)$	$j+1$	42

In most practical digital filtering applications, the input signal samples are defined by a number of significant bits comprised between seven bits and 20 bits. This means that, in order to avoid overflow, the transforms must operate on word lengths approximately double that of input words, comprising between 15 bits and about 50 bits.

The various possibilities for p nonprime and odd q are listed in Table 1. The case of q prime ($q = 23, 29, 37, 41, 43, 47$) corresponds to conventional Mersenne Transforms. When q is not a prime, the corresponding transforms have a very short length and their roots are not powers of 2.

In order to achieve maximum effectiveness in computing convolutions by means of pseudo-Mersenne Transforms, it would be desirable to have relatively long transforms with a number of terms highly factorizable. This does not seem possible with transforms computed modulo $p = 2^q - 1$. We note, however, that when p is not a prime, with the prime factorization of p defined by (13), we can define transforms modulo $p/p_i^{d_i}$ having power-of-2 roots and such that the number of terms is large and highly factorizable. These transforms can be defined by

$$A_k = \left(\sum_{n=0}^{8M-1} a_n (j+1)^{wnk} \right)_{\mathrm{mod}\; p/p_i^{d_i}}$$

$$k = 0, 1 \cdots 8M - 1$$
$$j = \sqrt{-1}. \qquad (16)$$

Various possibilities for such transforms are listed in Table 2. It can be seen that the maximum number of terms is both large (40 to 392 terms) and highly factorable, thereby leading to efficient FFT-type computation with a minimum number of operations.

It would seem, however, that these advantages are offset by the fact that the various operations are performed modulo $(2^q - 1)/(p_i^{d_i})$. The corresponding arithmetic circuits are obviously much more complex than arithmetic circuits modulo $2^q - 1$.

This difficulty can be circumvented by noticing that as $p = p_i^{d_1} \cdot p_i^{d_2} \cdots p_i^{d_i}$, we can compute the convolution modulo $p = 2^q - 1$ as with conventional Mersenne Transforms and obtain the final result by performing a last operation modulo $p/p_i^{d_i}$ on the convolutions computed modulo p,

$$z_m \bmod p/p_i^{d_i} = (z_m \bmod p) \bmod p/p_i^{d_i}. \qquad (17)$$

By proceeding in this fashion, relatively long convolutions can be computed efficiently by means of FFT-type algorithms with all but the last operation performed with easily implemented arithmetic circuits operating modulo $(2^q - 1)$. This technique is similar to what was proposed in [1] to compute Fermat Number Transforms with Mersenne arithmetic. The only drawback of this approach is that all the operations modulo $(2^q - 1)$ are done on words longer than that of the final result. It can be seen however, from Table 2 that many transforms with a highly factorable number of terms can be defined for which $p_i^{d_i}$ is small compared to p, so that the penalty in word length increase incurred when operating modulo p instead of modulo $p/p_i^{d_i}$ is only of the order of 20 percent.

The most interesting transforms are those which have a large, highly factorable number of terms combined with a useful word length as close as possible to q. Among these, the 200-point, 56-point, and 392-point Complex pseudo-Mersenne Transforms (corresponding, respectively, to q equal to 25, 35, and 49) seem particularly well adapted for digital filtering applications. Taking as an example the case of transforms defined by $q = 25$, one can see from Table 1 that the maximum odd length for real transforms computed modulo $p = 2^{25} - 1$ is 15 terms and that the corresponding roots are not powers of two. By operating modulo $(2^{25} - 1)/31$, it is possible to define real transforms having power-of-two roots with a maximum odd length increased to 25 terms. The maximum length is then expanded to 200 terms by using complex roots. Such a transform can be computed very efficiently by means of an FFT-type algorithm with a three-stage radix 2 decomposition, followed by a two-stage radix 5 decomposition.

One limitation of conventional Mersenne Transforms is the rigid relationship between word length and transform length. In this respect, pseudo-Mersenne Transforms provide a significant improvement, because their maximum number of terms $M_{\max}$ is highly composite and any transform length submultiple of $M_{\max}$ can be selected. It is even possible to have several transforms of identical length and defined modulo integers $P_1, \cdots P_i$ that are relatively prime. The convolution can than be computed separately modulo $P_1, \cdots P_i$ and the final result obtained modulo $(P_1 \cdot P_2 \cdots P_i)$ by the Chinese remainder theorem. This approach could, for instance, be used to compute a 40-term convolution with an approximate word length of 32 bits by means of transforms defined by modulo $(2^{15} - 1)/7$ and $(2^{25} - 1)/31$.

The computation of a time-variant convolution by means of Complex pseudo-Mersenne Transforms is shown in Fig. 1. When a time-invariant convolution is to be evaluated, the various samples A_k of the transform of $\{a_n\}$ are constant. They are usually precomputed and stored in a memory. In this case, minor additional savings on multiplication cost can be achieved by storing A_k Modulo $p/p_i^{d_i}$ instead of A_k modulo p.

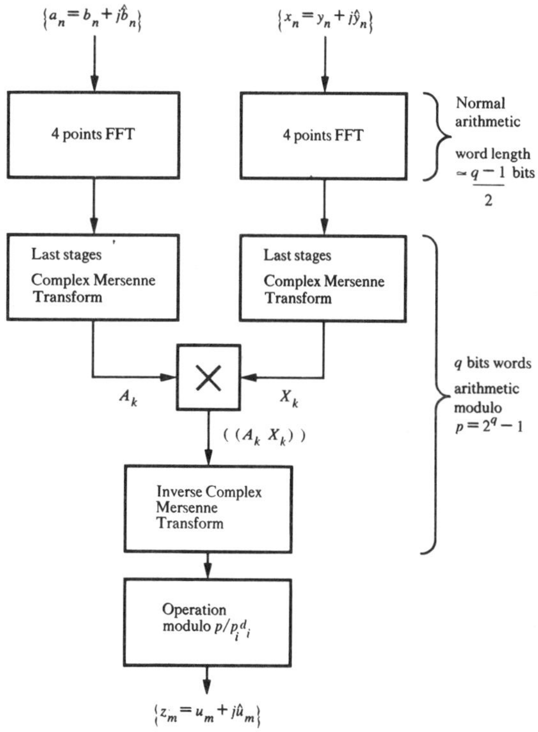

Figure 1 Computation of complex circular convolutions by means of Complex pseudo-Mersenne Transforms.

Concluding remarks

We have discussed Complex Mersenne Transforms that can be computed without multiplications. These transforms are very promising for computing convolutions because they can be partly computed with FFT-type algorithms and some of the operations can be performed on words of reduced length. Complex Mersenne Transforms also have the advantage of permitting operation on transform length and convolution lengths up to four times longer than is possible with conventional Mersenne Transforms.

These results have then been extended to cover the case of transforms operating in a ring modulo a pseudo-Mersenne number or a submultiple of such a number. It has been shown that some of these transforms have a highly composite transform length and therefore can be computed with an efficient FFT-type algorithm.

Complex Mersenne Transforms can be used for implementing digital filters in the same manner as Discrete Fourier Transforms. They have the advantage over DFT of permitting exact calculations, without round-off errors, and of being computationally much more efficient because they can be calculated without multiplications.

References

1. C. M. Rader, "Discrete Convolutions via Mersenne Transforms," *IEEE Trans. Computers* **C-21,** 1269 (1972).
2. R. C. Agarwal and C. S. Burrus, "Fast Convolution using Fermat Number Transforms with Applications to Digital Filtering," *IEEE Trans. Acoustics, Speech and Signal Processing* **ASSP-22,** 87 (1974).

3. R. C. Agarwal and C. S. Burrus, "Number Theoretic Transforms to Implement Fast Digital Convolution," *Proc. IEEE* **63,** 550 (1975).
4. B. Gold, C. M. Rader, A. V. Oppenheim, and T. G. Stockham, *Digital Processing of Signals*, McGraw-Hill Book Company, Inc., New York, 1969, Ch. 7, pp. 203-213.
5. J. W. Cooley and J. W. Tukey, "An Algorithm for Machine Calculation of Complex Fourier Series," *Math. Comp.* **19,** 297 (1966).
6. D. E. Knuth, *The Art of Computer Programming: Seminumerical Algorithms*, Addison-Wesley Publishing, Co., Inc., Reading, MA, 1969, Vol. 2, p. 356.
7. R. C. Agarwal and C. S. Burrus, "Fast One-Dimensional Digital Convolution by Multidimensional Techniques," *IEEE Trans. Acoustics, Speech and Signal Processing* **ASSP-22,** 1 (1974).
8. R. D. Gitlin, E. Y. Ho, and J. E. Mazo, "Passband Equalization of Differentially Phase-Modulated Data Signals," *Bell System Tech. J.* **52,** 219 (1973).
9. H. J. Nussbaumer, "Complex Convolutions via Fermat Number Transforms," *IBM J. Res. Develop.* **20,** 282 (1976).
10. I. S. Reed and T. K. Truong, "The Use of Finite Fields to Compute Convolutions," *IEEE Trans. Inf. Theory* **IT-21,** 208 (1975).
11. I. S. Reed and T. K. Truong, "Complex Integer Convolutions Over a Direct Sum of Galois Fields," *IEEE Trans. Inf. Theory* **IT-21,** 657 (1975).
12. H. J. Nussbaumer, "Dispositif Générateur de Fonction de Convolution Discrète et Filtre Numérique Incorporant Ledit Dispositif," French Patent Application No. 7,512,557, April 1975.
13. L. M. Leibowitz, "Fast Convolution by Number Theoretic Transforms," NRL Report 7924, Naval Research Laboratory, Washington, D. C., September 1975.
14. Gold, Rader, Oppenheim, and Stockham, *op. cit.*, Ch. 6, pp. 174-196.
15. J. M. Pollard, "The Fast Fourier Transform in a finite Field," *Math. Comp.* **25,** 365 (1971).

Received December 22, 1975

The author is located at Compagnie IBM, France, Centre d'Etudes et Recherches, 06610 La Gaude, France.

MATHEMATICAL SYSTEMS THEORY **10**, 169–180 (1977)

Some Bilinear Forms Whose Multiplicative Complexity Depends on the Field of Constants

by

S. WINOGRAD

IBM Thomas J. Watson Research Center
Yorktown Heights, New York

ABSTRACT

In this paper we consider the system of bilinear forms which are defined by a product of two polynomials modulo a third P. We show that the number of multiplications depend on how the field of constants used in the algorithm splits P. If $P = \prod_{i=1}^{k} P_i^{l_i}$ then $2 \cdot \deg(P) - k$ multiplications are needed. (We assume that P_i is irreducible.)

I. Introduction

The framework for many of the results on the number of multiplications necessary to compute algebraic functions is as follows: We start with a field G, called the field of constants, and then consider the functions to be computed as elements of $F = G(x_1, x_2, \ldots, x_n)$—the extension of G by the indeterminates $x_1, \ldots, x_n$. The lower bounds which are then derived on the number of multiplications needed to compute the functions are valid even if multiplications by a fixed element $g \in G$ is not counted as a multiplication. (See, for example, [1] and [2].)

This assumption, that a multiplication by a $g \in G$ is not counted, is valid if g is an integer, for then this multiplication may be replaced by several additions. In many applications the assumption may still be valid if g is a rational number, but if $G = Q(\sqrt{2})$, why assume that multiplication by $\sqrt{2}$ is not counted?

Many of the results in the literature are insensitive to the exact nature of G. The results of [1] and [2], for example, are of this nature, but as we will see in the next section the minimum number of multiplications may depend on the choice of G. We will give an example which requires 3 multiplications if $G = Q$, the rationals, but only 2 if $G = \mathbb{C}$, the complex number. We will then use the algorithm for $G = \mathbb{C}$ to construct an efficient algorithm to solve a related problem for the case $G = Q$.

In the following section we will analyze the dependence of the number of multiplications on the field G for a certain class of systems of bilinear forms. Fiduccia and Zalcstein [3] obtained the result of Theorem 3 for the case $l = 1$. They also obtained the result of the first half of Corollary 1 when the field G splits P and P has no repeated roots.

II. Examples

Consider the following system of bilinear forms:

$$\psi_1 = x_1 y_1 - x_2 y_2 \tag{1}$$
$$\psi_2 = x_2 y_1 + x_1 y_2.$$

This system is the real and imaginary part of the product of two complex numbers. It can be computed using 3 multiplications, for example:

$$m_1 = x_1(y_1+y_2) \qquad m_2 = (x_1+x_2)y_2 \qquad m_3 = (-x_1+x_2)y_1 \tag{2}$$
$$\psi_1 = m_1 - m_2 \qquad \psi_2 = m_1 + m_3.$$

It is known [4] that 3 multiplications is minimal if we take the field of constants to be Q. However, if we take $G = Q(i)$ then we have the following algorithm, which uses only 2 multiplications:

$$P_1 = \frac{x_1+ix_2}{2}(y_1+iy_2) \qquad P_2 = \frac{x_1-ix_2}{2}(y_1-iy_2) \tag{3}$$
$$\psi_1 = P_1 + P_2 \qquad \psi_2 = -iP_1 + iP_2.$$

The next example will illustrate how the second algorithm can be used to derive an efficient algorithm when $G = Q$.

Consider the system of bilinear forms

$$\begin{pmatrix}\psi_1\\ \psi_2\\ \psi_3\\ \psi_4\end{pmatrix} = \begin{pmatrix} x_1 & -x_2 & -x_3 & x_4\\ x_2 & x_1 & -x_4 & -x_3\\ x_3 & -x_4 & x_1 & -x_2\\ x_4 & x_3 & x_2 & x_1\end{pmatrix}\begin{pmatrix}y_1\\ y_2\\ y_3\\ y_4\end{pmatrix}. \tag{4}$$

This system is easily recognizable as the direct product of the complex algebra with itself, that is, if we denote:

$$X_1 = (x_1+ix_2), \quad X_2 = (x_3+ix_4), \quad Y_1 = (y_1+iy_2), \quad Y_2 = (y_3+iy_4)$$

then:

$$\psi_1 = \mathrm{Re}(X_1Y_1 - X_2Y_2) \tag{5}$$
$$\psi_2 = \mathrm{Im}(X_1Y_1 - X_2Y_2)$$
$$\psi_3 = \mathrm{Re}(X_2Y_1 + X_1Y_2)$$
$$\psi_3 = \mathrm{Im}(X_2Y_1 + X_1Y_2).$$

It may be seen that $9 = 3^2$ multiplications are necessary to compute (4) when $G = Q$, however, using algorithm (3) we can obtain an algorithm for computing (4) using only 6 multiplications. Using algorithm (3) we define

$$P_1 = \frac{X_1 + iX_2}{2}(Y_1 + iY_2) = \frac{(x_1 - x_4) + i(x_2 + x_3)}{2}((y_1 - y_4) + i(y_2 + y_3))$$

(6) $$P_2 = \frac{X_1 - iX_2}{2}(Y_1 - iY_2) = \frac{(x_1 + x_4) + i(x_2 - x_3)}{2}((y_1 + y_4) + i(y_2 - y_3))$$

$$X_1 Y_1 - X_2 Y_2 = P_1 + P_2$$
$$X_2 Y_1 + X_1 Y_2 = -iP_1 + iP_2,$$

and therefore:

(7) $$\begin{aligned}\psi_1 &= \mathrm{Re}(P_1) + \mathrm{Re}(P_2)\\ \psi_2 &= \mathrm{Im}(P_1) + \mathrm{Im}(P_2)\\ \psi_3 &= \mathrm{Im}(P_1) - \mathrm{Im}(P_2)\\ \psi_4 &= -\mathrm{Re}(P_1) + \mathrm{Re}(P_2).\end{aligned}$$

Using algorithm (2) twice we can compute $\mathrm{Re}(P_1)$, $\mathrm{Im}(P_1)$, $\mathrm{Re}(P_2)$, $\mathrm{Im}(P_2)$ in 6 multiplications, that is:

(8) $$\begin{aligned} m_1 &= \frac{x_1 - x_4}{2}(y_1 - y_4 + y_2 + y_3) & m_2 &= \frac{x_1 - x_4 + x_2 + y_3}{2}(y_2 + y_3)\\ m_3 &= \frac{-x_1 + x_4 + x_2 + x_3}{2}(y_1 - y_4) & m_4 &= \frac{x_1 + x_4}{2}(y_1 + y_4 + y_2 - y_3)\\ m_5 &= \frac{x_1 + x_4 + x_2 - x_3}{2}(y_2 - y_3) & m_6 &= \frac{-x_1 - x_4 + x_2 - x_3}{2}(y_1 + y_4)\end{aligned}$$

$$\mathrm{Re}(P_1) = m_1 - m_2 \qquad \mathrm{Im}(P_1) = m_1 + m_3$$
$$\mathrm{Re}(P_2) = m_4 - m_5 \qquad \mathrm{Im}(P_2) = m_4 + m_6.$$

Substituting (8) into (7) we obtain:

(9) $$\begin{aligned}\psi_1 &= m_1 - m_2 + m_4 - m_5 & \psi_2 &= m_1 + m_4 + m_3 + m_6\\ \psi_3 &= m_1 + m_3 - m_4 - m_6 & \psi_4 &= -m_1 + m_2 + m_4 - m_5.\end{aligned}$$

In the next section we will see that 6 multiplications is minimal for the system (4).

It should be clear that the construction used above is quite general. Using this construction one can show, for example, that if there is an algorithm for computing two $n_0 \times n_0$ matrices in k multiplications using any field $G \leq \mathbb{C}$, then it is possible to multiply two $n \times n$ in $An^{\log_{n_0} k}$ operations.

III. Results

In this paper we will write a system of bilinear forms as $A(\underline{x})\underline{y}$ where $A(\underline{x})$ is a

$t \times m$ matrix whose entries are linear forms (over G) of the indeterminates $\{x_1, x_2, \ldots, x_n\}$, and $\underline{y}$ is the (column) vector $(y_1, y_2, \ldots, y_m)^T$. We will need the following two results from [2]:

THEOREM 1: *Let $A(\underline{x})\underline{y}$ be a system of bilinear forms. If $A(x)$ has at least s columns such that no non-trivial linear combination of them (with coefficients in G) yields the column $\underline{0}$ then any algorithm for computing $A(\underline{x})\underline{y}$ requires at least s multiplications.*

THEOREM 2: *Let $A(\underline{x})\underline{y}$ be a system of bilinear forms, and let s be the minimum number of multiplications needed to compute $A(\underline{x})\underline{y}$, then there exists 2s linear forms (of x's and y's with coefficients in G) $L_1, L_2, \ldots, L_s, L_1', L_2', \ldots, L_s'$ such that $A(\underline{x})\underline{y} = U\underline{m}$ where U is a matrix with entries in G, m is the (column) vector $(m_1, m_2, \ldots, m_s)^T$ and $m_i = L_i \cdot L_i'$ $i = 1, 2, \ldots, s$.*

Let $R(z) = \sum_{i=0}^{a} x_i z^i$ and $S(z) = \sum_{i=0}^{b} y_i z^i$ be two polynomials with indeterminates as coefficients, and let $T(z) = R(z) \cdot S(z)$. The $a+b+1$ coefficients of T are a system of bilinear forms, denoted by $\tilde{T}$. It is known [3] that the minimum number of coefficients needed to compute $\tilde{T}$ is $a+b+1$.

One way of computing $\tilde{T}$ using $a+b+1$ multiplications starts by choosing $a+b+1$ distinct elements of G, $\alpha_0, \alpha_1, \ldots, a_{a+b}$. Since $\deg(T) = a+b$ we have

$$R(z) \cdot S(z) = R(z) \cdot S(z) \bmod \prod_{i=0}^{a+b} (z-\alpha_i). \tag{10}$$

By the Chinese Remainder Theorem $R(z) \cdot S(z) \bmod \prod_{i=0}^{a+b} (z-\alpha_i)$ can be obtained by computing $R(z) \cdot S(z) \bmod (u-\alpha_i) = R(\alpha_i) \cdot S(\alpha_i)$ for $i = 0, 1, \ldots, a+b$, and then using only multiplications by elements of G we obtain $R(z) \cdot S(z) \bmod \prod_{i=0}^{a+b} (z-\alpha_i)$. The $a+b+1$ multiplications are $R(\alpha_i) \cdot S(\alpha_i)$ $i = 0, 1, \ldots, a+b$. This method is essentially the one described in [6].

A second method starts by choosing $a+b$ distinct elements of G, $\beta_1, \beta_2, \ldots, \beta_{a+b}$, and uses the identity

$$R(z) \cdot S(z) = R(z) \cdot S(z) \bmod \prod_{i=1}^{a+b} (z-\beta_i) + x_a y_b \prod_{i=1}^{a+b} (z-\beta_i). \tag{11}$$

As before $R(z) \cdot S(z) \bmod \prod_{i=1}^{a+b} (z-\beta_i)$ is computed by using the Chinese Remainder Theorem using $a+b$ multiplications, and the $(a+b+1)^{\text{st}}$ multiplication is $x_a \cdot y_b$.

Of course we could replace the product $R(\alpha_i) \cdot S(\alpha_i)$ by the product $(g \cdot R(\alpha_i)) \cdot (h \cdot S(\alpha_i))$ for any $g, h \in G$ $(g, h \neq 0)$. If we agree to call two algorithms *essentially the same* if they differ only by this kind of changes then we obtain:

THEOREM 3: *Any algorithm for computing $\tilde{T}$ in $a+b+1$ multiplications is essentially the same as either the one derived from* (10) *or the one derived from* (11).

Proof: Let $m_0, m_1, \ldots, m_{a+b}$ be the $a+b+1$ multiplications. Since

$$
(12) \qquad T \equiv \begin{pmatrix}
x_0 & 0 & \cdots & 0 \\
x_1 & x_0 & \cdots & \vdots \\
\vdots & x_1 & & 0 \\
x_a & \vdots & & x_0 \\
0 & x_a & & x_1 \\
\vdots & 0 & & \vdots \\
0 & 0 & & x_a
\end{pmatrix}
\begin{pmatrix} y_0 \\ y_1 \\ \vdots \\ y_b \end{pmatrix} \equiv X\underline{y}
$$

we obtain that there exists an $(a+b+1)\times(a+b+1)$ matrix U with entries from G such that

$$
(13) \qquad X\underline{y} = U\begin{pmatrix} m_0 \\ m \\ m_{a+b} \end{pmatrix}
$$

Since all the rows of X are linearly independent (over G) we obtain that U is non-singular. Let $W = U^{-1}$. Consequently,

$$
(14) \qquad (WX)\underline{y} = \begin{pmatrix} m_0 \\ m_1 \\ m_{a+b} \end{pmatrix}
$$

Let $(w_0^i, w_1^i, \ldots, w_{a+b}^i)$ denote the ith row of W, then from (14) we obtain

$$
(15) \qquad \left(\sum_{j=0}^{a} w_j^i x_j, \sum_{j=0}^{a} w_{j+1}^i x_j, \ldots, \sum_{j=0}^{a} w_{j+b}^i x_j\right)\begin{pmatrix} y_0 \\ y_1 \\ y_b \end{pmatrix} = m_i,
$$

that is the bilinear form (15) can be computed in only one multiplication, and by Theorem 1 all of the forms $\sum_{j=0}^{a} w_{j+b}^i x_j$ $k = 0, 1, \ldots, b$ are multiples of one of them. That is the matrix

$$
(16) \qquad \begin{pmatrix}
w_0^i & w_1^i & \cdots & w_a^i \\
w_1^i & w_2^i & \cdots & w_{a+1}^i \\
 & & & \\
w_b^i & w_{b+1}^i & \cdots & w_{a+b}^i
\end{pmatrix}
$$

has rank 1. It is easily verified that this can happen under only two situations:

either $0 = w_0^i = w_1^i = w_2^i = \ldots = w_{a+b-1}^i$ and $w_{a+b}^i \neq 0$, or there exists α_i such that $w_j^i = (\alpha_i)^j w_0^i$ $j = 0, 1, \ldots, a+b$. Since W is non-singular at most one row can be of the second type, and if two rows are of the first kind their α's are different.

If W has no row of the second type, the algorithm is essentially the same as (10), and otherwise it is essentially the same as (11). This proves the theorem.

In the rest of the paper we will assume $a = b = n-1$.

Let $P = z^n + \sum_{i=0}^{n-1} a_i z^i$ be a polynomial with coefficients in a field F. Let $R = \sum_{i=0}^{n-1} x_i z^i$ and $S = \sum_{i=0}^{n-1} y_i z^i$ be two polynomials with indeterminates as coefficients. The coefficients of the polynomial $T(z) = R(z) \cdot S(z) \bmod P$ are a system of bilinear forms, denoted by $\tilde{T}_p$. For example, if $P = z^2 + 1$, the system $\tilde{T}_p$ is

$$(17) \qquad \begin{array}{c} x_0 y_0 - x_1 y_1 \\ x_1 y_0 + x_0 y_1, \end{array}$$

that is, the same as (1).

The results in this section will describe the minimum number of multiplications needed to compute $\tilde{T}_p$, and its dependence on G.

THEOREM 4: *Let $P = \bar{P}^l$ where $\bar{P}$ is irreducible (over G), and let $n = deg(P)$. The minimum number of multiplications needed to compute $\tilde{T}_p$ is $2n-1$.*

Proof. Let C_p be the companion matrix of P, and let $V_p = \{v \in G^n \mid \exists$ polynomial $q \neq 0$, $\deg(q) < n$ and $vq(C_p) = 0\}$. Since $\bar{P}$ is irreducible, $q(C_p)$ is non-singular whenever $\bar{P} \nmid q$. If $v \in V_p$ then $vq(C_p) = 0$ then $q = \bar{P}^r \cdot \bar{q}$ for some $1 > r > 0$ and $\bar{q}$ relatively prime to $\bar{P}$. Therefore $0 = vq(C_p) = v\bar{P}^r(C_p)\bar{q}(C_p)$, and since $\bar{q}(C_p)$ is non-singular $v\bar{P}^r(C_p) = 0$ and consequently $v\bar{P}^{l-1}(C_p) = 0$. So $V_p = \{v \in G^n \mid v\bar{P}^{l-1}(C_p) = 0\}$. We thus deduce that V_p is a subspace of G^n and $\dim(V_p) < n$.

Let $\tilde{T}p = A(\underset{\sim}{x})\underset{\sim}{y}$ then $A(\underset{\sim}{x})$ is

$$(18) \qquad A(\underset{\sim}{x}) = (\underset{\sim}{x} \mid C_p\underset{\sim}{x} \mid C_p^2\underset{\sim}{x} \mid \ldots \mid C_p^{n-1}\underset{\sim}{x})$$

where $\underset{\sim}{x}$ is the (column) vector $(x_0, x_1, \ldots, x_{n-1})^T$. (This follows from the observation that the coefficients of the polynomial $z \sum_{i=0}^{n-1} t_i z^i \bmod P$ are $C_p \cdot t$ where t^T is the vector $(t_0, t_1, \ldots, t_{n-1})$). Let t be the minimum number of multiplications needed to compute $\tilde{T}_p$, then by Theorem 2, $A(\underset{\sim}{x})\underset{\sim}{y} = U\underset{\sim}{m}$ where U is an $n \times t$ matrix with entries in G and $\underset{\sim}{m} = (m_1, m_2, \ldots m_t)^T$. Since for all non-zero (row) vectors) $w \in G^n$ $wA(\underset{\sim}{x}) \neq 0$ we obtain that rank $(U) = n$, and therefore U has n linearly independent columns. Assume with no loss of generality that the first n columns of U are linearly independent (otherwise we can permute the columns of U and the m_i's). There exists, therefore, an $n \times n$ non-singular matrix W such that $WU = (I \mid U')$. Since W is non-singular its rows span G^n and therefore there exists a row of W which is not in V_p. Assume with no loss of generality that it is the first row, and denote it by w. Since $WA(\underset{\sim}{x})\underset{\sim}{y} = (I \mid U')\underset{\sim}{m}$, we obtain $wA(\underset{\sim}{x})\underset{\sim}{y} = (1, 0, \overset{n}{\ldots}, 0, u_1', u_2', \ldots, u_{t-n}')\underset{\sim}{m}$, i.e., the bilinear form $wA(\underset{\sim}{x})\underset{\sim}{y}$ can be computed using $t-n+1$ multiplications. We claim that no non-trivial linear combination of the columns of $wA(\underset{\sim}{x})$ is 0. Assume

$$0 = \sum_{i=0}^{n-1} wC_p^i\underset{\sim}{x} \cdot \alpha_i = w\left(\sum_{i=0}^{n-1} \alpha_i C_p^i\right)\underset{\sim}{x} \text{ then } w \sum_{i=0}^{n-1} \alpha_i C_p^i = 0,$$

but $w \notin V_p$ and therefore $\alpha_i = 0$ $i = 0, 1, \ldots, n-1$. By Theorem 1 at least n multiplications are needed to compute $wA(\underline{x})\underline{y}$, that is, $t-n+1 \geq n$, and $t \geq 2n-1$ which proves the theorem. (Note that since $\mathcal{T}$ can be computed in $2n-1$ multiplications so can $\mathcal{T}_p$.)

Let $P = P_1^{l_1} \cdot P_2^{l_2}$ where P_1, P_2 are irreducible and $(P_1, P_2) = 1$. By the Chinese Remainder Theorem the system $\mathcal{T}_p$ can be computed by computing $\mathcal{T}_{p^{l_1}}$ and $\mathcal{T}_{p_2{}^{l_2}}$ and then combining the results using only additions and multiplications by elements of G.

Definition: Let $A(\underline{x})\underline{y}$ and $B(\underline{\xi})\underline{\eta}$ be two systems of bilinear forms on disjoint sets of indeterminates. The *disjoint sum of A and B*, denoted by $A+B$, is the system

$$\begin{pmatrix} A(x) & 0 \\ 0 & B(\underline{\xi}) \end{pmatrix} \begin{pmatrix} \underline{y} \\ \underline{\eta} \end{pmatrix}.$$

An *algorithm for computing $A \oplus B$ is said to be disjoint* if its matrix U is

$$U = \begin{pmatrix} U_1 & 0 \\ 0 & U_2 \end{pmatrix},$$

that is, if it can be viewed as computing $A(\underline{x})\underline{y}$ and $B(\underline{\xi})\underline{\eta}$ separately.

Conjecture: The minimum number of multiplications needed to compute $A \oplus B$ is $m(A)+m(B)$ where $m(A)$ denotes the minimum number of multiplications needed to compute $A(\underline{x})\underline{y}$, and similarly for $m(B)$. Moreover, every algorithm which computes $A \oplus B$ in the minimum number of multiplications is disjoint.

We cannot prove the conjecture in general but the next theorem proves it in a special case.

THEOREM 5: *Let $P_i = \bar{P}_i^{l_i}$ $i = 1, 2, \ldots, k$, where $\bar{P}_i$ is irreducible and $deg(P_i) = n_i$, and let $n = \sum_{i=1}^{k} n_i$. The minimum number of multiplications required to compute $\mathcal{T} = \mathcal{T}_{p_1} \oplus \mathcal{T}_{p_2} \oplus \ldots \oplus \mathcal{T}_{p_k}$ is $2n-k$. Moreover, every algorithm for computing this system in $2n-k$ multiplications is disjoint.*

Proof: Let $\mathcal{T}_{p_i}$ be $A_i(\underline{x}^i)\underline{y}^i$ and $\mathcal{T}$ be $A(\underline{x})\underline{y}$ where

$$A(\underline{x}) = \begin{pmatrix} A_1(\underline{x}^1) & & 0 \\ & A_2(\underline{x}^2) & \\ 0 & & A_k(\underline{x}^k) \end{pmatrix} \qquad y = \begin{pmatrix} y^1 \\ y^2 \\ \\ y^k \end{pmatrix} \tag{19}$$

Let $A(\underline{x})\underline{y} = U_{n \times t}\underline{m}$, (where t denotes the number of multiplications), then since all the rows of $A(\underline{x})$ are linearly independent, the rank of U is n. Therefore there exists an $n \times n$ non-singular matrix W such that $WU = (I \mid U')$. Partition W as $W = (W^1 \mid W^2 \mid \ldots \mid W^k)$ where W^1 is the first n_1 columns of W, W^2 the next n_2 columns of W, etc. Since W is non-singular $\text{rank}(W^i) = n_i$ $i = 1, 2, \ldots, k$. Therefore each W^i has a row which is not in V_{p_i}. Let w_j^i denote the jth row of W^i, and let $p(j)$ denote the cardinality of the set $\{i | w_j^i \notin V_{p_i}\}$.

LEMMA 1: *If for some j, $p(j) = s$ then $t \geq 2n-k+s-1$.*

Proof: Assume with no loss of generality that $p(1) = s$ and that $w_1^i \notin V_{P_i}$ $i = 1, 2, \ldots, s$. Let $j_{s+1}, j_{s+2}, \ldots, j_k$ be such that $w_{j_r}^r \notin V_{P_{j_r}}$ $r = s+1, s+2, \ldots, k$, then there exists a row vector β with non-zero coefficients only in positions $1, j_{s+1}, j_{s+2}, \ldots, j_k$ such that if $\gamma = \beta W = (\gamma_1 \gamma_2 \ldots \gamma_k)$ then $\gamma_i \notin V_{P_i}$ $i = 1, 2, \ldots, k$. Consider the bilinear form $\beta W A(\underline{x})\underline{y} = \beta(I \ U')\underline{m}$. We claim that at least n multiplications are needed to compute it. Consider any nontrivial linear combinations of the columns of $\beta W A(\underline{x}) = \gamma A(\underline{x})$. Substituting $C_{P_i}^j \underline{x}^i$ for the jth column of $A_i(\underline{x}^i)$ we obtain that this linear combination is

$$\sum_{i=1}^{k} \gamma_i \left(\sum_{j=0}^{n_j-1} \alpha_{i,j} C_{P_i}^j \right) \underline{x}^i.$$

This linear combination vanishes if and only if $\gamma_i \sum_{j=0}^{n_i-1} \alpha_{i,j} C_{P_i}^j = 0$ for $i = 1, 2, \ldots, k$. But $\gamma_i \notin V_{P_i}$ and therefore $\alpha_{i,j} = 0$ for all i and j. This establishes the claim. By Theorem 1 at least n multiplications are needed to compute $\gamma A(\underline{x})\underline{y}$, but $\beta(I \ U')$ has at most $k-s+1+t-n$ non-zero coefficients, and therefore $k-s+1+t-n \geq n$ or $t \geq 2n-k+s-1$. This proves the lemma.

Since $s \geq 1$ we have also proved that every algorithm for computing $\mathcal{T}$ requires at least $2n-k$ multiplications. Since it can be trivially computed in $2n-k$ multiplications, this proves the first half of the theorem.

To prove the second half of the theorem, consider an algorithm for computing $\mathcal{T}$ in $2n-k$ multiplications, and write it as $\mathcal{T} = Um$. Let W be as before, then as a consequence of Lemma 1 we obtain $p(j) \leq 1$ for $j = 1, 2, \ldots, n$ i.e., we can partition the rows of W into $k+1$ disjoint sets $N_0, N_1, N_2, \ldots N_k$ such that if $j \in N_0$ $w_j^i \in V_{P_i}$ for $i = 1, 2, \ldots, k$, and if $j \in N_r$ $(r \neq 0)$ then $w_j^r \notin V_{P_r}$ and $w_j^i \in V_{P_i}$ for $i \neq r$. With no loss of generality we may assume that $N_1 = \{1, 2, \ldots, m_1\}$, $N_2 = \{m_1+1, m_1+2, \ldots, m_1+m_2\}$, etc. and that $N_0 = \{\sum_{i=1}^k m_i+1, \sum_{i=1}^k m_i+2, \ldots, n\}$. (If this is not the case we can permute the rows of W, to bring it to this form.) As before we have $WU = (I \ U')$ and denote the i, j entry of U' by $u'_{i,j}$.

LEMMA 2: *If $i_1 \in N_{j_1}$ and $i_2 \in N_{j_2}$ $j_1 \neq j_2$ $(j_1, j_2 \neq 0)$ then for all j, $u'_{i_1,j} \cdot u'_{i_2,j} = 0$.*

Proof: Assume $u'_{i_1,j} \neq 0$ and $u'_{i_2,j} \neq 0$. Assume with no loss of generality, that $j_1 = 1$, $i_1 = 1$, $j_2 = 2$, $i_2 = m_1+1$, and $j = 1$. Let M be the matrix

$$M = \begin{pmatrix} 1/u'_{1,1} & 0 & 0 & \ldots & 0 \\ -\dfrac{u'_{2,1}}{u'_{1,1}} & 1 & 0 & & 0 \\ -\dfrac{u'_{3,1}}{u'_{1,1}} & 0 & 1 & & 0 \\ -\dfrac{u'_{n,1}}{u'_{1,1}} & 0 & 0 & & 1 \end{pmatrix} \tag{20}$$

Then $MWA(\underline{x})\underline{y} = \bar{W}^{A(\underline{x})}\underline{y} = M(I \mid U')\underline{m} = (I \mid \bar{U})\underline{m}'$ where $\underline{m}'$ is $\underline{m}$ with the first and $(n+1)^{\text{st}}$ entries exchanged, and $(I \mid \bar{U}) = M(I \mid U')$ with the first and $(n+1)^{\text{st}}$ columns exchanged. The $(m_1+1)^{\text{st}}$ row of $\bar{W}$ is

$$\left(w^1_{m_1+1} - \frac{u'_{m+1,1}}{u'_{1,1}} w_1^{\prime} \,\middle|\, w^2_{m_1+1} - \frac{u'_{m+1,1}}{u'_{1,1}} w_1^2 \,\middle|\, \ldots \,\middle|\, w^k_{m+1} - \frac{u'_{m+1,1}}{u'_{1,1}} w_1^k \right).$$

By construction $w^1_{m_1+1} \in V_{P_1}$ and $w^1_1 \notin V_{P_1}$ and therefore

$$w^1_{m_1+1} - \frac{u'_{m+1,1}}{u'_{1,1}} w^1_1 \notin V_{P_1},$$

similarly

$$w^2_{m_1+1} - \frac{u'_{m+1,1}}{u'_{1,1}} w^2_1 \notin V_{P_2},$$

and by Lemma 1 the algorithm uses at least $2n-k+1$ multiplications. Contradiction!

Lemma 2 states that the $n-k$ columns of U' can be partitioned into $k+1$ disjoint sets $M_0, M_1, \ldots, M_k$ such that if $i \in N_{j_1}$ $(j_1 \neq 0)$ and $j \in M_{j_2}$ $(j_2 \neq j_1)$ then $u'_{i,j} = 0$. Schematically then U' can be written as

$$U' = \begin{array}{c} \\ N_1 \\ N_2 \\ \\ N_k \\ N_0 \end{array} \left(\begin{array}{c|c|c|c|c} M_1 & M_2 & & M_k & M_0 \\ U'_1 & 0 & \cdots\cdots & 0 & \\ \hline 0 & U'_2 & \cdots\cdots & 0 & \\ \hline \vdots & \vdots & \vdots & \vdots & \\ 0 & 0 & \cdots\cdots & U'_k & \\ \hline & & U'_0 & & \end{array} \right) \tag{21}$$

LEMMA 3: (1) *For each $\dot{s}$ the cardinality of $M_{\dot{s}}$ is $n_{\dot{s}}-1$ (and therefore $M_0 = \emptyset$).*

(2) *If $i \in N_r$ and $j \in M_r$ $(r \neq 0)$ then $u'_{i,j} \neq 0$.*

(3) *If $i \in N_r$ then $w^j_i = 0$ for $j \neq r$.*

Proof: Let i be in $N_r (r \neq 0)$ then since $WA(\underline{x})\underline{y} = (I \mid U')\underline{m}$ we obtain that the bilinear form $(w^1_i, w^2_i, \ldots, w^k_i)A(\underline{x})y = (0, 0, \cdot\, 0, \overset{i}{1}, 0, \ldots \overset{n}{0}, 0 \ldots 0, \overset{M_r}{u^r_{r,i}}\, 0, \ldots, 0)\, \underline{m} = u_i m$. Since $w^r_i \notin V_{P_r}$ we obtain from Theorem 1 as in the proof of Theorem 4, that at least n_r multiplications are needed to compute it, and that therefore u_j has at least n_r non-zero entries. But that means that the cardinality of M_r is at least $n_r - 1$. Since U' has $n-k = \sum_{s=1}^k (n_s - 1)$ columns the cardinality of

M_r is exactly $n_r - 1$, and u^j has exactly n_r non-zero entries. To prove the third part of the lemma, observe that if $w_i^j \neq 0$ for $j \neq r$ then $(w_i^1 \mid w_i^2 \mid \ldots \mid w_i^k)A(\underline{x})$ has at least $n_r + 1$ linearly independent columns, namely the n_r columns $\sum_{a=1}^{r-1} n_a + 1, \sum_{a=1}^{r-1} n_a + 2, \ldots, \sum_{a=1}^{r} n_a$ (which are linearly independent since $w_i^r \notin V_{P_r}$) and without loss of generality, the column $\sum_{a=1}^{j-1} n_a + 1$ which is not zero since $w_i^j \neq 0$ and cannot depend on the other n_r columns since it has the indeterminates $\{x^j\}$ while the other n_r columns are linear combinations of the indeterminates $\{x^r\}$. But u_i has exactly n_r non-zero entries, contradicting Theorem 1.

LEMMA 4: *Let $i \in N_0$ and $j \in M_r$, and assume $u'_{i,j} \neq 0$, then:*

(1) *If $s \notin M_r$ then $u'_{i,s} = 0$.*

(2) *If $a \neq r$ then $w_i^a = 0$.*

Proof: Assume with no loss of generality that $j = 1 \in M_1$. Let M be as in the proof of Lemma 2 (that is, M as in (20)). Note that by Lemma 3, $u'_{1,1} \neq 0$. Let $\bar{W} = MW$ and let $M(I\ U')\underline{m} = (I\ \bar{U})\underline{m}'$ as in the proof of Lemma 2. Let $\bar{N}_0, \bar{N}_1, \ldots, \bar{N}_k$ be the partitioning of the rows of $\bar{W}$, then $i \in \bar{N}_1$, since

$$\bar{w}_i^1 = w_i^1 - \frac{u'_{i,1}}{u'_{1,1}} w_1^1 \notin V_{P_1}.$$

But by Lemma 2, if $s \notin M_r$ then $\bar{u}_{i,s} = u'_{i,s}$, and applying Lemma 2 again, we obtain $0 = \bar{u}_{i,s}$. This proves the first part of the lemma. By Lemma 3 (third part) $\bar{w}_i^s = w_i^s$, and again applying Lemma 3 we obtain $\bar{w}_i^s = 0$. This proves the lemma.

Lemma 4 states that the rows of W can be partitioned into k disjoint sets $\Pi_1, \Pi_2, \ldots, \Pi_k$, where $i \in \Pi_r$ if and only if $w_i^r \neq 0$. Since W is non-singular, the cardinality of Π_r is n_r, i.e.,

$$W = \begin{array}{c} \Pi_1 \\ \Pi_2 \\ \\ \\ \Pi_k \end{array} \left(\begin{array}{c|c|c|c} W_1 & 0 & \cdots\cdots & 0 \\ \hline 0 & W_2 & \cdots\cdots & 0 \\ \hline & & \ddots & \\ \hline 0 & 0 & \cdots\cdots & W_k \end{array}\right) \tag{22}$$

and similarly

$$U' = \begin{array}{c} \\ \Pi_1 \\ \Pi_2 \\ \\ \\ \Pi_k \end{array} \begin{array}{c} \begin{array}{cccc} M_1 & M_2 & \cdots & M_k \end{array} \\ \left(\begin{array}{c|c|c|c} U'_1 & 0 & \cdots & 0 \\ \hline 0 & U'_2 & \cdots & 0 \\ \hline & & \ddots & \\ \hline 0 & 0 & \cdots & U'_k \end{array}\right) \end{array} \tag{23}$$

and therefore the columns of $(I \mid U')$ can be permuted by a permutation P such that

$$(I\ U')P = \begin{pmatrix} \tilde{U}_1 & 0 & \cdots & 0 \\ 0 & \tilde{U}_2 & \cdots & 0 \\ \vdots & \vdots & \ddots & \\ 0 & 0 & & U_k \end{pmatrix}. \tag{24}$$

But $WA(\underline{x})\underline{y} = ((I \mid U')P)P^{-1}\underline{m}$ and therefore $A(\underline{x})\underline{y} = ((W^{-1}(I \mid U')P)P^{-1}\underline{m}$, that is, the columns of the matrix U could be permuted such that

$$U = \begin{pmatrix} U_1 & 0 & \cdots & 0 \\ 0 & U_2 & & \vdots \\ \vdots & \vdots & \ddots & \vdots \\ 0 & 0 & \cdots & U_k \end{pmatrix} \tag{25}$$

and therefore the algorithm is disjoint, and we have proved the second half of the theorem.

COROLLARY 1: *Let k be the number of irreducible factors of a polynomial P (not counting multiplicity) then the minimum number of multiplications needed to compute $\tilde{T}_p$ is $2 \cdot deg(P) - k$. Moreover, if P has no repeated roots and it is split by G then the minimal algorithms is unique.*

Proof: Let $P = \prod_{i=1}^{k} P_i^{l_i}$, by the Chinese Remainder Theorem multiplication modulo P is the same as multiplication modulo the $P_i^{l_i}$'s. The second part of the corollary follows from the observation that if P is linear the minimal algorithm for computing $\tilde{T}_p$ is unique.

COROLLARY 2: *Let $G = \mathbb{R}$, the reals. The minimal number of multiplications needed to compute the system (4) of bilinear forms is 6.*

Proof: Define $\xi_1 = x_1 + x_4$, $\xi_2 = x_2 - x_3$, $\xi_3 = x_1 - x_4$, $\xi_4 = x_2 + x_3$, $\eta_1 = y_1 + y_4$, $\eta_2 = y_2 - y_3$, $\eta_3 = y_1 - y_4$, $\eta_4 = y_2 + y_3$, then we obtain

$$\begin{pmatrix} \psi_1 + \psi_4 \\ \psi_2 - \psi_3 \\ \psi_1 - \psi_4 \\ \psi_2 + \psi_3 \end{pmatrix} = \begin{pmatrix} \xi_1 & -\xi_2 & 0 & 0 \\ \xi_2 & \xi_1 & 0 & 0 \\ 0 & 0 & \xi_3 & -\xi_4 \\ 0 & 0 & \xi_4 & \xi_3 \end{pmatrix} \begin{pmatrix} \eta_1 \\ \eta_2 \\ \eta_3 \\ \eta_4 \end{pmatrix}. \tag{26}$$

System (19 is clearly equivalent to (4), and (19) is $\tilde{T}_p \oplus \tilde{T}_p$ where $P = z^2 + 1$.

REFERENCES

[1] PAN, V. YA., Methods of computing values of polynomials, *Russian Mathematical Surveys*, **21** (1966) 105–136.

[2] WINOGRAD, S., On the number of multiplications necessary to compute certain functions, *Comm. Pure and Appl. Math.*, **23** (1970) 165–179.

[3] FIDUCCIA, C. M. and Y. ZALCSTEIN, *Algebras having linear multiplicative complexities*, Technical Report 46, Department of Computer Science, State University of New York at Stony Brook, (August 1975).

[4] WINOGRAD, S., On multiplication of 2×2 matrices, *Linear Alg. Appl.*, **4** (1971) 381–388.

[5] STRASSEN, V., Vermeidung von Divisionen, *J. für reine und angewandte Mathematik*, **264** (1973), 184–202.

[6] TOOM, A. L., The Complexity of a Scheme of Functional Elements, Soviet Mathematics, Translations of Doklady Akademie, *Nauk. S.S.S.R.*, **4,** (June 1963), 714–717.

Received December 3, 1975 and in revised form April 8, 1976 and June 16, 1976

Part Three

IX

complex general-N Winograd Fourier transform algorithm (WFTA)

Purpose

This program computes the discrete Fourier transform (DFT) of sequences whose length N is a product of relatively prime factors taken from the set {2, 3, 4, 5, 7, 8, 9, 16}. The method in this program for computing the DFT is a general-N implementation of Winograd's algorithm [4], referred to as the WFTA [113, 114].

Method

In order to set up a common notational framework we first summarize the basic form of the WFTA. Consider the computation of the N-point DFT of the sequence $x[n]$, $n = 0, 1, \ldots, N-1$. Suppose the transform length factors into the product of μ mutually prime integers; $N = N_1 \times N_2 \times \ldots \times N_\mu$. The WFTA consists of five steps as illustrated in Figure 9.1 for the case $N = 15$.

Step 1: The one-dimensional sequence $x[n]$ is mapped into a μ-dimensional array $s[n_1, n_2, \ldots, n_\mu]$ via the Sino correspondence

$$n = \left\langle \sum_{i=1}^{\mu} \left(\frac{N}{N_i}\right) n_i \right\rangle_N \tag{9.1}$$

Step 2: The *pre-weave* [70] modules are implemented, one for each factor N_i. Each pre-weave module contains only additions and subtractions. The size of the data array may expand when passing through the pre-weave elements. The order of the modules can be chosen to minimize the total number of additions in the algorithm.

Step 3: The data array is multiplied (point-by-point) by an array of real constants† derived from the multipliers of the small-N DFT algorithms for N_i, $i = 1, 2, \ldots, \mu$. These constants depend solely on the complex exponentials of the DFT, W_N, and are the only multiplications required in the algorithm.

†Note that the constants are not allowed to be imaginary as proposed by others; rather the post-weave modules contain "multiplications" by j to effectively obtain imaginary constants. Thus the need for a flag array is eliminated.

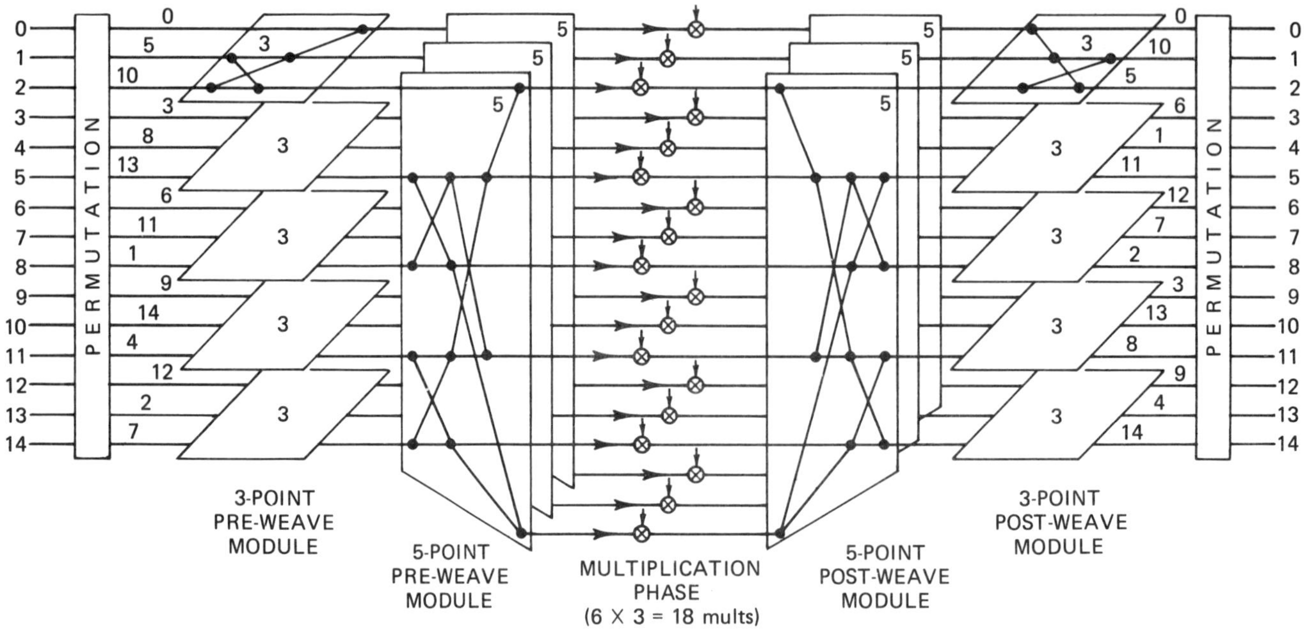

Figure 9-1 Fifteen-point DFT algorithm decomposed using Good's mapping to obtain a two-dimensional (3 × 5)-point DFT and Winograd's algorithms for 3-point and 5-point DFTs. The algorithm requires 18 = 6 × 3 multiplications which are "nested" between the pre-weave and post-weave modules.

Step 4: The *post-weave* [70] modules are implemented for each of the μ dimensions. Each of these modules contains only additions, subtractions, and multiplications by j. A different module is required for each dimension. After passing through the post-weave section, the data array $S[k_1, k_2, \dots, k_\mu]$ shrinks back to its original size, $N_1 \times N_2 \times \dots \times N_\mu$. At this point in the computation, we have obtained the μ-dimensional $N_1 \times N_2 \times \dots \times N_\mu$ DFT of the array $s[n_1, n_2, \dots, n_\mu]$.

Step 5: A mapping of the μ-dimensional array $S[k_1, k_2, \dots, k_\mu]$ is performed in accordance with the Chinese Remainder theorem to obtain the correct one-dimensional DFT, $X[k]$. This correspondence is

$$k = \langle s_1k_1 + s_2k_2 + \dots + s_\mu k_\mu \rangle_N \tag{9.2}$$

where the integers s_i satisfy

$$s_i \equiv \delta_{ij} \bmod N_j; \quad i, j = 1, 2, \dots, \mu \tag{9.3}$$

Initialization

In order to have an efficient WFTA program several computations need be performed only once when DFT's of the same length are to be calculated over and over. First, the factors of N must be determined. Since the WFTA is composed of small-N DFT algorithms that have been developed only for certain lengths, a check against the allowable values of N is made. Next, the constant multiplier array is formed by multiplying together the pre-stored constants for the individual small-N factors. Finally, permutation vectors for the input and output mappings are prepared based on the factors of N. This speeds up the actual mapping process because the congruences of the Chinese

Remainder Theorem are solved during initialization. The expense, of course, is the storage to hold the mapping vectors.

Program Description

The WFTA program consists of seven subroutines, each of which is described in the following material. The flow of control in the algorithm is given in Figure 9.2. Documentation of the individual computational modules (pre-weave and post-weave) is presented using the annotated flowgraphs of Figures 9.3–9.10.

The WFTA is invoked by calling the subroutine WFTA with the arguments:

- N transform length
- XR(N) real part of array to be transformed
- XI(N) imaginary part of array to be transformed. The real and imaginary parts of the transform are returned in XR and XI.
- INIT a flag to specify whether the call to WFTA requires initialization. INIT = 0 calls for initialization; INIT ≠ 0 skips the initialization phase. When many DFT's of the same length are to be performed, initialization is needed only for the first DFT but not on succeeding calls. See subroutine INISHL for details.
- IERR is an integer variable that contains, upon completion of a DFT, an error code. If the DFT was done successfully, IERR = 0; if an error occurred, IERR = −1 or −2. There are two causes of the error: (1) the transform length is illegal (that is, it contains factors outside the set {2, 3, 4, 5, 7, 8, 9, 16}), (IERR = −1), or (2) the program has not been initialized for the particular value of the transform length specified (IERR = −2).

The subroutine WFTA calls in turn the subroutines INISHL (if needed), PERM1, WEAVE1, MULT, WEAVE2, and PERM2.

The subroutine INISHL performs those calculations, for a fixed value of N, that need not be done on each pass of the algorithm. These include factoring the transform length N into its mutually prime factors, calculating the constant coefficients to be used in the multiplication step (subroutine MULT), and calculating the mapping vectors for the pre- and post-permutations of the data to be used in subroutines PERM1 and PERM2. The transform length N is factored into mutually prime factors taken from the set {2, 3, 4, 5, 7, 8, 9, 16}. Since N can have at most four factors from this set, the factors are assigned to the integer variables NA, NB, NC, and ND as follows:

NA is assigned the largest value from the set {1, 2, 4, 8, 16} that divides N; NB the largest divisor of N from the set {1, 3, 9}; NC the largest from {1, 7}; and ND the largest from {1, 5}. The product NA*NB*NC*ND should equal N, if the transform length is permissible. In the following, NA will be referred to as the *first* (or outermost) factor, NB the *second*, NC the *third*, and ND the *fourth* (or innermost). This reference is with respect to the order of pre-weave modules in the calculations (that is, NA is the size of the first pre-weave module and last post-weave module).†

From the factors of N we can now determine via Table 9-1 the amount of data expansion that will occur in the pre-weave phase and, equivalently, the number of

†This order of pre-weave and post-weave modules will minimize the number of additions in the algorithm in all cases except N = 48, 240, 336, and 1680. The increases in additions for these cases is less than 4 per cent compared with the optimal strategy [113]. The benefit of the order adopted here is a simpler program structure.

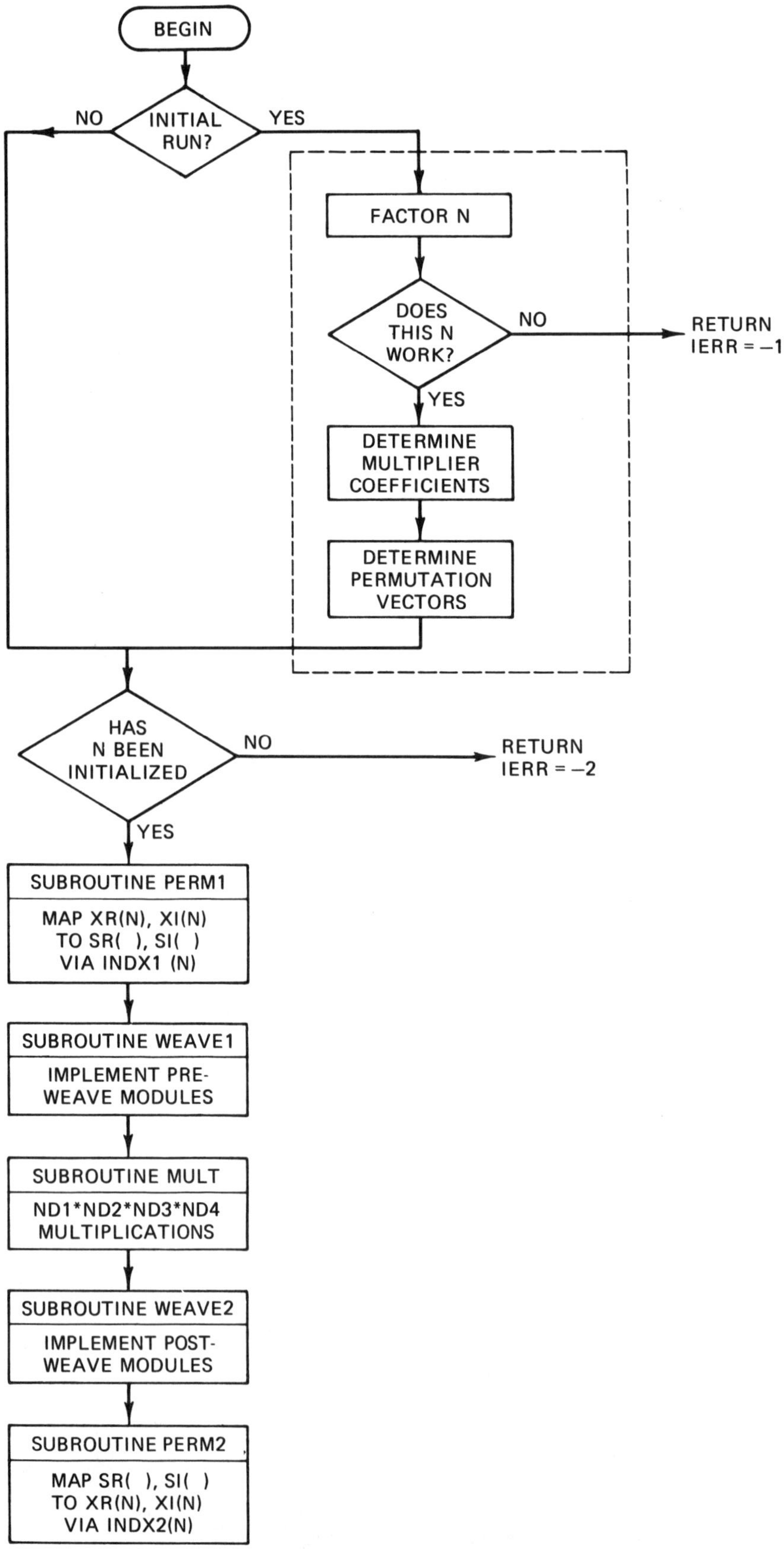

Figure 9-2 Flowchart for Winograd Fourier Transform Algorithm Program.

multiplications that will be performed in the WFTA. For each of the four variables (NA, NB, NC, and ND) the number of multiplies is assigned to the variables ND1, ND2, ND3 and ND4, respectively. In order to accommodate the data expansion of the WFTA, the working array SR(·) and SI(·) must be dimensioned as large as ND1*ND2*ND3*ND4. If the present upper limit on the transform size of 5040 is to be reduced, the size of the arrays can be reduced using Table 9.1 to determine the data expansion for the largest allowable transform size.

Table 9.1. Number of Multiplications, $M(N)$, Needed for N-point WFTA

N	$M(N)$
2	2
3	3
4	4
5	6
7	9
8	8
9	11
16	18

NOTE: if $N = r_1 \times r_2 \times \ldots \times r_\mu$
then $M(N) = M(r_1)M(r_2)\ldots M(r_\mu)$

The second step of the INISHL subroutine is to calculate the constant coefficients for the multiplication routine. These are generated from the coefficients of the individual factors as

$$\text{COEF}(n_1, n_2, n_3, n_4) = \text{COA}(n_1)*\text{COB}(n_2)*\text{COC}(n_3)*\text{COD}(n_4)$$

where

$$n_1 = 1, 2, \ldots, \text{ND1}$$
$$n_2 = 1, 2, \ldots, \text{ND2}$$
$$n_3 = 1, 2, \ldots, \text{ND3}$$
$$n_4 = 1, 2, \ldots, \text{ND4}$$

The array COEF is actually treated as a one-dimensional array in the FORTRAN program, but can also be thought of as a multidimensional array in a mixed-radix representation with the first index varying the most rapidly. Thus, the following addresses are equivalent

$$\text{COEF}(n_1, n_2, n_3, n_4) = \text{COEF}(n_1 + \text{ND1}*n_2 + \text{ND1}*\text{ND2}*n_3 + \text{ND1}*\text{ND2}*\text{ND3}*n_4)$$

This method of data addressing will also apply to the data arrays SR(·) and SI(·).

The third step of the INISHL subroutine is the calculation of the mapping vectors INDX1(·) and INDX2(·) for the pre- and post-permutations of the data. The pre-permutation of the data is done according to the Sino correspondence of (9.1). The mapping is implemented using the mapping vector INDX1(·) to obtain the highest speed possible. The post-permutation requires the inverse mapping of the Chinese Remainder Theorem, as in (9.2) and(9.3). Again, for maximum speed a mapping vector INDX2(·) is constructed.

The three functions of the INISHL subroutine need be performed only once for a given value of transform length N. The remaining five subroutines form the computational core of the algorithm. The subroutine PERM1 maps the data input arrays XR and XI into the working arrays SR and SI using the pre-permutation vector INDX1. The arrays SR and SI are treated as one-dimensional arrays of size ND1*ND2*ND3*ND4 in subroutine PERM1. Note that the WFTA is NOT in-place because of the nature of the permutations.

Next the subroutine WEAVE1 is invoked to perform the pre-weave modules. The first pre-weave module performed is of length NA. If NA = 1 the next factor, NB, is tested. Each factor NB, NC, and ND is checked in turn. In WEAVE1 the arrays SR and SI are treated as one-dimensional arrays to gain speed in execution. There is no 4-point pre-weave module because the 4-point DFT is implemented entirely as a post-weave module. Likewise, the 2-point pre-weave module is a 2-point DFT and there is no 2-point post-weave module. Documentation of the code for each pre-weave module can be found in the flowgraphs of Figures 9.3–9.10, and Table 9.2.

Following the WEAVE1 subroutine the data are multiplied point-by-point by the real array COEF in the subroutine MULT. Then the post-weave modules are implemented in the subroutine WEAVE2. The module of length ND (=5 or 1) is done first, then NC, NB, and finally NA. The flowgraphs of Figures 9.3–9.10 give the post-weave algorithms for the various factors. The last subroutine called is PERM2 which performs the post-permutation by mapping SR and SI to XR and XI via the mapping vector INDX2(·). This completes the WFTA calculation.

Description of Parameters

CDA, CDB, CDC, CDD: real arrays that are assigned coefficients of the factors of N during initial call (CDA: for outermost factor, CDD: for innermost factor).

CO3, CO4, CO8, CO9, CO16: real arrays of short-DFT coefficients for factors 3, 4, 8, 9, 16 respectively.

COEF: real array of size ND1*ND2*ND3*ND4 to which the N-point DFT multiply coefficients are assigned during initial run.

INDX1, INDX2: integer arrays of size N for storing initial and final permutation mappings respectively.

INIT: integer variable passed from user program to WFTA. It must be zero on the initial run and nonzero otherwise.

IERR: integer variable that returns error code indicating successful completion (IERR = 0) or abnormal termination (IERR = −1 or −2).

N: transform length

NA, NB, NC, ND: integer variables for four factors of N (NA: outermost, ND: innermost).

ND1, ND2, ND3, ND4: integer variables representing the "expansion" of factors NA, NB, NC, ND, respectively. See Table 9.1.

NMULT: integer variable equal to product of ND1, ND2, ND3, and ND4. This is the number of multiplications.

SR, SI: real arrays of size ND1*ND2*ND3*ND4 to which XR, XI are transferred during initial permutation.

XR, XI: real and imaginary parts respectively of input data array. The transformed data are also placed in these arrays by the final permutation.

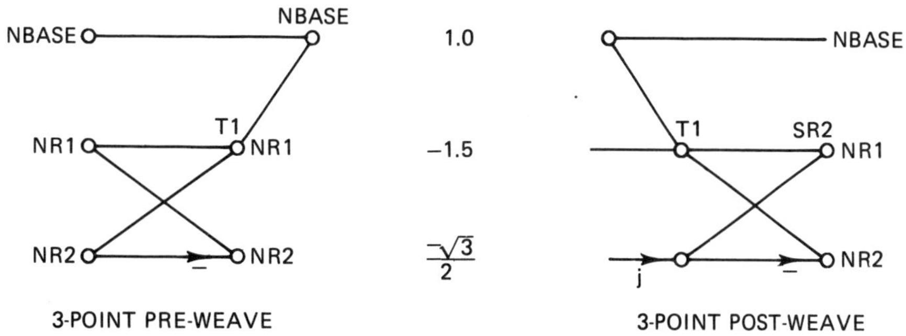

Figure 9-3 Three-point DFT algorithm using Winograd's decomposition. Labels correspond to storage locations in FORTRAN program.

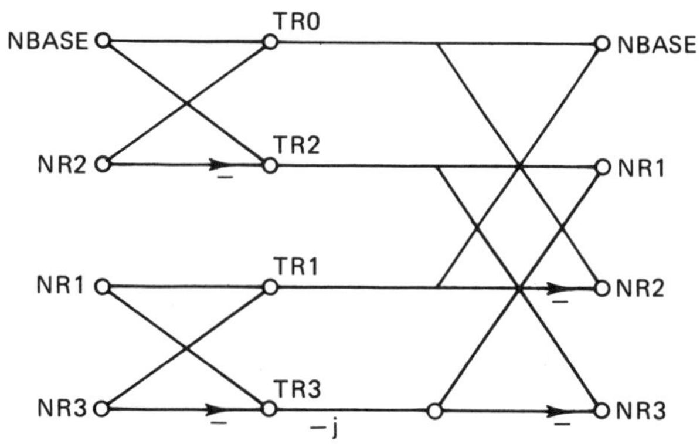

Figure 9-4 Four-point DFT which is also used entirely as a post-weave module in WFTA program.

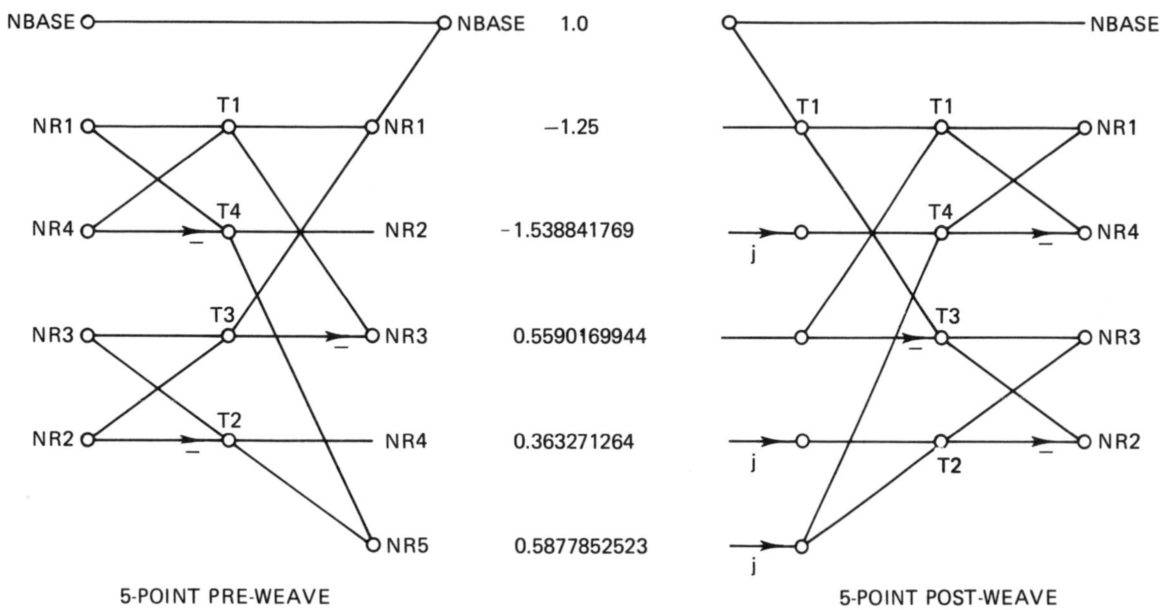

Figure 9-5 Five-point DFT algorithm using Winograd's decomposition.

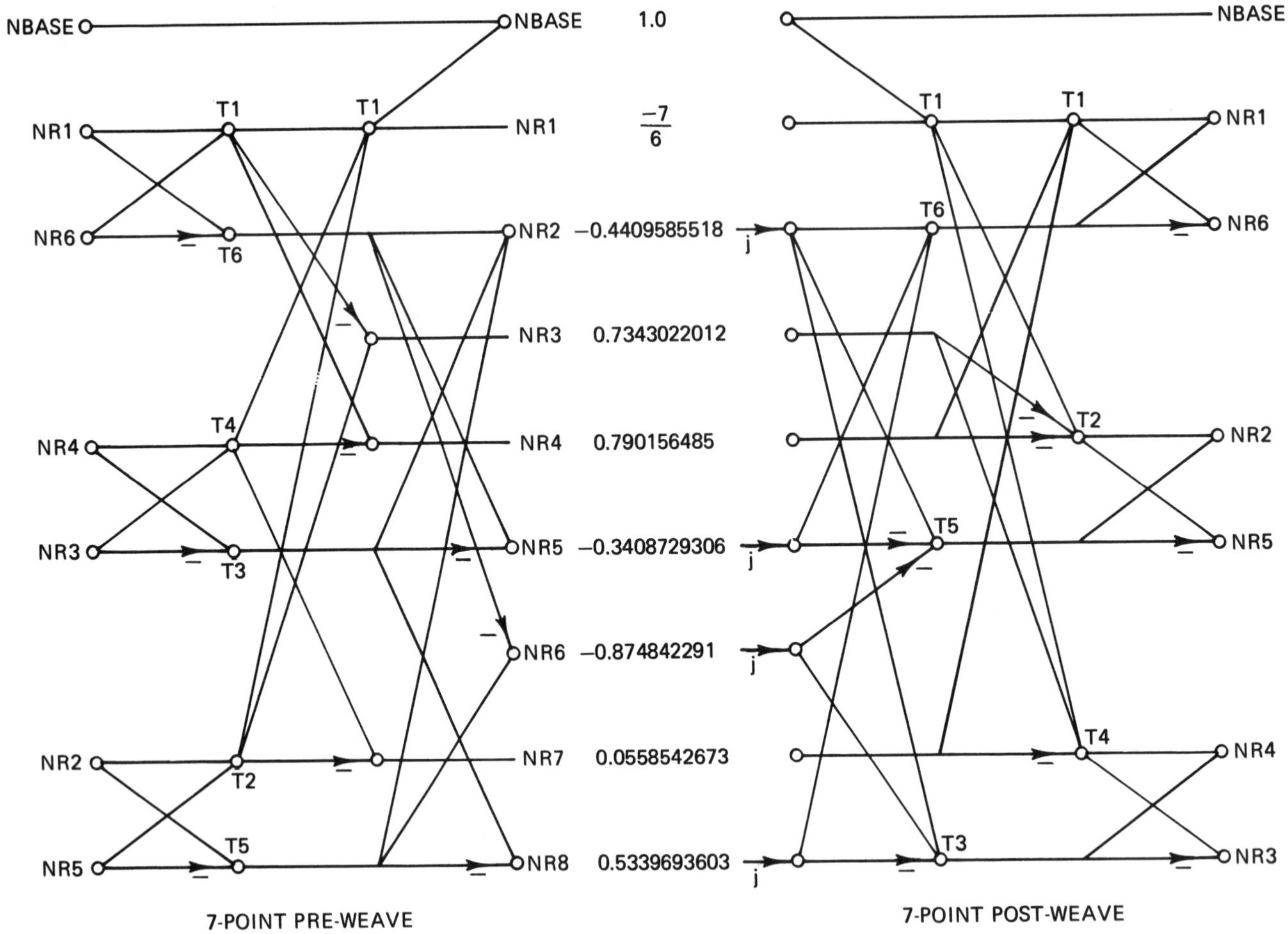

Figure 9-6 Seven-point DFT using Winograd's decomposition. Nine multiplications are required.

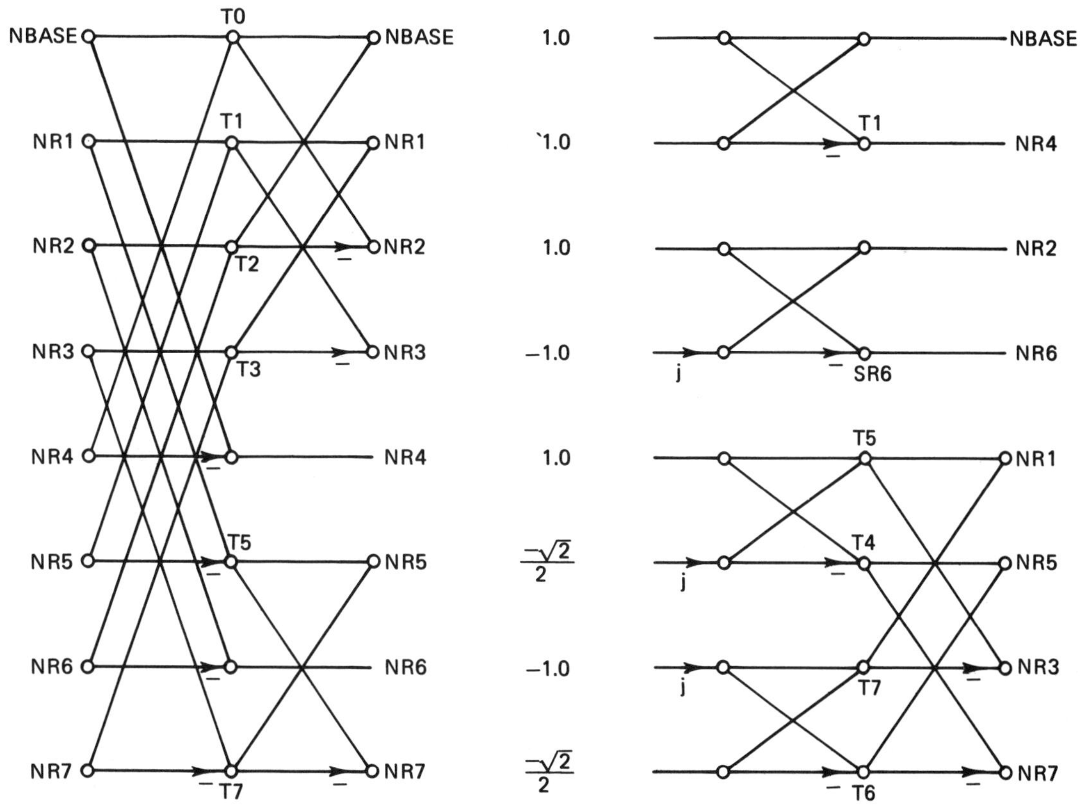

8-POINT PRE-WEAVE

8-POINT POST-WEAVE

Figure 9-7 Eight-point DFT algorithm using Winograd's decomposition.

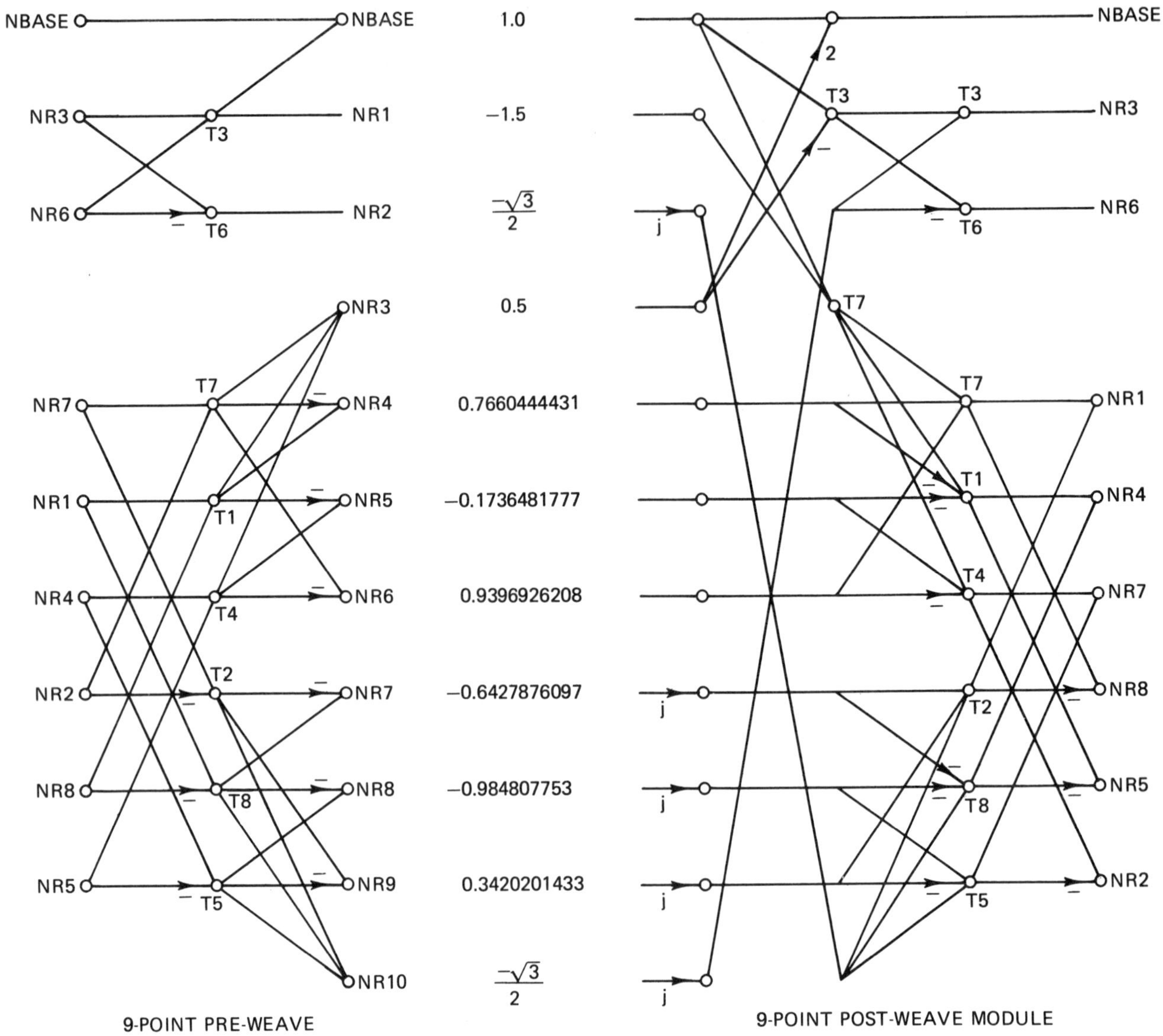

Figure 9-8 Nine-point DFT using Winograd's decomposition. Eleven multiplications are required.

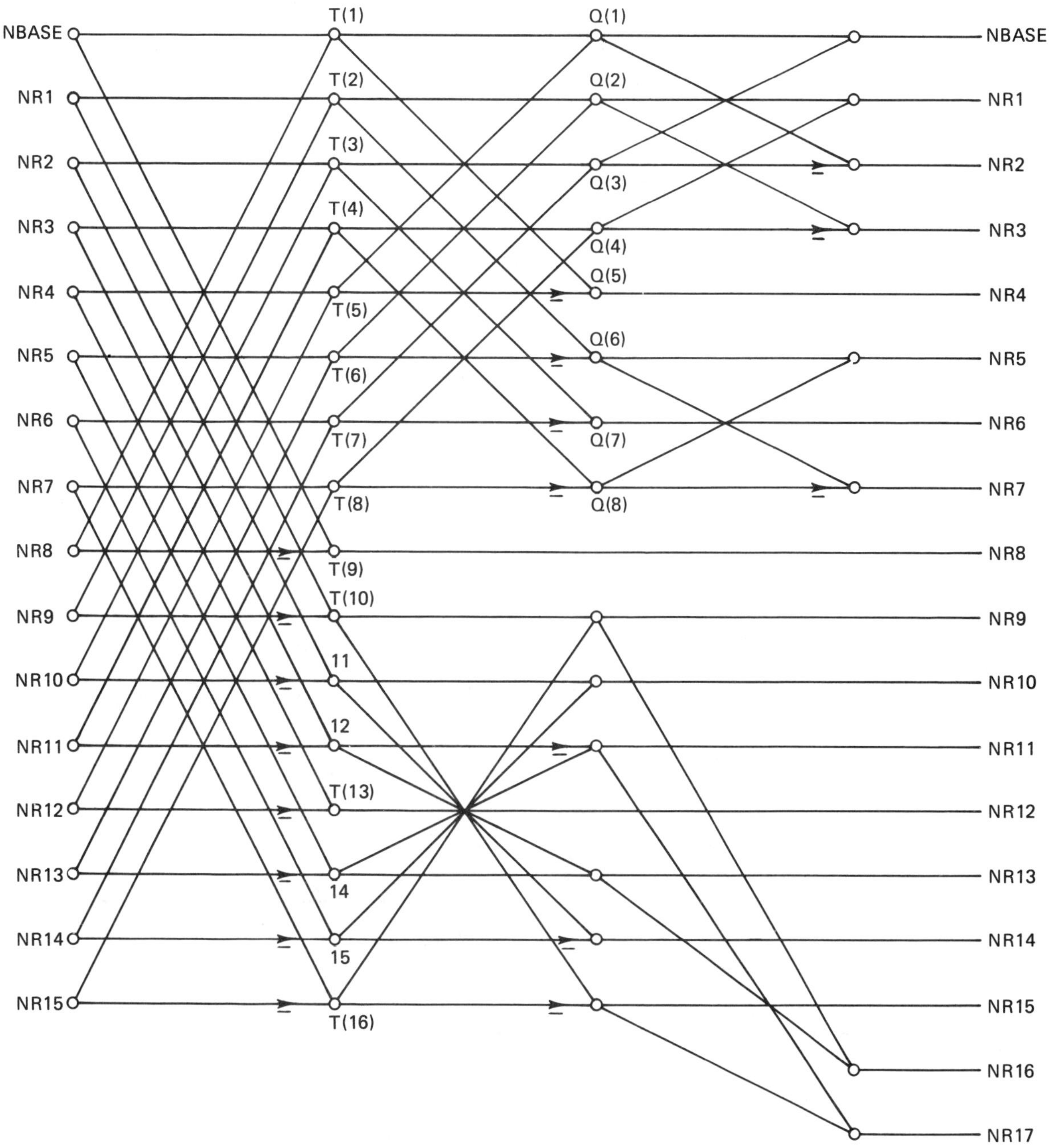

Figure 9-9 Sixteen-point pre-weave module.

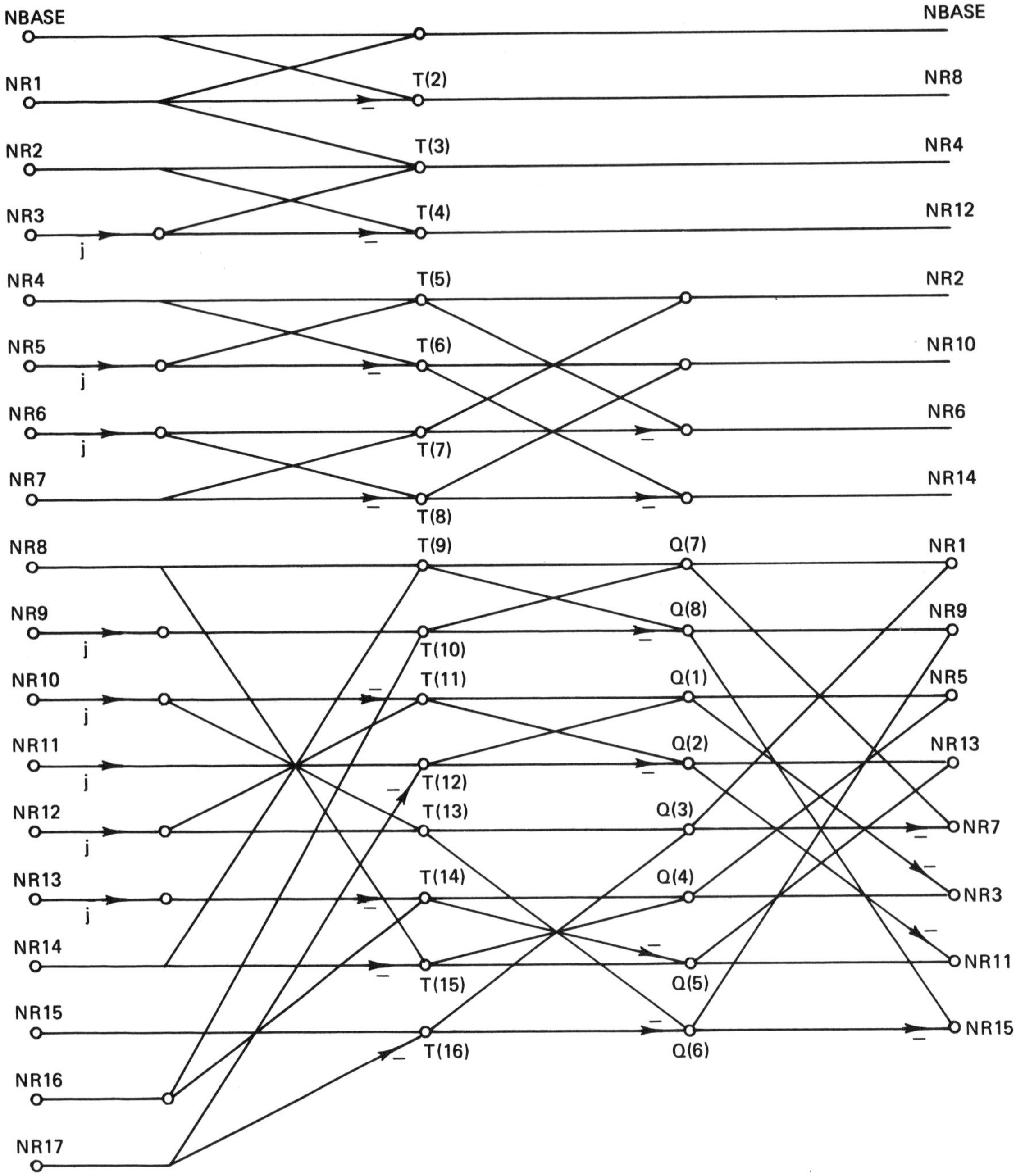

Figure 9-10 Sixteen-point post-weave module.

Table 9.2. Multipliers for 16-point DFT Algorithm

i	CD16(i)
1	1.0
2	1.0
3	1.0
4	−1.0
5	1.0
6	$-\frac{\sqrt{2}}{2}$
7	−1.0
8	$\frac{\sqrt{2}}{2}$
9	1.0
10	0.5411961001
11	$-\frac{\sqrt{2}}{2}$
12	−0.5411961001
13	−1.0
14	−1.306562965
15	$\frac{\sqrt{2}}{2}$
16	1.306562965
17	−0.9238795325
18	0.3826834324

Summary of User Requirements

1. Specify input sequences XR and XI and parameters cap N, INIT, and IERR.
2. Call WFTA. For initial call use INIT = 0.
3. Output is obtained in XR and XI. The error code IERR is zero for successful completion and −1 or −2 if the value of cap N is not permissible or was not initialized.
4. After initial call, use INIT $\neq$ 0 for all other calls as long as cap N is not changed.

Test Example †

A main program is provided to generate a test waveform whose DFT is known analytically. For this purpose a linear frequency modulated (LFM) "chirp" signal was chosen. If the transform length N is even, the signal is

$$x[n] = \exp\left[j\frac{\pi}{N}n^2\right]$$

†This test program measures the computational accuracy of a working WFTA program. However, it should not be used to determine whether or not the WFTA program code is thoroughly debugged. The symmetry of a chirp sequence persists even after the sequence is mapped into a multidimensional array. A preweave operation, therefore, frequently involves subtraction of equal quantities, giving intermediate results that are equal to zero. However, many possible program coding errors could also give rise to zero-valued intermediate variables, and such errors would not be detected by this simple test program.

If the transform length is odd, the test signal is taken to be

$$x[n] = \exp\left[j\frac{2\pi}{N}n^2\right]$$

In both cases the DFT is known to have a constant magnitude equal to $\sqrt{N}$. Thus a check of the WFTA is performed by comparing the magnitude of all DFT values with $\sqrt{N}$ and printing out the maximum percentage deviation.

Summary

A FORTRAN program for calculating the DFT using Winograd's "nested" algorithm has been presented. An effort has been made to provide a computationally efficient program, but some compromises have been made to simplify the program structure. For example, the multiplications of the WFTA are all performed in a separate subroutine, MULT. A slightly more efficient implementation of the algorithm would provide code so that the innermost module of the pre-weave and post-weave computations is combined with the multiplications. For example, in Figure 9.1 the innermost module is the 5-point factor. The advantage of such an approach is that it avoids some memory transfers (load and stores) that are required in our implementation. Furthermore, it reduces slightly the amount of memory required for the arrays SR and SI. The disadvantage, of course, is that an additional block of code must be provided for each factor in the set {2, 3, 4, 5, 7, 8, 9, 16} to allow for the possibility of any factor being the innermost. This could conceivably double the amount of program code.

Another possible and helpful improvement in the WFTA would be the development of input and output permutations that could be done "in-place." If this were possible, the arrays SR and SI could share memory with XR and XI and only an additional (ND1*ND2*ND3*ND4 − N) locations would be needed to accommodate the data expansion of the WFTA.

If one requires only an algorithm to compute the DFT of real data, a rather simple modification of the existing program will exploit the real-valued nature of the input. The subroutines PERM1, WEAVE1, and MULT should be modified to remove all calculations involving the array SI, which must be zero until subroutine WEAVE2 is begun. This will result in a slight improvement in the running time of the algorithm for real data.

As a final comment, this program is only an initial attempt to produce an efficient WFTA program; it is quite likely that future developments related to Winograd's algorithm will lead to more efficient implementations.

```
C      TEST PROGRAM TO EXERCISE THE WFTA SUBROUTINE
C      THE TEST WAVEFORM IS A CHIRP WHOSE MAGNITUDE WHEN TRANSFORMED
C      SHOULD BE A CONSTANT EQUAL TO THE SQRT(N)
C
C      ************************************************************
C
        COMMON NA,NB,NC,ND,ND1,ND2,ND3,ND4
        DOUBLE PRECISION PI,PIN,XN,XJ,XT
        REAL*4 XR(5040),XI(5040)
C
C------  ENTER WFTA LENGTH AND
C------  NUMBER OF TIMES WFTA IS TO BE CALLED
C
50      WRITE(6,100)
100     FORMAT(1X,'LENGTH,NO.TIMES')
        READ(5,*)N,NOTIME
        WRITE(6,150) N
150     FORMAT(' LENGTH = ',I5)
        IF(N .LE. 0) CALL EXIT
        PI=3.141592653589793D0
        XN=FLOAT(N)
        PIN=PI/XN
C
C------  GENERATE A CHIRP TEST SIGNAL
C------  THE CHIRP IS DIFFERENT FOR N EVEN AND N ODD
C
        IF(2*(N/2).NE.N) PIN=2.0*PIN
        INIT=0
        DO 200 J=1,N
        XJ=(J-1)*(J-1)
        XJ=XJ*PIN
        XT=DCOS(XJ)
        XR(J)=XT
        XT=DSIN(XJ)
        XI(J)=XT
200     CONTINUE
C
C------  CALL THE WFTA
C
        DO 400 LL=1,NOTIME
        CALL WFTA(N,XR,XI,INIT,IERR)
        IF(IERR.LT.0) GO TO 50
C
C------  CHECK THE MAGNITUDE OF THE
C------  TRANSFORMED CHIRP
C
        DEVMAX=0.0
        DO 350 J=1,N
        XMAG=XR(J)*XR(J)+XI(J)*XI(J)
        DEV=(XMAG-XN)/XN
        DEV=100.0*ABS(DEV)
        IF(DEV.GT.DEVMAX)DEVMAX=DEV
350     CONTINUE




C
C------  PRINT OUT THE MAXIMUM DEVIATION (%)
C
        WRITE(6,300) DEVMAX
300     FORMAT(1X,'MAXIMUM DEVIATION = ',F10.4,' PERCENT')
        INIT=1
        XN=N*XN
400     CONTINUE
        GO TO 50
        END
```

```
      SUBROUTINE WFTA(N,XR,XI,INIT,IERR)
      REAL*4 XR(1),XI(1)
C
C   **************************************
C     THE FOLLOWING TWO CARDS MAY BE CHANGED IF THE MAXIMUM
C     DESIRED TRANSFORM LENGTH IS LESS THAN 5040
C     NOTE THAT 6*9*11*18 = 10692 (EXPANSION)
C   **************************************
      REAL*4 SR(10692),SI(10692),COEF(10692)
      INTEGER INDX1(5040),INDX2(5040)
C   **************************************
C
      COMMON NA,NB,NC,ND,ND1,ND2,ND3,ND4
C
C-------   TEST FOR INITIAL RUN
C
      IF(INIT.EQ.0) CALL INISHL(N,COEF,XR,XI,INDX1,INDX2,IERR)
C
      IF(IERR.LT.0) RETURN
      M=NA*NB*NC*ND
      IF(M.EQ.N) GO TO 100
      IERR=-2
      RETURN
C
C-------   PROGRAM NOT INITIALIZED FOR THIS VALUE OF N
C
100   MULT=ND1*ND2*ND3*ND4
C
C-------   PERMUTE THE INPUT DATA
C
      CALL PERM1(SR,SI,XR,XI,INDX1)
C
C-------   DO THE PRE-WEAVE MODULES
C
      CALL WEAVE1(SR,SI)
C
C-------   DO THE NESTED MULTIPLICATIONS
C
      CALL MULT(SR,SI,COEF,NMULT)
C
C-------   DO THE POST-WEAVE MODULES
C
      CALL WEAVE2(SR,SI)
C
C-------   PERMUTE THE OUTPUT DATA
C
      CALL PERM2(SR,SI,XR,XI,INDX2)
      RETURN
      END

C     THIS ROUTINE INITIALIZES THE WFTA PROGRAM
C     SO THAT SUCCESSIVE CALLS TO THE TRANSFORM
C     CALCULATION WILL BE AS EFFICIENT AS POSSIBLE.
C     FOR A FIXED VALUE OF THE TRANSFORM LENGTH, N,
C     THIS SUBROUTINE WILL BE EXECUTED ONLY ON THE
C     FIRST CALL TO THE WFTA ROUTINE.
C
C   ************************************************
C
      SUBROUTINE INISHL(N,COEF,XR,XI,INDX1,INDX2,IERR)
      REAL*4 COEF(1),XR(1),XI(1)
      INTEGER S1,S2,S3,S4,INDX1(1),INDX2(1),P1
      REAL*4 CO3(3),CO4(4),CO8(8),CO9(11),CO16(18),CDA(18),CDB(11),
     1 CDC(9),CDD(6)
      COMMON NA,NB,NC,ND,ND1,ND2,ND3,ND4
C
C-------  DATA STATEMENTS ASSIGN SHORT DFT COEFFICIENTS.
C
      DATA CO3/1.0,-1.5,-0.8660254038/
      DATA CO4/1.0,1.0,1.0,1.0/
      DATA CO8/1.0,1.0,1.0,-1.0,1.0,-0.7071067812,
     1 -1.0,0.7071067812/
      DATA CO9/1.0,-1.5,-0.8660254038,0.5,0.7660444431,
     1 -0.1736481777,0.9396926208,-0.6427876097,
     2 -0.984807753,0.3420201433,-0.8660254038/
      DATA CO16/1.0,1.0,1.0,-1.0,1.0,-0.7071067812,-1.0,
     1 0.7071067812,1.0,0.5411961001,-0.7071067812,
     2 -0.5411961001,-1.0,-1.306562965,0.7071067812,
     3 1.306562965,-0.9238795325,0.3826834324/
```

```
      DATA CDA/18*1./
      DATA CDB/11*1./
      DATA CDC/1.0,-1.1666666667,-0.4409585518,0.7343022012,
     1 0.7901564685,-0.3408729306,-0.874842291,
     2 0.0558542673,0.5339693603/
      DATA CDD/1.0,-1.25,-1.538841769,0.5590169944,0.363271264,
     1 0.5877852523/
C
C------  FOLLOWING SEGMENT DETERMINES FACTORS OF N AND CHOOSES
C------  THE APPROPRIATE SHORT DFT COEFFICIENTS.
C
      IERR=0
      ND1=1
      NA=1
      NB=1
      ND2=1
      NC=1
      ND3=1
      ND=1
      ND4=1
      IF(N.LE.0) GO TO 190
      IF(16*(N/16).EQ.N) GO TO 30
      IF(8*(N/8).EQ.N) GO TO 40
      IF(4*(N/4).EQ.N) GO TO 50
      IF(2*(N/2).NE.N) GO TO 70
      ND1=2

      NA=2
      CDA(2)=1.0
      GO TO 70
30    ND1=18
      NA=16
      DO 31 J=1,18
31    CDA(J)=CO16(J)
      GO TO 70
40    ND1=8
      NA=8
      DO 41 J=1,8
41    CDA(J)=CO8(J)
      GO TO 70
50    ND1=4
      NA=4
      DO 51 J=1,4
51    CDA(J)=CO4(J)
70    IF(3*(N/3).NE.N) GO TO 120
      IF(9*(N/9).EQ.N) GO TO 100
      ND2=3
      NB=3
      DO 71 J=1,3
71    CDB(J)=CO3(J)
      GO TO 120
100   ND2=11
      NB=9
      DO 110 J=1,11
110   CDB(J)=CO9(J)
120   IF(7*(N/7).NE.N) GO TO 160
      ND3=9
      NC=7
160   IF(5*(N/5).NE.N) GO TO 190
      ND4=6
      ND=5
190   M=NA*NB*NC*ND
      IF(M.EQ.N) GO TO 250
      WRITE(6,210)
210   FORMAT(' THIS N DOES NOT WORK')
      IERR=-1
      RETURN
C
C------  NEXT SEGMENT GENERATES THE DFT COEFFICIENTS
C------  THE FLAG ARRAY.
C
250   J=1
      DO 300 N4=1,ND4
      DO 300 N3=1,ND3
      DO 300 N2=1,ND2
      DO 300 N1=1,ND1
      COEF(J)=CDA(N1)*CDB(N2)*CDC(N3)*CDD(N4)
      J=J+1
300   CONTINUE
```

```
C
C------    FOLLOWING SEGMENT GENERATES INPUT
C------    INDEXING MAP
C
          S1=0
          S2=0
          S3=0
          S4=0
          S5=0
          J=1
          NU=NB*NC*ND
          NV=NA*NC*ND
          NW=NA*NB*ND
          NY=NA*NB*NC
          K=1
          DO 440 N4=1,ND
          DO 430 N3=1,NC
          DO 420 N2=1,NB
          DO 410 N1=1,NA
405       IF(K.LE.N) GO TO 408
          K=K-N
          GO TO 405
408       INDX1(J)=K
          J=J+1
410       K=K+NU
420       K=K+NV
430       K=K+NW
440       K=K+NY
C
C------   FOLLOWING SEGMENT GENERATES OUTPUT
C------   INDEXING MAP
C
          M=1
          IF(NA.EQ.1) GO TO 530
520       P1=M*NU-1
          IF((P1/NA)*NA.EQ.P1) GO TO 510
          M=M+1
          GO TO 520
510       S1=P1+1
530       IF(NB.EQ.1) GO TO 540
          M=1
550       P1=M*NV-1
          IF((P1/NB)*NB.EQ.P1) GO TO 560
          M=M+1
          GO TO 550
560       S2=P1+1
540       IF(NC.EQ.1) GO TO 630
          M=1
620       P1=M*NW-1
          IF((P1/NC)*NC.EQ.P1) GO TO 610
          M=M+1
          GO TO 620
610       S3=P1+1
630       IF(ND.EQ.1) GO TO 660
          M=1



640       P1=M*NY-1
          IF((P1/ND)*ND.EQ.P1) GO TO 650
          M=M+1
          GO TO 640
650       S4=P1+1
660       J=1
          DO 810 N4=1,ND
          DO 810 N3=1,NC
          DO 810 N2=1,NB
          DO 810 N1=1,NA
          INDX2(J)=S1*(N1-1)+S2*(N2-1)+S3*(N3-1)+S4*(N4-1)+1
900       IF(INDX2(J).LE.N) GO TO 910
          INDX2(J)=INDX2(J)-N
          GO TO 900
910       J=J+1
810       CONTINUE
          RETURN
          END
```

```
C       SUBROUTINE TO PERFORM PERMUTATION OF THE
C       INPUT DATA ACCORDING TO THE
C       CHINESE REMAINDER THEOREM
C       (SINO-CORRESPONDENCE)
C
C     *******************************************
C
        SUBROUTINE PERM1(SR,SI,XR,XI,INDX1)
        COMMON NA,NB,NC,ND,ND1,ND2,ND3,ND4
        REAL*4 XR(1),XI(1),SR(1),SI(1)
        INTEGER INDX1(1)
        J=1
        K=1
        INC1=ND1-NA
        INC2=ND1*(ND2-NB)
        INC3=ND1*ND2*(ND3-NC)
        DO 40 N4=1,ND
        DO 30 N3=1,NC
        DO 20 N2=1,NB
        DO 10 N1=1,NA
        SR(J)=XR(INDX1(K))
        SI(J)=XI(INDX1(K))
        K=K+1
10      J=J+1
20      J=J+INC1
30      J=J+INC2
40      J=J+INC3
        RETURN
        END
C
C
C      SUBROUTINE TO PERMUTE THE
C      TRANSFORMED DATA ACCORDING TO THE
C      MAPPING OF THE CHINESE REMAINDER THEOREM
C
C     *******************************************
C
        SUBROUTINE PERM2(SR,SI,XR,XI,INDX2)
        COMMON NA,NB,NC,ND,ND1,ND2,ND3,ND4
        REAL*4 XR(1),XI(1),SR(1),SI(1)
        INTEGER INDX2(1)
        J=1
        K=1
        INC1=ND1-NA
        INC2=ND1*(ND2-NB)
        INC3=ND1*ND2*(ND3-NC)
        DO 40 N4=1,ND
        DO 30 N3=1,NC
        DO 20 N2=1,NB
        DO 10 N1=1,NA
        XR(INDX2(K))=SR(J)
        XI(INDX2(K))=SI(J)
        K=K+1
10      J=J+1
20      J=J+INC1
30      J=J+INC2
40      J=J+INC3
        RETURN
        END




C       ALL MULTIPLICATIONS OF THE WFTA
C       ARE PERFORMED IN THIS ROUTINE
C       COMPLEX DATA ARE MULTIPLIED BY
C       REAL COEFFICIENTS
C
C     *************************************
C
        SUBROUTINE MULT (SR,SI,COEF,NMULT)
        COMMON NA,NB,NC,ND,ND1,ND2,ND3,ND4
        REAL*4 SR(1),SI(1),COEF(1)
        DO 10 J=1,NMULT
        SR(J)=SR(J)*COEF(J)
        SI(J)=SI(J)*COEF(J)
  10    CONTINUE
        RETURN
        END
```

```
      SUBROUTINE WEAVE1(SR,SI)
      COMMON NA,NB,NC,ND,ND1,ND2,ND3,ND4
      REAL*4 Q(8),T(16)
      REAL*4 SR(1),SI(1)
      IF(NA.EQ.1) GO TO 300
      IF(NA.NE.2) GO TO 800
C
C ******************************************************
C
C     THE FOLLOWING CODE IMPLEMENTS THE 2 POINT PRE-WEAVE MODULE
C
C ******************************************************
C
      NLUP2=2*(ND2-NB)
      NLUP23=2*ND2*(ND3-NC)
      NBASE=1
      DO 240 N4=1,ND
      DO 230 N3=1,NC
      DO 220 N2=1,NB
      NR1=NBASE+1
      T0=SR(NBASE)+SR(NR1)
      SR(NR1)=SR(NBASE)-SR(NR1)
      SR(NBASE)=T0
      T0=SI(NBASE)+SI(NR1)
      SI(NR1)=SI(NBASE)-SI(NR1)
      SI(NBASE)=T0
220   NBASE=NBASE+2
230   NBASE=NBASE+NLUP2
240   NBASE=NBASE+NLUP23
800   IF(NA.NE.8) GO TO 1600
C
C ******************************************************
C
C     THE FOLLOWING CODE IMPLEMENTS THE 8 POINT PRE-WEAVE MODULE
C
C ******************************************************
C
      NLUP2=8*(ND2-NB)
      NLUP23=8*ND2*(ND3-NC)
      NBASE=1
      DO 840 N4=1,ND
      DO 830 N3=1,NC
      DO 820 N2=1,NB
      NR1=NBASE+1
      NR2=NR1+1
      NR3=NR2+1
      NR4=NR3+1
      NR5=NR4+1
      NR6=NR5+1
      NR7=NR6+1
      T3=SR(NR3)+SR(NR7)
      T7=SR(NR3)-SR(NR7)
      T0=SR(NBASE)+SR(NR4)
      SR(NR4)=SR(NBASE)-SR(NR4)
      T1=SR(NR1)+SR(NR5)



      T5=SR(NR1)-SR(NR5)
      T2=SR(NR2)+SR(NR6)



      SR(NR6)=SR(NR2)-SR(NR6)
      SR(NBASE)=T0+T2
      SR(NR2)=T0-T2
      SR(NR1)=T1+T3
      SR(NR3)=T1-T3
      SR(NR5)=T5+T7
      SR(NR7)=T5-T7
```

```
      T3=SI(NR3)+SI(NR7)
      T7=SI(NR3)-SI(NR7)
      T0=SI(NBASE)+SI(NR4)
      SI(NR4)=SI(NBASE)-SI(NR4)
      T1=SI(NR1)+SI(NR5)
      T5=SI(NR1)-SI(NR5)
      T2=SI(NR2)+SI(NR6)
      SI(NR6)=SI(NR2)-SI(NR6)
      SI(NBASE)=T0+T2
      SI(NR2)=T0-T2
      SI(NR1)=T1+T3
      SI(NR3)=T1-T3
      SI(NR5)=T5+T7
      SI(NR7)=T5-T7
820   NBASE=NBASE+8
830   NBASE=NBASE+NLUP2
840   NBASE=NBASE+NLUP23
1600  IF(NA.NE.16) GO TO 300
C
C ******************************************************
C
C    THE FOLLOWING CODE IMPLEMENTS THE 16 POINT PRE-WEAVE MODULE
C
C ******************************************************
C
      NLUP2=18*(ND2-NB)
      NLUP23=18*ND2*(ND3-NC)
      NBASE=1
      DO 1640 N4=1,ND
      DO 1630 N3=1,NC
      DO 1620 N2=1,NB
      NR1=NBASE+1
      NR2=NR1+1
      NR3=NR2+1
      NR4=NR3+1
      NR5=NR4+1
      NR6=NR5+1
      NR7=NR6+1
      NR8=NR7+1
      NR9=NR8+1
      NR10=NR9+1
      NR11=NR10+1
      NR12=NR11+1
      NR13=NR12+1
      NR14=NR13+1
      NR15=NR14+1
      NR16=NR15+1
      NR17=NR16+1
      JBASE=NBASE
      DO 1645 J=1,8
      T(J)=SR(JBASE)+SR(JBASE+8)
      T(J+8)=SR(JBASE)-SR(JBASE+8)
      JBASE=JBASE+1
1645  CONTINUE
      DO 1650 J=1,4
      Q(J)=T(J)+T(J+4)
      Q(J+4)=T(J)-T(J+4)
1650  CONTINUE
      SR(NBASE)=Q(1)+Q(3)
      SR(NR2)=Q(1)-Q(3)
      SR(NR1)=Q(2)+Q(4)
      SR(NR3)=Q(2)-Q(4)
      SR(NR5)=Q(6)+Q(8)
      SR(NR7)=Q(6)-Q(8)
      SR(NR4)=Q(5)
      SR(NR6)=Q(7)
      SR(NR8)=T(9)
      SR(NR9)=T(10)+T(16)
      SR(NR15)=T(10)-T(16)
      SR(NR13)=T(14)+T(12)
      SR(NR11)=T(14)-T(12)
      SR(NR17)=SR(NR11)+SR(NR15)
      SR(NR16)=SR(NR9)+SR(NR13)
      SR(NR10)=T(11)+T(15)
      SR(NR14)=T(11)-T(15)
      SR(NR12)=T(13)
```

```
      JBASE=NBASE
      DO 1745 J=1,8
      T(J)=SI(JBASE)+SI(JBASE+8)
      T(J+8)=SI(JBASE)-SI(JBASE+8)
      JBASE=JBASE+1
1745  CONTINUE
      DO 1750 J=1,4
      Q(J)=T(J)+T(J+4)
      Q(J+4)=T(J)-T(J+4)
1750  CONTINUE
      SI(NBASE)=Q(1)+Q(3)
      SI(NR2)=Q(1)-Q(3)
      SI(NR1)=Q(2)+Q(4)
      SI(NR3)=Q(2)-Q(4)
      SI(NR5)=Q(6)+Q(8)
      SI(NR7)=Q(6)-Q(8)
      SI(NR4)=Q(5)
      SI(NR6)=Q(7)
      SI(NR8)=T(9)
      SI(NR9)=T(10)+T(16)
      SI(NR15)=T(10)-T(16)
      SI(NR13)=T(14)+T(12)
      SI(NR11)=T(14)-T(12)
      SI(NR17)=SI(NR11)+SI(NR15)
      SI(NR16)=SI(NR9)+SI(NR13)



      SI(NR10)=T(11)+T(15)
      SI(NR14)=T(11)-T(15)
      SI(NR12)=T(13)
1620  NBASE=NBASE+18
1630  NBASE=NBASE+NLUP2
1640  NBASE=NBASE+NLUP23
300   IF(NB.EQ.1) GO TO 700
      IF(NB.NE.3) GO TO 900
C
C ******************************************************
C
C     THE FOLLOWING CODE IMPLEMENTS THE 3 POINT PRE-WEAVE MODULE
C
C ******************************************************
C
      NLUP2=2*ND1
      NLUP23=3*ND1*(ND3-NC)
      NBASE=1
      NOFF=ND1
      DO 340 N4=1,ND
      DO 330 N3=1,NC
      DO 310 N2=1,ND1
      NR1=NBASE+NOFF
      NR2=NR1+NOFF
      T1=SR(NR1)+SR(NR2)
      SR(NBASE)=SR(NBASE)+T1
      SR(NR2)=SR(NR1)-SR(NR2)
      SR(NR1)=T1
      T1=SI(NR1)+SI(NR2)
      SI(NBASE)=SI(NBASE)+T1
      SI(NR2)=SI(NR1)-SI(NR2)
      SI(NR1)=T1
310   NBASE=NBASE+1
330   NBASE=NBASE+NLUP2
340   NBASE=NBASE+NLUP23
```

```
900     IF(NB.NE.9) GO TO 700
C
C ****************************************************
C
C     THE FOLLOWING CODE IMPLEMENTS THE 9 POINT PRE-WEAVE MODULE
C
C ****************************************************
C
        NLUP2=10*ND1
        NLUP23=11*ND1*(ND3-NC)
        NBASE=1
        NOFF=ND1
        DO 940 N4=1,ND
        DO 930 N3=1,NC
        DO 910 N2=1,ND1
        NR1=NBASE+NOFF
        NR2=NR1+NOFF
        NR3=NR2+NOFF
        NR4=NR3+NOFF
        NR5=NR4+NOFF
        NR6=NR5+NOFF
        NR7=NR6+NOFF
        NR8=NR7+NOFF
        NR9=NR8+NOFF
        NR10=NR9+NOFF
        T3=SR(NR3)+SR(NR6)
        T6=SR(NR3)-SR(NR6)
        SR(NBASE)=SR(NBASE)+T3
        T7=SR(NR7)+SR(NR2)
        T2=SR(NR7)-SR(NR2)
        SR(NR2)=T6
        T1=SR(NR1)+SR(NR8)
        T8=SR(NR1)-SR(NR8)
        SR(NR1)=T3
        T4=SR(NR4)+SR(NR5)
        T5=SR(NR4)-SR(NR5)
        SR(NR3)=T1+T4+T7
        SR(NR4)=T1-T7
        SR(NR5)=T4-T1
        SR(NR6)=T7-T4
        SR(NR10)=T2+T5+T8
        SR(NR7)=T8-T2
        SR(NR8)=T5-T8
        SR(NR9)=T2-T5
        T3=SI(NR3)+SI(NR6)
        T6=SI(NR3)-SI(NR6)
        SI(NBASE)=SI(NBASE)+T3
        T7=SI(NR7)+SI(NR2)
        T2=SI(NR7)-SI(NR2)
        SI(NR2)=T6
        T1=SI(NR1)+SI(NR8)
        T8=SI(NR1)-SI(NR8)
        SI(NR1)=T3
        T4=SI(NR4)+SI(NR5)
        T5=SI(NR4)-SI(NR5)
        SI(NR3)=T1+T4+T7
        SI(NR4)=T1-T7
        SI(NR5)=T4-T1
        SI(NR6)=T7-T4
        SI(NR10)=T2+T5+T8
        SI(NR7)=T8-T2
        SI(NR8)=T5-T8
        SI(NR9)=T2-T5
910     NBASE=NBASE+1
930     NBASE=NBASE+NLUP2
940     NBASE=NBASE+NLUP23
```

```
700     IF(NC.NE.7) GO TO 500
C
C ********************************************************
C
C     THE FOLLOWING CODE IMPLEMENTS THE 7 POINT PRE-WEAVE MODULE
C
C ********************************************************
C
        NOFF=ND1*ND2
        NBASE=1
        NLUP2=8*NOFF
        DO 740 N4=1,ND
        DO 710 N1=1,NOFF
        NR1=NBASE+NOFF
        NR2=NR1+NOFF
        NR3=NR2+NOFF
        NR4=NR3+NOFF
        NR5=NR4+NOFF
        NR6=NR5+NOFF
        NR7=NR6+NOFF
        NR8=NR7+NOFF
        T1=SR(NR1)+SR(NR6)
        T6=SR(NR1)-SR(NR6)
        T4=SR(NR4)+SR(NR3)
        T3=SR(NR4)-SR(NR3)
        T2=SR(NR2)+SR(NR5)
        T5=SR(NR2)-SR(NR5)
        SR(NR5)=T6-T3
        SR(NR2)=T5+T3+T6
        SR(NR6)=T5-T6
        SR(NR8)=T3-T5
        SR(NR3)=T2-T1
        SR(NR4)=T1-T4
        SR(NR7)=T4-T2
        T1=T1+T4+T2
        SR(NBASE)=SR(NBASE)+T1
        SR(NR1)=T1
        T1=SI(NR1)+SI(NR6)
        T6=SI(NR1)-SI(NR6)
        T4=SI(NR4)+SI(NR3)
        T3=SI(NR4)-SI(NR3)
        T2=SI(NR2)+SI(NR5)
        T5=SI(NR2)-SI(NR5)
        SI(NR5)=T6-T3
        SI(NR2)=T5+T3+T6
        SI(NR6)=T5-T6
        SI(NR8)=T3-T5
        SI(NR3)=T2-T1
        SI(NR4)=T1-T4
        SI(NR7)=T4-T2
        T1=T1+T4+T2
        SI(NBASE)=SI(NBASE)+T1
        SI(NR1)=T1
710     NBASE=NBASE+1
740     NBASE=NBASE+NLUP2
```

```
500     IF(ND.NE.5) RETURN

C
C ******************************************************
C
C     THE FOLLOWING CODE IMPLEMENTS THE 5 POINT PRE-WEAVE MODULE
C
C ******************************************************
C
        NOFF=ND1*ND2*ND3
        NBASE=1
        DO 510 N1=1,NOFF
        NR1=NBASE+NOFF
        NR2=NR1+NOFF
        NR3=NR2+NOFF
        NR4=NR3+NOFF
        NR5=NR4+NOFF
        T4=SR(NR1)-SR(NR4)
        T1=SR(NR1)+SR(NR4)
        T3=SR(NR3)+SR(NR2)
        T2=SR(NR3)-SR(NR2)
        SR(NR3)=T1-T3
        SR(NR1)=T1+T3
        SR(NBASE)=SR(NBASE)+SR(NR1)
        SR(NR5)=T2+T4
        SR(NR2)=T4
        SR(NR4)=T2
        T4=SI(NR1)-SI(NR4)
        T1=SI(NR1)+SI(NR4)
        T3=SI(NR3)+SI(NR2)
        T2=SI(NR3)-SI(NR2)
        SI(NR3)=T1-T3
        SI(NR1)=T1+T3
        SI(NBASE)=SI(NBASE)+SI(NR1)
        SI(NR5)=T2+T4
        SI(NR2)=T4
        SI(NR4)=T2
510     NBASE=NBASE+1
        RETURN
        END
```

```
      SUBROUTINE WEAVE2(SR,SI)
      REAL*4 SR(1),SI(1)
      COMMON NA,NB,NC,ND,ND1,ND2,ND3,ND4
      REAL*4 Q(8),T(16)
      IF(ND.NE.5) GO TO 700
C
C ****************************************************
C
C     THE FOLLOWING CODE IMPLEMENTS THE 5 POINT POST-WEAVE MODULE
C
C ****************************************************
C
      NOFF=ND1*ND2*ND3
      NBASE=1
      DO 510 N1=1,NOFF
      NR1=NBASE+NOFF
      NR2=NR1+NOFF
      NR3=NR2+NOFF
      NR4=NR3+NOFF
      NR5=NR4+NOFF
      T1=SR(NBASE)+SR(NR1)
      T3=T1-SR(NR3)
      T1=T1+SR(NR3)
      T4=SI(NR2)+SI(NR5)
      T2=SI(NR4)+SI(NR5)
      SR(NR1)=T1-T4
      SR4=T1+T4
      SR2=T3+T2
      SR(NR3)=T3-T2
      T1=SI(NBASE)+SI(NR1)
      T3=T1-SI(NR3)
      T1=T1+SI(NR3)
      T4=SR(NR2)+SR(NR5)
      T2=SR(NR4)+SR(NR5)
      SI(NR1)=T1+T4
      SI(NR4)=T1-T4
      SI(NR2)=T3-T2
      SI(NR3)=T3+T2
      SR(NR2)=SR2
      SR(NR4)=SR4
510   NBASE=NBASE+1
```

```
700     IF(NC.NE.7) GO TO 300
C
C *****************************************************
C
C     THE FOLLOWING CODE IMPLEMENTS THE 7 POINT POST-WEAVE MODULE
C
C *****************************************************
C
        NOFF=ND1*ND2
        NBASE=1
        NLUP2=8*NOFF
        DO 740 N4=1,ND
        DO 710 N1=1,NOFF
        NR1=NBASE+NOFF
        NR2=NR1+NOFF
        NR3=NR2+NOFF
        NR4=NR3+NOFF
        NR5=NR4+NOFF
        NR6=NR5+NOFF
        NR7=NR6+NOFF
        NR8=NR7+NOFF
        T1=SR(NR1)+SR(NBASE)
        T2=T1-SR(NR3)-SR(NR4)
        T4=T1+SR(NR3)-SR(NR7)
        T1=T1+SR(NR4)+SR(NR7)
        T6=SI(NR2)+SI(NR5)+SI(NR8)
        T5=SI(NR2)-SI(NR5)-SI(NR6)
        T3=SI(NR2)+SI(NR6)-SI(NR8)
        SR(NR1)=T1-T6
        SR6=T1+T6
        SR2=T2-T5
        SR5=T2+T5
        SR(NR4)=T4-T3
        SR(NR3)=T4+T3
        T1=SI(NR1)+SI(NBASE)
        T2=T1-SI(NR3)-SI(NR4)
        T4=T1+SI(NR3)-SI(NR7)
        T1=T1+SI(NR4)+SI(NR7)
        T6=SR(NR2)+SR(NR5)+SR(NR8)
        T5=SR(NR2)-SR(NR5)-SR(NR6)
        T3=SR(NR2)+SR(NR6)-SR(NR8)
        SI(NR1)=T1+T6
        SI(NR6)=T1-T6
        SI(NR2)=T2+T5
        SI(NR5)=T2-T5
        SI(NR4)=T4+T3
        SI(NR3)=T4-T3
        SR(NR2)=SR2
        SR(NR5)=SR5
        SR(NR6)=SR6
710     NBASE=NBASE+1
740     NBASE=NBASE+NLUP2
```

```
300     IF(NB.EQ.1) GO TO 400
        IF(NB.NE.3) GO TO 900



C
C ******************************************************
C
C     THE FOLLOWING CODE IMPLEMENTS THE 3 POINT POST-WEAVE MODULE
C
C ******************************************************
C
        NLUP2=2*ND1
        NLUP23=3*ND1*(ND3-NC)
        NBASE=1
        NOFF=ND1
        DO 340 N5=1,ND
        DO 330 N4=1,NC
        DO 310 N2=1,ND1
        NR1=NBASE+NOFF
        NR2=NR1+NOFF
        T1=SR(NBASE)+SR(NR1)
        SR(NR1)=T1-SI(NR2)
        SR2=T1+SI(NR2)
        T1=SI(NBASE)+SI(NR1)
        SI(NR1)=T1+SR(NR2)
        SI(NR2)=T1-SR(NR2)
        SR(NR2)=SR2
310     NBASE=NBASE+1
330     NBASE=NBASE+NLUP2
340     NBASE=NBASE+NLUP23
900     IF(NB.NE.9) GO TO 400
C
C ******************************************************
C
C     THE FOLLOWING CODE IMPLEMENTS THE 9 POINT POST-WEAVE MODULE
C
C ******************************************************
C
        NLUP2=10*ND1
        NLUP23=11*ND1*(ND3-NC)
        NBASE=1
        NOFF=ND1
        DO 940 N4=1,ND
        DO 930 N3=1,NC
        DO 910 N2=1,ND1
        NR1=NBASE+NOFF
        NR2=NR1+NOFF
        NR3=NR2+NOFF
        NR4=NR3+NOFF
        NR5=NR4+NOFF
        NR6=NR5+NOFF
        NR7=NR6+NOFF
        NR8=NR7+NOFF
        NR9=NR8+NOFF
        NR10=NR9+NOFF
        T3=SR(NBASE)-SR(NR3)
        T7=SR(NBASE)+SR(NR1)
        SR(NBASE)=SR(NBASE)+SR(NR3)+SR(NR3)
        T6=T3+SI(NR10)



        SR(NR3)=T3-SI(NR10)
        T4=T7+SR(NR5)-SR(NR6)
        T1=T7-SR(NR4)-SR(NR5)
        T7=T7+SR(NR4)+SR(NR6)
        SR(NR6)=T6
```

```
      T8=SI(NR2)-SI(NR7)-SI(NR8)
      T5=SI(NR2)+SI(NR8)-SI(NR9)
      T2=SI(NR2)+SI(NR7)+SI(NR9)
      SR(NR1)=T7-T2
      SR8=T7+T2
      SR(NR4)=T1-T8
      SR(NR5)=T1+T8
      SR7=T4-T5
      SR2=T4+T5
      T3=SI(NBASE)-SI(NR3)
      T7=SI(NBASE)+SI(NR1)
      SI(NBASE)=SI(NBASE)+SI(NR3)+SI(NR3)
      T6=T3-SR(NR10)
      SI(NR3)=T3+SR(NR10)
      T4=T7+SI(NR5)-SI(NR6)
      T1=T7-SI(NR4)-SI(NR5)
      T7=T7+SI(NR4)+SI(NR6)
      SI(NR6)=T6
      T8=SR(NR2)-SR(NR7)-SR(NR8)
      T5=SR(NR2)+SR(NR8)-SR(NR9)
      T2=SR(NR2)+SR(NR7)+SR(NR9)
      SI(NR1)=T7+T2
      SI(NR8)=T7-T2
      SI(NR4)=T1+T8
      SI(NR5)=T1-T8
      SI(NR7)=T4+T5
      SI(NR2)=T4-T5
      SR(NR2)=SR2
      SR(NR7)=SR7
      SR(NR8)=SR8
910   NBASE=NBASE+1
930   NBASE=NBASE+NLUP2
940   NBASE=NBASE+NLUP23
400   IF(NA.EQ.1) RETURN
      IF(NA.NE.4) GO TO 800
C
C *******************************************************
C
C     THE FOLLOWING CODE IMPLEMENTS THE 4 POINT POST-WEAVE MODULE
C
C *******************************************************
C
      NLUP2=4*(ND2-NB)
      NLUP23=4*ND2*(ND3-NC)
      NBASE=1
      DO 440 N4=1,ND
      DO 430 N3=1,NC
      DO 420 N2=1,NB
      NR1=NBASE+1
      NR2=NR1+1
      NR3=NR2+1
      TR0=SR(NBASE)+SR(NR2)
      TR2=SR(NBASE)-SR(NR2)
      TR1=SR(NR1)+SR(NR3)
      TR3=SR(NR1)-SR(NR3)
      TI1=SI(NR1)+SI(NR3)
      TI3=SI(NR1)-SI(NR3)
      SR(NBASE)=TR0+TR1
      SR(NR2)=TR0-TR1
      SR(NR1)=TR2+TI3
      SR(NR3)=TR2-TI3
      TI0=SI(NBASE)+SI(NR2)
      TI2=SI(NBASE)-SI(NR2)
      SI(NBASE)=TI0+TI1
      SI(NR2)=TI0-TI1
      SI(NR1)=TI2-TR3
      SI(NR3)=TI2+TR3
420   NBASE=NBASE+4
430   NBASE=NBASE+NLUP2
440   NBASE=NBASE+NLUP23
```

```
800    IF(NA.NE.8) GO TO 1600
C
C *****************************************************
C
C      THE FOLLOWING CODE IMPLEMENTS THE 8 POINT POST-WEAVE MODULE
C
C *****************************************************
C
       NLUP2=8*(ND2-NB)
       NLUP23=8*ND2*(ND3-NC)
       NBASE=1
       DO 840 N4=1,ND
       DO 830 N3=1,NC
       DO 820 N2=1,NB
       NR1=NBASE+1
       NR2=NR1+1
       NR3=NR2+1
       NR4=NR3+1
       NR5=NR4+1
       NR6=NR5+1
       NR7=NR6+1
       T1=SR(NBASE)-SR(NR1)
       SR(NBASE)=SR(NBASE)+SR(NR1)
       SR6=SR(NR2)+SI(NR3)
       SR(NR2)=SR(NR2)-SI(NR3)
       T4=SR(NR4)-SI(NR5)
       T5=SR(NR4)+SI(NR5)
       T6=SR(NR7)-SI(NR6)
       T7=SR(NR7)+SI(NR6)
       SR(NR4)=T1
       SR(NR1)=T4+T6
       SR3=T4-T6
       SR5=T5-T7
       SR(NR7)=T5+T7
       T1=SI(NBASE)-SI(NR1)


       SI(NBASE)=SI(NBASE)+SI(NR1)
       T3=SI(NR2)-SR(NR3)
       SI(NR2)=SI(NR2)+SR(NR3)
       T4=SI(NR4)+SR(NR5)
       T5=SI(NR4)-SR(NR5)
       SI(NR6)=T3
       T6=SR(NR6)+SI(NR7)
       T7=SR(NR6)-SI(NR7)
       SI(NR4)=T1
       SI(NR1)=T4+T6
       SI(NR3)=T4-T6
       SI(NR5)=T5+T7
       SI(NR7)=T5-T7
       SR(NR3)=SR3
       SR(NR5)=SR5
       SR(NR6)=SR6
820    NBASE=NBASE+8
830    NBASE=NBASE+NLUP2
840    NBASE=NBASE+NLUP23
```

```
1600    IF(NA.NE.16) RETURN
C
C *****************************************************
C
C     THE FOLLOWING CODE IMPLEMENTS THE 16 POINT PRE-WEAVE MODULE
C
C *****************************************************
C
        NLUP2=18*(ND2-NB)
        NLUP23=18*ND2*(ND3-NC)
        NBASE=1
        DO 1640 N4=1,ND
        DO 1630 N3=1,NC
        DO 1620 N2=1,NB
        NR1=NBASE+1
        NR2=NR1+1
        NR3=NR2+1
        NR4=NR3+1
        NR5=NR4+1
        NR6=NR5+1
        NR7=NR6+1
        NR8=NR7+1
        NR9=NR8+1
        NR10=NR9+1
        NR11=NR10+1
        NR12=NR11+1
        NR13=NR12+1
        NR14=NR13+1
        NR15=NR14+1
        NR16=NR15+1
        NR17=NR16+1
        T(2)=SR(NBASE)-SR(NR1)
        SR(NBASE)=SR(NR1)+SR(NBASE)
        T(4)=SR(NR2)+SI(NR3)
        T(3)=SR(NR2)-SI(NR3)
        T(6)=SR(NR4)+SI(NR5)
        T(5)=SR(NR4)-SI(NR5)
        T(8)=-SI(NR6)-SR(NR7)
        T(7)=-SI(NR6)+SR(NR7)
        T(9)=SR(NR8)+SR(NR14)
        T(15)=SR(NR8)-SR(NR14)
        T(13)=-SI(NR10)-SI(NR12)
        T(11)=SI(NR10)-SI(NR12)
        T(16)=SR(NR15)-SR(NR17)
        T(12)=SR(NR11)-SR(NR17)
        T(10)=-SI(NR9)-SI(NR16)
        T(14)=-SI(NR16)+SI(NR13)
        SR(NR2)=T(5)+T(7)
        SR6=T(5)-T(7)
        SR10=T(6)+T(8)
        SR(NR14)=T(6)-T(8)
        Q(7)=T(9)+T(10)
        Q(8)=T(9)-T(10)
        Q(1)=T(11)+T(12)
        Q(2)=T(11)-T(12)
        Q(4)=T(14)+T(15)
        Q(5)=T(15)-T(14)
        Q(3)=T(13)+T(16)
        Q(6)=T(13)-T(16)
        SR(NR1)=Q(3)+Q(7)
        SR(NR7)=Q(7)-Q(3)
        SR9=Q(8)+Q(6)
        SR(NR15)=Q(8)-Q(6)
        SR5=Q(1)+Q(4)
        SR3=Q(4)-Q(1)
        SR13=Q(2)+Q(5)
        SR11=Q(5)-Q(2)
        SR(NR8)=T(2)
        SR(NR4)=T(3)
        SR12=T(4)
```

```
      T(2)=SI(NBASE)-SI(NR1)
      SI(NBASE)=SI(NR1)+SI(NBASE)
      T(4)=SI(NR2)-SR(NR3)
      T(3)=SI(NR2)+SR(NR3)
      T(6)=SI(NR4)-SR(NR5)
      T(5)=SI(NR4)+SR(NR5)
      T(8)=SR(NR6)-SI(NR7)
      T(7)=SR(NR6)+SI(NR7)
      T(9)=SI(NR8)+SI(NR14)
      T(15)=SI(NR8)-SI(NR14)
      T(13)=SR(NR10)+SR(NR12)
      T(11)=SR(NR12)-SR(NR10)
      T(16)=SI(NR15)-SI(NR17)
      T(12)=SI(NR11)-SI(NR17)
      T(10)=SR(NR9)+SR(NR16)
      SI(NR2)=T(5)+T(7)
      SI(NR6)=T(5)-T(7)
      SI(NR10)=T(6)+T(8)
      SI(NR14)=T(6)-T(8)
      Q(7)=T(9)+T(10)
      Q(8)=T(9)-T(10)


      Q(1)=T(11)+T(12)
      Q(2)=T(11)-T(12)
      Q(4)=T(14)+T(15)
      Q(5)=T(15)-T(14)
      Q(3)=T(13)+T(16)
      Q(6)=T(13)-T(16)
      SI(NR1)=Q(3)+Q(7)
      SI(NR7)=Q(7)-Q(3)
      SI(NR9)=Q(8)+Q(6)
      SI(NR15)=Q(8)-Q(6)
      SI(NR5)=Q(1)+Q(4)
      SI(NR3)=Q(4)-Q(1)
      SI(NR13)=Q(2)+Q(5)
      SI(NR11)=Q(5)-Q(2)
      SI(NR8)=T(2)
      SI(NR4)=T(3)
      SI(NR12)=T(4)
      SR(NR3)=SR3
      SR(NR5)=SR5
      SR(NR6)=SR6
      SR(NR9)=SR9
      SR(NR10)=SR10
      SR(NR11)=SR11
      SR(NR12)=SR12
      SR(NR13)=SR13
 1620 NBASE=NBASE+18
 1630 NBASE=NBASE+NLUP2
 1640 NBASE=NBASE+NLUP23
      RETURN
      END
```

Bibliography

Reprinted Papers

[1] Kolba, D. P., and T. W. Parks. "A Prime Factor FFT Algorithm Using High Speed Convolution." *IEEE Trans. on Acoustics, Speech, and Signal Processing* ASSP-25, August 1977): 281–94.

[2] Rader, C. M. "Discrete Fourier Transforms When the Number of Data Samples is Prime." *Proc. IEEE* 56 (June, 1968): 1107–1108.

[3] Agarwal, R. C., and J. W. Cooley. "New Algorithms for Digital Convolution." *IEEE Trans. on Acoustics, Speech, and Signal Processing* ASSP-25, No. 5 (October 1977): 392–409.

[4] Winograd, S. "On Computing the Discrete Fourier Transform." *Mathematics of Computation* 32 (January 1978): 175–199.

[5] Good, I. J. "The Relationship between Two Fast Fourier Transforms." *IEEE Trans. on Computers* C-20 (March 1971): 310–17.

[6] Pollard, J. M. "The Fast Fourier Transform in a Finite Field." *Mathematics of Computation* 25 (April 1971): 365–74.

[7] Agarwal, R. C., and C. S. Burrus. "Fast Convolution Using Fermat Number Transforms with Applications to Digital Filtering." *IEEE Trans. on Acoustics, Speech, and Signal Processing* ASSP-22 (April 1974): 87–97.

[8] ———, and ———. "Fast One-Dimensional Convolution by Multidimensional Techniques." *IEEE Trans. on Acoustics, Speech, and Signal Processing* ASSP-22 (February 1974): 1–10.

[9] McClellan, J. H. "Hardware Realization of a Fermat Number Transform." *IEEE Trans. on Acoustics, Speech, and Signal Processing* ASSP-24 (June 1976): 216–25.

[10] Leibowitz, L. M. "A Simplified Binary Arithmetic for the Fermat Number Transform." *IEEE Trans. on Acoustics, Speech, and Signal Processing* ASSP-24 (October 1976): 356–59.

[11] Reed, I. S., and T. K. Truong. "The Use of Finite Fields to Compute Convolutions." *IEEE Trans. on Information Theory* IT-21 (March 1975): 208–213.

[12] Nussbaumer, H. J. "Digital Filtering Using Complex Mersenne Transforms." *IBM Journal of Research and Development* 20 (September 1976): 498–504.

[13] Winograd, S. "Some Bilinear Forms Whose Multiplicative Complexity Depends on the Field of Constants." *Mathematical Systems Theory* 10 (September 1977): 169–80.

General References

[14] Agarwal, R. C., and C. S. Burrus. "Fast Digital Convolutions Using Fermat Transforms." in *Southwest IEEE Conference Record* Houston, Texas: April 1973. pp. 538–43.

[15] ———. "On the Realization of Digital Filters." Ph.D. dissertation, Rice University, 1973.

[16] ———, and C. S. Burrus. "Number-Theoretic Transforms to Implement Fast Digital Convolution." *Proc. IEEE* 63 (April 1975): 556–60.

[17] ———, and J. W. Cooley. "Some New Algorithms for Fast Convolution by Multidimensional Techniques." In *IEEE Arden House Workshop on Digital Signal Processing*. Harriman, N.Y.: February 1976.

[18] ———, and ———. "Efficient Implementation of One-Dimensional Convolutions by Multidimensional Techniques." *Proc. EASCON-76*, September 1976, 156.

[19] ———, and ———. "New Algorithms for Digital Convolution." In *IEEE Int. Conf. on Acoustics, Speech, and Signal Processing*, pp. 360–62. Hartford, Conn.: May 9–11, 1977.

[20] Allwright, J. C. "Real Factorization of Noncyclic Convolution Operators with Applications to Fast Convolution." *Electronics Letters* 7 (December 1971): 718–19.

[21] Bailey, D. "Winograd's Algorithm Applied to Number Theoretic Transforms." *Electronics Letters* 13, (September 1, 1977): 548–49.

[22] Bergland, G. D. "A FFT Algorithm Using Base-8 Iterations." *Mathematics of Computation* 22 (April 1968): 275–79.

[23] Berlekamp, E. R. *Algebraic Coding Theory*. New York: McGraw-Hill, 1968.

[24] Bernstein, R. "Schnelle Faltung mit der Rader Transformation." Diplomarbeit, Institut für Nachrichtentechnik, Universität Erlangen-Nurnberg, 1974.

[25] Birkhoff, G., and S. Maclaine. *A Survey of Modern Algebra*. New York: Macmillan, 1953, pp. 370–78.

[26] Bluestein, L. I. "A Linear Filtering Approach to the Computation of Discrete Fourier Transform." *IEEE Trans. on Audio and Electroacoustics*, 18 (1970): 451–56.

[27] Bonneau, R. J. "A Class of Finite Computation Structures Supporting the Fast Fourier Transform." Project MAC, Memo #31, March 1973.

[28] Brenner, N. "Algorithm 467; Matrix Transposition in Place (F1)." *Communications ACM* 16 (November 1973): 692–94.

[29] Cohn-Sfetcu, S., and J. E. Gibbs. "Harmonic Differential Calculus and Filtering in Galois Fields." In *IEEE Int'l. Conference on Acoustics, Speech, and Signal Processing*, pp. 148–53, Philadelphia: April, 1976.

[30] ———. "Multiplicative Convolutions and Fourier Transforms." In *IEEE Int'l. Symposium on Information Theory*, p. 94. Cornell University, Ithaca, N.Y.: October, 1977.

[31] Cooley, J. W., and J. W. Tukey. "An Algorithm for the Machine Computation of Complex Fourier Series." *Mathematics of Computation* (April 1965): 297–301.

[32] ———, P. A. W. Lewis, and P. D. Welch. "Historical Notes on the FFT." *IEEE Trans. on Audio and Electroacoustics* AU-15 (June 1967): 76–79.

[33] Conway, J. H., and M. J. T. Guy. "A Tabulation of Some Information Concerning Finite Fields." In R. F. Churchhouse and J. C. Herz, eds., *Computers in Mathematical Research*, pp. 37–50. North Holland: 1968.

[34] Davis, W. F. "A Class of Efficient Convolution Algorithms." In *Proc. Symp. Applications of Walsh Functions*, pp. 318–29. Washington, D.C.: March 1972.

[35] Dickson, L. E. *History of the Theory of Numbers*, p. 376. Washington, D.C.: Carnegie Institute, 1919.

[36] ———. *Introduction to the Theory of Numbers*. New York: Dover, 1957.

[37] Dubois, E., and A. N. Venetsanopoulos. "Fast Convolution Using the Discrete Fourier Transform over Commutative Rings with Identity." In *20th Midwest Symposium on Circuits and Systems*. Lubbock, Texas: August 15–17, 1977.

[38] Erdelsky, P. J. "Fermat Number Transform Hardware Configuration." *Rept. No.* TN-1975-18. San Diego, Calif.: Naval Undersea Center, 1975.

[39] Fiduccia, C. M., and Y. Zalcstein. "Algebras having Linear Complexities." State University of New York at Stony Brook Dept. of Computer Science *Technical Report 46*. Stony Brook, N.Y., August 1975.

[40] GETHÖFFER, H. "Fast Transform Digital Filtering Using Non-Diagonal Spectral Operators." In *IEEE Int'l. Conf. on Acoustics, Speech, and Signal Processing*, pp. 356–59. Hartford, Conn.: May 9–11, 1977.

[41] GOLD, B., and C. M. RADER. *Digital Processing of Signals*. New York: McGraw-Hill, 1969.

[42] GOLOMB, S. W. "Properties of the Sequence $3 \cdot 2^n + 1$." *Mathematics of Computation* (July 1976): 657–63.

[43] ———, I. S. REED, and T. K. TRUONG. "Integer Convolutions over the Finite Field $GF(3 \cdot 2^n + 1)$." *SIAM Journal on Applied Math.* 32 (March 1977): 356–65.

[44] GOOD, I. J. "On the Inversion of Circulant Matrices." *Biometrika* 37 (1950): 185–86.

[45] ———. "A New Finite Series for Legendre Polynominals." *Proc. Camb. Phil. Soc.* 51 (1958): 385–88.

[46] ———. "The Interaction Algorithm and Practical Fourier Series." *J. Royal Stat. Soc.*, ser. B, 20 (1958): 361–72; *Addendum* 22: (1960): 372–75.

[47] ———. "Polynomial Algebra: An Application of the Fast Fourier Transform." *Nature* 222: 1302.

[48] HARDY, G. H., and E. M. WRIGHT. *The Theory of Numbers*. Oxford: Oxford University Press, 1960.

[49] ———, and ———. *An Introduction to the Theory of Numbers*. London: Oxford University Press, 1968.

[50] HERSTEIN, I. N. *Topics in Algebra*, 2d ed. Lexington, Mass.: Xerox College Publishing, 1975.

[51] HOPCROFT, J. E., and J. MUSINSKI. "Duality Applied to the Complexity of Matrix Multiplication and Other Bilinear Forms." *SIAM Journal on Computing* 2 (1973): 159–73.

[52] IBM Corp. "Procedure Library-Mathematics (PL-MATH) Program Description." White Plains, N.Y.: IBM Progr. Product SH20-0985-0, 1971.

[53] JENKINS, W. K., and B. J. LEON. "The Use of Residue Coding in the Design of Hardware for Nonrecursive Digital Filters." In *Eighth Asilomar Conference on Circuits, Systems, and Computers*, pp. 458–62, December 1974.

[54] ———. "Composite Number-theoretic Transforms for Digital Filtering." In *Ninth Asilomar Conference on Circuits, Systems, and Computers*, pp. 421–24, November 1975.

[55] ———, and B. J. LEON. "The Use of Residue Number Systems in the Design of Finite Impulse Response Digital Filters." *IEEE Trans. on Circuits and Systems* CAS-24 (April 1977): 191–201.

[56] KIBLER, D., and R. C. AGARWAL. "Necessary and Sufficient Conditions for the Existence of the Modular Fourier Transform: Comments on 'Number Theoretic Transforms to Implement Fast Digital Convolution'." *Proc. IEEE* 65 (February 1977): 265–67.

[57] KNUTH, D. E. *The Art of Computer Programming*, vol. 2, *Seminumerical Algorithms*. Reading, Mass.: Addison-Wesley, 1969.

[58] KOLBA, D. P., and T. W. PARKS. "Convolution Algorithms for the DFT." Rice University, Elec. Eng. Dept. *Technical Report No. 7609*. Houston, Texas: October 1976.

[59] KRIZ, T. A., and D. F. BACHMAN. "Computational Efficiency of Number-Theoretic Transform Implemented Finite Impulse Response Filters." *IEEE Int'l. Symposium on Information Theory*, p. 93. Ithaca, N.Y.: October 1977.

[60] LEIBOWITZ, L. M. "Fast Convolution by Number Theoretic Transforms." *NRL Report* 7924. Washington, D.C.: Naval Research Laboratory, September 1975.

[61] ———. "A Binary Arithmetic for the Fermat Number Transform." *NRL Report* 7971. Washington, D.C.: Naval Research Laboratory, March 1976.

[62] LIU, K. Y., I. S. REED, and T. K. TRUONG. "Fast Number-Theoretic Transforms for Digital Filtering." *Electronics Letters* 12 (November 1976): 1644–46.

[63] MAKAROV, O. M. "On the Relations Between the Fast Fourier and Hadamard Transform Algorithms and the Algorithms of Karatsuba, Winograd and Strassen." 15 *Zh. Vychisl. Mat. Fiz.* [USSR] (July 1975): 1095–1105.

[64] MATHEWS, G. B. *Theory of Numbers*. New York: Chelsea, 1961.

[65] MCCLELLAN, J. H. "Hardware for the Fermat Number Transform." *Technical Note* ESD-TR-75-149. M.I.T., Lincoln Lab., 1975.

[66] MCCOY, N. H. *The Theory of Numbers*. London: McMillan, 1965.

[67] MELHUISH, P. "Fermat-Transform Implementation by a Minicomputer." *Electronics Letters* 11 (March 1975): 109–111.

[68] MICHELSON, A. M. "A Fast Transform in Galois Fields and an Application to Decoding Reed-Solomon Codes." Presented at *IEEE Intl. Symposium on Information Theory*. Ronneby, Sweden, June 1976.

[69] MORGERA, S. D. "Synthesizing Large DFTs Using the Mutual Prime Factor Cyclic Algorithm." In *IEEE Int'l. Symposium on Information Theory*, p. 93. Ithaca, N.Y.: October, 1977.

[70] MORRIS, L. R. "A Comparative Study of Time-Efficient FFT and WFTA Programs for General Purpose Computers." *IEEE Trans. on Acoustics, Speech, and Signal Processing*. ASSP-26 (April 1978): 141–150.

[71] MURAKAMI, H., and I. S. REED. "Recursive Realization of Finite Impulse Filters Using Finite Field Arithmetic." *IEEE Trans. on Information Theory* IT-23 (March 1977): 232–42.

[72] ———, ———, and L. R. WELCH. "A Transform Decoder for Reed-Solomon Codes in Multiple User Communication Systems." *IEEE Trans. on Information Theory* IT-23 (November 1977): 675–83.

[73] NAGELL, T. *Introduction to Number Theory*. New York: Chelsea, 1964.

[74] NAWAB, H. "Parallelism and Code Optimization Issues in FFT and WFT Algorithms." S.M. thesis, Massachusetts Institute of Technology, September 1978.

[75] NEVIN, R. L. "Complex Convolutions Using Radix-2 Number-Theoretic Transforms." Ph.D. dissertation, Syracuse University, October 1976.

[76] ———. "Application of the Rader-Brenner FFT Algorithm to Number-Theoretic Transforms." *IEEE Trans. on Acoustics, Speech, and Signal Processing* ASSP-25 (April 1977): 196–98.

[77] NICHOLSON, P. J. "Algebraic Theory of the Finite Fourier Transform." Ph.D. dissertation, Stanford University, 1969.

[78] ———. "Algebraic Theory of Finite Fourier Transforms." *Journal of Computer and Systems Science* 5 (1971): 524–47.

[79] NIVEN, I., and H. S. ZUCKERMAN. *An Introduction to the Theory of Numbers*. 3d ed. New York: Wiley, 1972.

[80] NUSSBAUMER, H. J. "Dispositif Generateur de Fonction de Convolution Discrete et Filtre Numerique Incorporant Ledit Dispositif." French Patent Application No. 7,512,557, April 1975.

[81] ———. "Complex Convolutions via Fermat Number Transforms." *IBM Journal of Research and Development* 20 (May 1976): 282–84.

[82] ———. "Digital Filtering Using Cotransforms in Finite Fields." *Electronics Letters* 12 (June 1976): 113–14.

[83] ———. "Digital Filtering Using Pseudo-Fermat Number Transforms." *IEEE Trans. on Acoustics, Speech, and Signal Processing* ASSP-25, (February 1977): 79–83.

[84] ———. "Digital Filtering Using Polynomial Transforms." *Electronics Letters* 13 (June 1977): 386–87.

[85] OLDS, C. D. *Continued Fractions*. New York: Random House, 1963.

[86] ORE, O. *Number Theory and Its History*. New York: McGraw-Hill, 1948.

[87] PALL, G., and E. A. SEIDEN, "A Problem in Abelian Groups, with Application to the Transposition of a Matrix on an Electronic Computer." *Mathematics of Computation* 14 (1960): 189–92.

[88] PATTERSON, R. W. "Fixed Point Error Analysis of the Winograd Fourier Transform Algorithm." S.M. thesis, Massachusetts Institute of Technology, September 1977.

[89] PITASSI, D. A. "Fast Convolution Using the Walsh Transforms. In *Proc. Conf. Applications of Walsh Functions*, pp. 130–33. Washington, D.C.: April 1971.

[90] POLLARD, J. M. "Implementation of Number-Theoretic Transforms," *Electronics Letters* 12 (June 1976): 378–79.

[91] PREPARATA, F. P., and D. V. SARWATE. "Computational Complexity of Fourier Transforms over Finite Fields." University of Illinois *Co-ord. Science Lab., Rept No.* R-745, UILU-ENG-76-2233, 1976.

[92] RABINER, L. R., R. W. SCHAFER, and C. M. RADER. "The Chirp z-Transform Algorithm and Its Application." *Bell Sys. Tech. J.*, 48 (May–June, 1969): 1249–92.

[93] ———, and RADER, eds. *Digital Signal Processing*. New York: IEEE Press, 1972.

[94] ———, and B. GOLD. *Theory and Application of Digital Signal Processing*, pp. 371–80. Englewood Cliffs, N.J.: Prentice-Hall, 1975.

[95] RADER, C. M. "The Number-Theoretic DFT and Exact Discrete Convolution." Presented at *IEEE Arden House Workshop on Digital Signal Processing*. Harriman, N.Y.: Jan. 11, 1972.

[96] ———. "Discrete Convolution Via Mersenne Transforms." *IEEE Trans. on Computers* C-21 (December 1972): 1269–73.

[97] ———. "A Note on Exact Discrete Fourier Transforms." *IEEE Trans. on Audio and Electroacoustics* AU-21 (December 1973): 558–59.

[98] ———. "On the Application of the Number-Theoretic Transforms of High Speed Convolution to Two-Dimensional Filtering." *IEEE Trans. on Circuits and Systems* CAS-22 (June 1975): 575.

[99] ———. "Convolution and Correlation Using Number-Theoretic Transforms," in *Theory and Application of Digital Signal Processing*, L. R. RABINER and B. GOLD Sect. 6.19. Englewood Cliffs, N.J.: Prentice-Hall, 1975.

[100] ———, and N. M. BRENNER. "A New Principle for Fast Fourier Transformation." *IEEE Trans. on Acoustics, Speech, and Signal Processing* ASSP-24 (April 1976): 264–66.

[101] ———. "Recovery of Under-Sampled Periodic Waveforms." *IEEE Trans. on ASSP* ASSP-25 (June 1977): 242–49.

[102] RAYNER, P. J. W. "A Fast Cyclic Convolution Algorithm." Presented at *Symp. Digital Filtering*, London: Imperial College, August 1971.

[103] ———. "Diagonalization of Circular Convolution Matrices by a Class of Finite Arithmetic Transforms." *Univ. Cambridge Report No. CUED/B-ELECT/TR19*, 1973.

[104] REED, I. S., and T. K. TRUONG. "Complex Integer Convolution over a Direct Sum of Galois Fields." *IEEE Trans. on Information Theory* IT-21 (November 1975): 657–61.

[105] ———, and ———. "Convolutions over Residue Classes of Quadratic Integers." *IEEE Trans. on Information Theory* IT-22 (July 1976): 468–75.

[106] ———, Y. S. KWOH, T. K. TRUONG, and E. L. HALL. "X-Ray Reconstruction by Finite Field Transforms." *Third International Joint Conference on Pattern Recognition*, pp. 339–43. Coronado, Calif.: November 1976.

[107] ———, K. Y. LIU, and T. K. TRUONG. "Fast Algorithm for Computing Complex Number-Theoretic Transforms." *Electronics Letters* 13 (May 1977): 278–80.

[108] ———, T. K. TRUONG, Y. S. KWOH, and E. L. HALL. "Image Processing by Transforms over a Finite Field." *IEEE Trans. on Computers* C-26 (September 1977): 874–81.

[109] RIVARD, G. E. "Direct Fast Fourier Transform of Bivariate Functions." *IEEE Trans. on Acoustics, Speech, and Signal Processing* ASSP-25 (June 1977): 250–52.

[110] SCHONHAGE, A., and V. STRASSEN. "Fast Multiplication of Large Numbers." *Comput.* (German) 7 (1971): 281–92.

[111] SELBY, S. M. ed. *CRC Standard Mathematical Tables*, 146–56. Cleveland, Ohio: The Chemical Rubber Co., 1973.

[112] SHANKS, D. *Solved and Unsolved Problems in Number Theory, vol. 1.* Washington, D.C.: Spartan Books, 1962.

[113] SILVERMAN, H. F. "An Introduction to Programming the Winograd Fourier Transform Algorithm (WFTA)." *IEEE Trans. on Acoustics, Speech, and Signal Processing* ASSP-25 (April 1977): 152–65.

[114] SILVERMAN, H. F. "A Method for Programming the Complex General-M Winograd Fourier Transform Algorithm." In *IEEE Int. Conf. on ASSP*, pp. 369–72. Hartford, Conn.: May, 9–11, 1977.

[115] SINGLETON, R. C. "A Short Bibliography on the Fast Fourier Transform." *IEEE Trans. on Audio and Electroacoustics* AU-15 (June 1967): 166–69.

[116] ———. "An ALGOL Procedure for the Fast Fourier Transform with Arbitrary Factors." *Commun. Ass. Comput. Mach.* 11 (November 1968): 776–79.

[117] ———. "An Algorithm for Computing the Mixed-Radix Fast Fourier Transform." *IEEE Trans. on Audio and Electroacoustics* AU-17 (June 1969): 93–103.

[118] STOCKHAM, T. G. "High Speed Convolution and Correlation." In *Spring Joint Comput. Conf., AFIPS Conf. Proc.* 28: 229–33. Washington, D.C.: Spartan, 1966.

[119] SZABO, N. S., and R. I. TANAKA. *Residue Arithmetic and Its Applications to Computer Technology*, pp. 118–22. New York: McGraw-Hill, 1967.

[120] TAKAHASI, H., and Y. ISHIBASHI. "A New Method for 'Exact Calculation' by a Digital Computer." *Inform. Process. Jap.* I (1962): 28–42.

[121] TERRELL, P. M., and P. J. W. RAYNER. "Improved Algorithm for Low-Order Digital Convolution." *Electronics Letters* 9 (1973): 618–619.

[122] ———, and ———. "Improved Algorithm for Low-Order Real Digital Circular Convolution." *Electronics Letters* 10 (1974): 224–25.

[123] THOMAS, L. H. "Using a Computer to Solve Problems in Physics." In *Applications of Digital Computers.* Boston: Ginn and Co. 1963.

[124] VANWORMHOUDT, M. C. "On Number-Theoretic Fourier Transforms in Residue Class Rings." *IEEE Trans. on Acoustics, Speech, and Signal Processing* ASSP-25 (December 1977): 585–86.

[125] VEGH, E., and L. M. LEIBOWITZ. "Discrete Convolution of Complex Integer Sequences." *IEEE Int'l. Conf. on Acoustics, Speech, and Signal Processing*, pp. 134–35. Philadelphia: April 1976.

[126] ———, and ———. "Fast Complex Convolution in Finite Rings." *IEEE Trans. Acoustics, Speech, and Signal Processing* ASSP-24 (August 1976): 343–44.

[127] WINOGRAD, S. "On the Number of Multiplications Necessary to Compute Certain Functions." *Communications on Pure and Applied Mathematics* 23 (1970): 165–79.

[128] ———. "Some Bilinear Forms Whose Multiplicative Complexity Depends on the Field of Constants." *IBM Research Report* RC5669, October 10, 1975.

[129] ———. "On Computing the Discrete Fourier Transform." *Proc. National Academy of Sciences, USA* 73 (April 1976): 1005–1006.

[130] ———. "On Computing the Discrete Fourier Transform." *IBM Research Report* RC-6291, November 1976; *Mathematics of Computation*, January 1978.

[131] ———. "A New Method for Computing the DFT." *IEEE. Int'l. Conf. on ASSP*, 77CH1197-3 (1977): 366–68.

[132] ZALCSTEIN, Y. "A Note on Fast Convolution." *IEEE Trans. on Computers*. C-20 (June 1971): 665–666.

Index

Page numbers in boldface give the page on which the entry is defined.